MATHEMATICAL THINKING
AND
PROBLEM SOLVING

STUDIES IN MATHEMATICAL THINKING AND LEARNING

A series of volumes edited by
Alan Schoenfeld

Carpenter/Fennema/Romberg • Rational
Numbers: An Integration of Research

Romberg/Fennema/Carpenter • Integrating Research
on the Graphical Representation of Functions

Schoenfeld • Mathematical Thinking
and Problem Solving

MATHEMATICAL THINKING AND PROBLEM SOLVING

Edited by
ALAN H. SCHOENFELD
University of California, Berkeley

LAWRENCE ERLBAUM ASSOCIATES, PUBLISHERS
1994   Hillsdale, New Jersey   Hove, UK

Copyright © 1994, by Lawrence Erlbaum Associates, Inc.
All rights reserved. No part of the book may be reproduced in
any form, by photostat, microform, retrieval system, or any other
means, without the prior written permission of the publishers.

Lawrence Erlbaum Associates, Inc., Publishers
365 Broadway
Hillsdale, New Jersey 07642

Library of Congress Cataloging-in-Publication Data

Mathematical thinking and problem solving / edited by Alan H.
Schoenfeld.
   p. cm.
   Includes bibliographical references and indexes.
   ISBN 0-8058-0989-9 (acid-free). — ISBN 0-8058-0990-2 (pbk. : acid
-free)
   1. Mathematics—Study and teaching—Congresses. I. Schoenfeld,
Alan H.
QA11.A1M2767   1994
510'.71—dc20   93-49480
                                                        CIP

Books published by Lawrence Erlbaum Associates are printed on acid-free paper, and their
bindings are chosen for strength and durability.

Printed in the United States of America
10 9 8 7 6 5 4 3 2 1

For Jane and Anna,
with love

Contents

Preface   xvii

1. **The Role of Research in Reforming Mathematics Education: A Different Approach**   1
   Judah L. Schwartz

   A Discussion of Judah Schwartz's Chapter   8
   Marcia Linn and Roy Pea

2. **Some Thoughts on Writing for the Putnam**   19
   Bruce Reznick

   Comments on Bruce Reznick's Chapter   30
   Loren C. Larson

   A Discussion of Bruce Reznick's Chapter   39
   Ingram Olkin and Alan H. Schoenfeld

3. **Reflections on Doing and Teaching Mathematics**   53
   Alan H. Schoenfeld

   A Discussion of Alan Schoenfeld's Chapter   71
   Leon Henkin and Judah L. Schwartz

4. **Democratizing Access to Calculus: New Routes to Old Roots**   77
   James J. Kaput

   Comments on James Kaput's Chapter   157
   Ed Dubinsky

   Comments on James Kaput's Chapter   172
   Jere Confrey and Erick Smith

5. **Making Calculus Students Think With Research Projects**   193
   Marcus Cohen, Arthur Knoebel, Douglas S. Kurtz, and David J. Pengelley

   A Discussion of Cohen, Knoebel, Kurtz, and Pengelley's Chapter   209
   Barbara Y. White and Ronald G. Douglas

6. **A Theory and Practice of Learning College Mathematics**   221
   Ed Dubinsky

   Comments on Ed Dubinsky's Chapter   244
   Ronald G. Wenger

   Comments on Ed Dubinsky's Chapter   248
   Andrea A. diSessa

7. **The Role of Proof in Problem Solving**   257
   Susanna S. Epp

   Comments on Susanna Epp's Chapter   270
   James G. Greeno

   A Discussion of Susanna Epp's Chapter   279
   John Addison

8. **Classroom Instruction That Fosters Mathematical Thinking and Problem Solving: Connections Between Theory and Practice**   287
   Thomas A. Romberg

Comments on Thomas Romberg's Chapter   305
Gaea Leinhardt

Comments on Thomas Romberg's Chapter   312
Robert B. Davis

Epilogue   323
Alan H. Schoenfeld

Author Index   327
Subject Index   333

Contributors

John Addison
Mathematics Department
University of California at Berkeley

Marcus Cohen
Department of Mathematics
New Mexico State University

Jere Confrey
Mathematics Education
Cornell University

Robert B. Davis
Graduate School of Education
Rutgers University

Andrea A. diSessa
Education in Mathematics, Science, and Technology
School of Education
University of California at Berkeley

Ronald G. Douglas
Dean of Physical Sciences and Mathematics
College of Arts and Sciences
State University of New York at Stony Brook

Ed Dubinsky
Department of Mathematics
Purdue University

Susanna S. Epp
Department of Mathematics
DePaul University

James G. Greeno
School of Education
Stanford University

Leon Henkin
Mathematics Department
University of California at Berkeley

James J. Kaput
Department of Mathematics
University of Massachusetts, Dartmouth

Arthur Knoebel
Department of Mathematics
New Mexico State University

Douglas S. Kurtz
Department of Mathematics
New Mexico State University

Loren C. Larson
Department of Mathematics
St. Olaf College

Gaea Leinhardt
Learning Research Development Center
University of Pittsburgh

Marcia Linn
Education in Mathematics, Science, and Technology
School of Education
University of California at Berkeley

Ingram Olkin
Department of Statistics
Stanford University

Roy Pea
School of Education
Northwestern University

David J. Pengelley
Department of Mathematics
New Mexico State University

Bruce Reznick
Department of Mathematics
University of Illinois

Thomas A. Romberg
Wisconsin Center for Education Research
School of Education
University of Wisconsin–Madison

Alan H. Schoenfeld
Education in Mathematics, Science, and Technology
School of Education
University of California at Berkeley

Judah L. Schwartz
Educational Technology Center
Harvard Graduate School of Education

Erick Smith
Mathematics Education
University of Illinois at Chicago

Ronald G. Wenger
Mathematical Sciences Teaching and Learning Center
University of Delaware

Barbara Y. White
Education in Mathematics, Science, and Technology
School of Education
University of California at Berkeley

Acknowledgments

This book is the product of multiple and overlapping communities: the community of scholars at large; the mathematicians, mathematics educators and cognitive scientists who contributed to our conference discussions about mathematical thinking and problem solving, and who contributed the chapters to this book; the local community at Berkeley, which took care of everything from arranging logistical details at the conference to making and transcribing tapes; and the myriad people involved in producing the book itself. I owe sincere thanks to the Sloan Foundation and the National Science Foundation, whose generous support made it possible for us to get together in the first place. I want to thank the many conference participants for all of their contributions. The conference and the follow-up couldn't have taken place without the help of the Functions Group at Berkeley, which is a consistent source of pride and pleasure for me. And special thanks go to Cathy Kessel, whose contributions to the production of the manuscript (including editing, producing figures, compiling indexes, and checking mathematical typesetting) have been innumerable and invaluable.

Preface

Not long ago, Leon Henkin asked a group of university mathematicians which of the following three items they thought was indicative of the greatest societal change to take place in the 20th century:

- differences in gender-related expectations and stereotypes;
- the break-up of the former Soviet Union; or
- the fact that mathematicians are discussing mathematics education at the annual mathematics meetings.

His question, asked only partly in jest, pointed to a series of extraordinary changes in the mathematical community. In the early 1980s there was virtually no serious communication among the groups that have major things to say about mathematics education: mathematicians, mathematics educators, classroom teachers, and cognitive scientists. Members of these groups came from different traditions, had different perspectives, and rarely if ever gathered in the same place to discuss issues of common interest. Part of the problem was that there was no common ground for the discussions: Given the disparate traditions and perspectives, there wasn't a common knowledge base to serve as a starting place for communication.

As one way of addressing this problem, the Sloan Foundation funded two conferences in the mid-1980s, bringing together members of the different communities in a ground-clearing effort, designed to establish a base for communication. In those conferences, interdisciplinary teams reviewed major topic areas and put together distillations of what was known about them. The product of their work, the edited volume *Cognitive Science and Mathematics Education* (Schoenfeld, 1987) helped to establish a platform for dialogue.

By the end of the 1980s we were ready for next steps: to have members of the different communities present and discuss current and proposed work, and to begin carving out an agenda for future collaborative endeavors.

Ingram Olkin, Lauren Resnick, Lynn Steen, and I served as an organizing committee for the next round. We designed a working conference in which various people involved in educational reform would present their work, and members of the broad communities gathered would comment on it. The focus of the conference would be largely, but not exclusively, on college mathematics; reform efforts are informed, of course, by developments in K-12 mathematics. The main issues would be mathematical thinking and problem solving. The conference, held in Berkeley at the dawn of the 1990s, was jointly sponsored by the Sloan Foundation and the National Science Foundation, for which we are most grateful. The written product of that conference—the chapters and reactions—is the volume you are now reading.

In Chapter 1, Judah Schwartz describes his approach to making change in education. Schwartz's software has had a significant effect on classroom practice. Here he discusses the idea of the "Trojan Mouse," a deceptively simple way of catalyzing change in classrooms. The discussion of his chapter was led by Marcia Linn and Roy Pea, who offer an integrated reaction.

Chapters 2 and 3 focus explicitly on problem solving. In the United States the primary test of undergraduate problem solving is the Putnam examination, a nationwide competition that has so strong a reputation that scoring in the top ten virtually guarantees admission, with a full fellowship, to the graduate school of one's choice. Bruce Reznick has played a major role in putting the exam together, and in Chapter 2 he reflects on his experiences. Loren Larson, a problemist from much the same tradition, offers some additional comments. The discussion of Reznick's paper was led by Ingram Olkin and myself, and our comments follow. In Chapter 3, I discuss aspects of my problem solving courses—courses that might be best conceived as "introductions to mathematical thinking for college students." I describe some of my goals for the courses, and the ways I go about achieving them. Reactions are provided by Leon Henkin and Judah Schwartz.

In Chapter 4 James Kaput offers a reconceptualization of calculus grounded in both historical and psychological analyses; he also suggests technological tools that may make the fundamental ideas of calculus more accessible to younger students. Ed Dubinsky, Jere Confrey and Erick Smith offer critiques and alternatives. Marcus Cohen, Arthur Knoebel, Douglas Kurtz, and David Pengelley continue the discussion of calculus reform in Chapter 5. The calculus project at New Mexico State University has its own Trojan Mouse to offer: ostensibly simple research projects for students to work on, which ultimately have the property of transforming the character of the whole courses. The commentary is by Barbara White and Ronald Douglas.

In Chapter 6 Ed Dubinsky discusses the dialectic between research and development in mathematics education, using his work in introductory college mathematics as a case in point. Ron Wenger and Andy diSessa critique the chapter. Issues of proof, which are introduced by Dubinsky in Chapter 6, are the focus of Susanna Epp's Chapter 7. Epp, like Dubinsky, found that her students had difficulty grasping the power and generality of mathematical argumentation. She presents her reflections on the issue and her attempts to deal with it. The commentary is by James Greeno, and the discussion is led by John Addison.

Thomas Romberg's Chapter 8 and its discussion bring us full circle, where we once again confront the issue of change. In discussing exemplary illustrations of current practice, Romberg raises the main issues to be faced when confronting prospects of new pedagogy, new content, and new technologies. In their reactions, Gaea Leinhardt and Robert Davis expand on this theme. Systemic change is hard to achieve, and we had best be aware of the obstacles we are likely to confront.

These last chapters also set the stage for our concluding discussions. Following the group of presentations, conference participants worked on an agenda describing "next steps"—what we need to know, and what we consider to be high priority issues as the field develops. The product of these discussions is given in the Epilogue.

The conference was challenging; it was fun. The various groups found they could talk to each other. They learned much from the interactions; and, although they represented quite diverse perspectives and backgrounds, they came to a strong and clear consensus regarding major issues, as reflected in the Epilogue. This volume reflects the work of the conference, making its deliberations public. Whichever hat or hats you wear—that of mathematician, mathematics educator, classroom teacher, or cognitive scientist, I hope it will enable you to profit from it, as the participants did.

Alan H. Schoenfeld

1 The Role of Research in Reforming Mathematics Education: A Different Approach

Judah L. Schwartz
Massachusetts Institute of Technology and Harvard Graduate School of Education

FIRST THINGS FIRST

This chapter is about a strategy for reform in mathematics education that mixes research, development, and implementation in somewhat unusual proportions and arranges their temporal sequence in a way that some will find puzzling and irreverent. Perhaps because of its untraditional nature, this approach to mathematics education reform has not found favor among the more traditional governmental sources of support for education reform. Nonetheless, I hope that the perplexed or offended reader may be persuaded that there is some pragmatic merit to the view that I present.

Before presenting the main line of argument of the chapter, it is necessary to set down the assumptions that underlie my thinking about reform in mathematics education. In particular, it is necessary to address explicitly the question, what might be considered successful mathematical education reform?

My formulation of the notion of successful education reform is not couched in terms of grade levels or performance on standardized tests. Certainly it is necessary for us to teach more and for our youngsters to learn more mathematics than they do now. Moreover, we and they must be able to demonstrate this greater knowledge in some suitable way. (We explicitly put aside here the question of the adequacy of the assessment instruments we normally use to demonstrate the extent of what we have succeeded in teaching and our students have succeeded in learning.) However, if we limit our aspirations to having the students "know more," and we succeed in achieving these aspirations, then all we will have succeeded in doing is having a somewhat larger segment of our population have the commonly held rote, formulaic, and ceremonial notion of the mathematical enterprise.

My notion of successful mathematics education and successful reform thereof includes changing habits of mind so that students become critical, in a constructive way, of all that is served up on their intellectual plates. Students who have been successfully educated mathematically are skeptical students, who look for evidence, example, counterexample, and proof, not simply because school exercises demand it, but because of an internalized compulsion to know and to understand.

In addition, my motion of successful mathematics education, or for that matter successful education in any discipline, has students coming to understand that intellectuality is not a spectator sport, and that they can and should be challenged to create in each and every discipline that they study. This is not a widespread idea in the minds of the American public when they think about schools, and it is often absent entirely when the public thinks about mathematics in schools. The popular notion of school mathematics is that it is mostly arithmetic, which must, above all else, be learned in a deadeningly rote and ununderstanding fashion before any thinking whatsoever is even contemplated. Even when one gets beyond "the facts" and "the tables," thinking is limited to understanding the inventions of the past and almost never to the inventing of new mathematics.

A DIFFERENT STRATEGY

In this chapter I present a strategy for education reform in mathematics that is drawn directly from the experience my colleagues and I have had with a recent powerful innovative reform and indirectly from other innovations that have managed to infiltrate and permeate the educational scene.

An overview of the strategy includes:

- Getting an important new idea
- Disguising the idea so that it appears nonthreatening
- Making curricular artifacts built around the idea
- Distributing the artifacts and building a social network of users
- Looking carefully at what, how, what to do differently

In order to make the discussion concrete, I illustrate these points with experiences drawn from the development, implementation, and research that surrounds the Geometric Supposers.[1] The Supposer materials include a series of four microcomputer software environments that run on widely available microcomputers, copious written materials for both students and teachers, as well as several videotapes.

---

[1] THE GEOMETRIC SUPPOSER SERIES, THE GEOMETRIC SUPERSUPPOSER, by Judah L. Schwartz & Michal Yerushalmy, 1983–1993, Sunburst Communications, Pleasantville, NY.

AN IMPORTANT IDEA

The teaching and learning of geometry in school is saddled with a conflict. There is a body of mathematics to be taught, learned, and made that deals with properties of shapes and relations among shapes. Clearly, there is a fair amount of generality in such statements as "the medians in any triangles are concurrent." To understand the importance of this statement one needs to draw a triangle and its medians. However, and here's the problem, one cannot draw *any* triangle. One can only draw a specific triangle. Aside from size, each triangle that one draws is unique. In this sense, the isosceles right triangle is no more special than the triangle with angles of 39.7, 82.1, and 58.2 degrees.

If the human mind indeed needs images to help it think about spatial matters, how do we avoid the inevitable specificity that images possess when we are trying to understand the generalities that characterize the objects we are exploring?

The central idea of the Geometric Supposer software is that the user can select or construct a primitive shape, such as a trapezoid or a scalene triangle, and make a sequence of Euclidean constructions such as parallels, perpendiculars, angle bisectors, and so forth, on this shape. The software preserves the construction as a procedure, which can then be carried out on any other shape of the same type. Measurements carried out on the original shape can be repeated in this way as well.

The Supposer software environments thus allow the decoupling of the inevitable specificity of the image from the generality of the construction that happens to have been carried out on it. This makes it both possible and inviting to make and explore conjectures about the generality of results discovered to be true in particular cases. Lowering the logistical and psychological costs of exploring conjectures can and often does have the dramatic consequence of bringing curiosity, inquiry, and research into the mathematics classroom.

At this point in the normal research paradigm, a host of researchers would descend on experimental subjects and try to ascertain how well students taught geometry with the making and exploring of conjectures central to their learning compare, by traditional assessment measures, to students taught in the traditional way. I find this notion absurd, because it makes an asymmetric assumption about quality and desirability. It places the "burden of proof" on the innovation, despite the weight of evidence against the desirability of the status quo.

"A TROJAN MOUSE"

If the important idea that one devises is to find its way into the way mathematics is taught, learned, and made in schools, then it is important that it not appear to threaten current practice. No individual teacher, principal, superintendent, or school board member likes to be threatened and told, even if only by implication, that what he or she has been doing is not adequate. Further, there are few institutions in our society more inertia-ridden than schools and boards of education, both state and local. This gives rise to a corresponding reluctance to change on an institutional level.

Thus, it is important to disguise the innovation in such a way that it does not appear to threaten. Moreover, depending on sensitivities, it is often important that the innovation does not appear to criticize. It should appear to augment, rather than replace.

In the case of the Geometric Supposer materials, we did this by insisting on the relevance of these materials to the standard tenth-grade geometry curriculum in Euclidean plane geometry and to the standard middle and junior high school geometry activities. We wrote written materials in the form of Projects & Problems Books that focused on familiar content. Our written materials always included sections for the teacher indicating something of the range of responses the materials were likely to elicit from their students. We did everything we could to make the enterprise as evolutionary as possible, eschewing the revolutionary.

What did we hope would happen? Because the Supposer software environments made the making and exploring of conjectures so easy and inviting, we hoped that students and teachers using the materials would be drawn into doing so. Indeed, in many classrooms, that is exactly what happened. When that happened, students and teachers came to realize, at first implicitly, and then explicitly, that geometry is not a finished subject, and that it still is possible to make new geometry. New knowledge of the subject can originate with teachers, but it can also originate with students. As a consequence, the bases for authority in a classroom have to be renegotiated, because authority by assertion rooted in the possession of knowledge is no longer an adequate basis for exercising authority.

CURRICULAR ARTIFACTS

"You can't fight something with nothing," a wise man once said. If you want to change existing practice, you must give people new artifacts to signify the change. In all domains of human endeavor, the artifacts people enjoy are emblematic of the undertaking. A new undertaking requires new emblems. In the case of education reform, there need to be new curricular artifacts that allow the users to mark the newness of their undertaking.

With respect to the question of making new artifacts to celebrate innovation, we live at a particularly interesting time, with all of the ambiguity of the adjective *interesting* intended. The microcomputer, the digital chameleon of our era, allows us to make an endless stream of artifacts, each encased in the increasingly familiar box with screen and keyboard. We, as a society, have developed an enthusiasm for the computer tools that we have learned to build, some of which are eminently sensible and useful. For the most part educators have shared this enthusiasm for new tools. As a consequence, those of us interested in education reform have another arrow in our quiver with which to attack the problem.

One must be careful in seizing this opportunity. Despite the enthusiasm of some, there are still many educators who are frightened of the newer technologies. In addition, financial constraints and inequities are such that not all classrooms and teachers have free and easy access to the use of the newer technologies.

As a consequence, one must be careful to implement microcomputer materials on those machines that are most likely to be found in schools and least likely to frighten inexperienced users, be they students or teachers. This constraint rules out writing software meant to be run on "cutting-edge" hardware. This constraint also rules out making the systems overly capable. The more features the system has, the harder it becomes to use and the less likely the educational system is to perceive it as an extension and augmentation of its current practice.

Who should develop these artifacts? There is no general solution to this problem. It seems clear that full-time teachers are too overburdened with the demands of the classroom from day to day as to be able to devote the time to such development. Further, I think it is difficult for most teachers to detach themselves from the curriculum they are teaching sufficiently to think about the material with the total freshness I am advocating. However, let it be clearly said and recognized, that any curricular development that takes place without the continuing central and focal involvement of teachers is almost certain to fall far short of its potential. The penchant that humans have for deluding themselves seems to be unbounded. Leaving curricular design totally in the hands of those who are not teaching seems to be a fatal error.

In similar fashion, mathematicians alone cannot be trusted to generate artifacts. They are far too prone to be captivated by a beauty and an aesthetic that they have not succeeded in getting others to appreciate. But clearly, no undertaking of the sort described here makes sense without them. In addition, the perspective of the cognitive scientist, particularly one who works in the field of mathematics learning and teaching is equally insufficient and equally necessary.

Finally, insofar as curricular artifacts that are based on the computer are concerned, the nature of the public's expectations has gotten to the point that smoothness, look, feel, and speed of software exceeds the capabilities of most nonprofessional programmers.

SOCIAL NETWORKS

Just as you cannot fight something with nothing, you cannot fight somebody with nobody. There must not only be artifacts. There must be people—more people than simply the enthusiastic intellectual parents of the innovation. There have to be teachers and students who feel themselves part of a movement that is trying to do something about changing mathematics education.

In many respects, the most straightforward way to develop a network of people using the materials is to publish them. This makes them widely available, a necessary but not sufficient condition for the building of a social network. We did this with the Supposer materials. Many of the people who were involved in the development of the computer and written materials spoke (and continue to speak) at both regional and national meetings of the National Council of Teachers of Mathematics (NCTM). We published articles on the curricular and instructional aspects of the innovation in several journals.

We also tried a variety of less common techniques to grow a social network of teachers who felt themselves and their classes to be part of an education reform movement. Under the auspices of the publisher we formed a nationwide pilot project, with two sites per state. At each of these sites, in exchange for a promise to use the Supposer materials as the basis of their geometry course, and to provide periodic feedback about their classes, teachers were supplied with the Supposer materials at no charge. We also formed a Geometric Supposer Society, in which membership was open to teachers and students who submitted some new insight, problem, or theorem they had devised.

The dissemination of materials and techniques is but one aspect of the social network. Less visible, but equally if not more important, is the role of the social network in legitimating the adoption of new habits of mind. It is difficult for students well into a school career in which mathematics has been an endless series of incompletely understood calculation and manipulation ceremonies to shift gears and to exercise in class their curiosity and inventiveness. It is difficult for teachers and principals, who will be held accountable to superintendents and school boards (through the medium of machine-scorable preanswered tests), to imagine mathematics classes in which mathematics is "discovered" rather than "covered." It is difficult for school boards to imagine that the mathematics they learned and the way they learned it is possibly not universal or eternal in its importance. All of these groups need support as they develop new "habits of mind." It should not be surprising that this support comes best from people in similar situations.

As the use of the Supposers grows, a variety of intertwined social networks of people has also grown. More importantly, these networks have devised ways of their own to grow. There are electronic mail conferencing systems dealing with the use of the Supposer. Virtually every regional NCTM meeting contains several talks by teachers about new ways to use the Supposer materials. Even reports from various federal agencies about the promise of technology in education now readily, albeit inappropriately, take credit for it.

RESEARCH

Now that there is a substantial number of students and teachers using the Supposer materials, it becomes reasonable to ask many questions about the nature and quality of the learning and teaching that takes place with these materials. Many of the questions are of a traditional sort and rely on the use of traditional instruments in order to obtain answers. For the most part, such questions and the answers to them are uninteresting.

However, there are many things to be learned about the nature of teaching and learning, generally, and teaching and learning mathematics, particularly, by trying to assess the enterprise in profoundly different ways. This chapter offers one illustration of a different approach to assessment. Suppose, for example, that we forswore forevermore the asking of any mathematical question or problem that had a *unique* correct answer or solution. For those who are poised to attack me for promoting a "loss of certainty," I hasten to stress that I am not proposing questions or problems without correct and incorrect answers.

By using such problems in order to do research on the quality of learning and teaching as well as to assess their extent, we would oblige the student to synthesize responses from universes of possible responses, thereby offering some insight into their modes of thought, styles, as well as their tastes and judgments.

AN AFTERWORD: IS THIS CASE A CASE OF ANYTHING OTHER THAN ITSELF?

Is the Supposer experience unique, or could it serve as a guide to future action? We are not certain but believe that with proper attention to the similarities and differences, we can use the Supposer experience to guide an effort to reform the learning and teaching of algebra. We are trying to hold on to what we think are those features of the Supposer experience that we believe contributed to its success, while trying to avoid the repetition of mistakes. There have been some small encouraging signs, but the road ahead looks difficult and lined with Cassandras.

A Discussion of Judah Schwartz's Chapter

Marcia Linn
Roy Pea

> **Marcia Linn,** *University of California, Berkeley*
> **Roy Pea,** *Northwestern University*
> **Judah Schwartz,** *Harvard Graduate School of Education*
> **Uri Treisman,** *University of Texas, Austin*
> **Alan H. Schoenfeld,** *University of California, Berkeley*
> **Andrea diSessa,** *University of California, Berkeley*
> **Jere Confrey,** *Cornell University*
> **Jim Kaput,** *University of Massachusetts, Dartmouth*
> **Ron Wenger,** *The University of Delaware*

Marcia Linn. The first question that Roy and I thought might be interesting to discuss is the research paradigm that Judah advocates in his chapter. Judah suggests a 5-step process. The main thrust of the approach is to create an artifact, put it out in schools, and see what happens. One thing I wonder about is the following: Does this approach deny participation in the artifact construction process to those who are ultimately going to use it? Is this approach taken for logistical reasons, or does it have a more philosophical underpinning? In my own experience, with the Computer as Lab Partner Project, I found that there are real advantages to having a joint constructive process involving teachers, researchers, and developers.

Roy Pea. There is a general trend in studies of workplace technology, influenced by a group in Denmark that works in participatory design, in which there are close observations of existing work practices. Here, analogous analyses might be concerned with what teachers and students are doing in the algebra classroom, taking a look at the kinds of resources that they have available and what kinds of tools they currently use, and then to engage in a co-design practice with them. Of course, this is intended to complement the good design ideas and instructional goals Judah starts with. Such ideas and goals would not have necessarily come out of conversations with teachers and students.

Judah Schwartz. Let me interject, because there is a very interesting issue of balance here. Obviously we do have teachers participating in our design efforts. But there's an interesting conundrum about observations of classroom practice, because if you ask teachers in the main "What do you want?," what you get are requests to support what they know how to do. In algebra, for example, where we are working on the design of some software called the Algebraic Proposers, you get requests for symbol manipulators and graphing packages.

Marcia Linn. I think that's absolutely right. Ask people what they think they want, and you get requests for electronic books. Accomplishing an order of magnitude change is clearly a difficult issue. On the other hand, I question whether merely putting an artifact out in classrooms will result in an order of magnitude change. In fact, I think you implied that wasn't necessarily true, that the Geometric Supposers were sometimes used for pretty retrograde purposes.

Judah Schwartz. Any tool can be used or abused. For that reason it's very important to get moles in the system. The way to get moles in the system is to use the teachers with whom you've been working in the development process as your front people. I don't talk about the Supposer to teachers. Richard Hood talks about the Supposer. Dan Chazan does; Mary Sapienza does. And that's very important, because then it's talked about to teachers by teachers who have had a hand in the development process. That's a key ingredient in the strategy.

Roy Pea. How do you go about establishing attitudinal changes in the teachers you work with? If they're having difficulties of a particular kind, it's not necessarily because they're not learning the technology well, but maybe because the technology could be redesigned in ways that more fit what they come to see as useful work practices with this new tool in their environment.

Judah Schwartz. You listen hard and you talk to people as honestly as you can, but you don't take an advocacy position. We work very hard to have the advocates be the teachers we've worked with in the development process. It's not as if the teachers don't talk to us, but we're not the advocates. The ones who are have lived through the process, in classrooms.

Marcia Linn. How do you help teachers become experimenters in their own classrooms? Do you feed them a few good ideas, not just about the artifact but about the instructional process, to facilitate a broader instructional impact?

Judah Schwartz. You make lots of written materials. For every one of the Supposers there's a thick book of problems and projects.

Marcia Linn. And that's not constraining to the open-ended experimental paradigm you advocate?

Judah Schwartz. No, because the projects are open-ended, the problems are open-ended.

Marcia Linn. There are some serious questions about what instructional lessons students can take from dynamic, interactive media. The Geometric Supposers represent one class of examples, the Function Proposers another. Dynamic representations often flash rapidly across the screen and overload the processing capacity of the student. For example, all of the boxes in the Function Proposers might be difficult to integrate and, instead of leading to a deeper understanding of the relationship between formalisms and functions, might simply lead to confusion. Or, they might spark interest in questions like "Let's see what kind of a picture I can put on the screen," rather than in the prediction and the reconciliation of outcomes. Let me mention some of our experience with the Computer as Lab Partner project. In that project we have real-time data collection in a science class. At first the students were wildly enthusiastic. However, we discovered that the students weren't taking the graph from the screen and relating it to what they thought was going to happen at all. When we made prediction a rigid constraint of the curriculum, it led to an order of magnitude increase in understanding. The difficulty is to integrate the proper use of the dynamic representation with the cognitive excitement it engenders.

Uri Treisman. When we talk about this wonderful tool, let's remember that it's only a tool. One has to also develop the settings in which this tool is going to be used. One of the problems with algebra is that students can manipulate quadratics, but they never learn when they would expect to see a quadratic. Or if you ask a teacher to describe a situation in which one might expect a quadratic function to arise, one may or may not get an answer that's productive in classroom terms. We have to help build the contexts of tool use, not just the tools themselves. That raises assessment issues as well. It's difficult to assess beautiful tools without having an understanding of the settings in which students might get to work with them.

Roy Pea. I think that's the kind of thing that we've been discussing, having a reciprocal evolution between the technology and the work practices. Indeed, that kind of reciprocal evolution applies as well to the research questions you would want to ask about the classroom environment: What are the kids learning? What does the teacher need to know to be able to pull off activities like this? The research questions that might have been asked by any of us before the Supposers were put into classrooms will change dramatically after we see what goes on in them. Seymour Papert had wonderful ideas about what you can do with Logo in the classroom, but different things happened in the classroom with a lot of teachers. The same will be true of the Supposers, of course. So, we have to keep questioning the assumptions about what the appropriate layout and functionality of the tool is, what the appropriate research questions are to ask, and what the appropriate work practices and curriculum are. And that leads to the topic of curriculum transformation.

Judah Schwartz. There are a number of things to respond to in what's just been said. First, I should stress that we aren't developing all-purpose tools; some tools are good for some things, and others serve different ends. Second, I can at least point to some existent proofs.

I'll tell you one story, because I think it's wonderful. A student of mine who is now teaching in high school is struggling in an undistinguished algebra two class. He's giving the kids functions to classify graphically, lots and lots of pictures of functions. He wants students to classify them, to invent ways of describing them. So the students are classifying them by their asymptotic behavior, by how many zeros they have, how many extrema, and so on. They have some sort of an ill-defined notion of slope. They know what slope means for a straight line, but they don't know what you do with all those wiggly things, where the slope obviously keeps changing. Then one kid says,

"I know, because the slope changes it has to have an x in it."

He says:

"In fact, I thought about it. Look, suppose you want the slope of [the function $y =$] x^3. If you look closely at x^3, it looks like a straight line, because everything looks like a straight line if you look closely. So the thing is, you gotta write x^3 the way you write a straight line, which is m times x plus b. So how do you write x^3 like that? You write it x^2 times x plus b, and x^2 plays the role of m."

That's not bad, not bad at all. That's one of my good stories. To counter that there are a lot of less good stories, of course.

Alan Schoenfeld. You said that you want to change the atmosphere in the classroom, to establish a different kind of classroom dynamic that would ultimately affect the students' habits of mind. The classroom reality reflected by your stories is that sometimes it doesn't work, sometimes it does, and there are some wonderful stories to tell when it does. The research question for me is: What makes the magic happen when it happens?

Judah Schwartz. I can tell you a little bit about that in the case of geometry. Part of what makes it happen in geometry has nothing to do with the software. It has to do with the organizational surroundings. Some of it has to do with the degree to which a principal is willing to empower teachers and say: "Yes, this is a good way to teach; yes, this is all right; yes, I will shield you from the school board when they say they have to take the same exams." (Never mind that when they do take the same exams they do equally well or better.)

In addition, you need to have networks of teachers who can talk to one another and say, "You know what I tried today," or "You know what some kid came up with today?" They don't necessarily have to be in the same school.

Those are some of the things that seem to help make the good things happen. They don't seem to be geometry specific, so when we come to the same point in algebra we will apply the same strategies as a first approximation and hope that works.

Roy Pea. One interesting thing about the anecdotes or "war stories" you just referred to is that they could be used in some really interesting and positive ways as resources for the Supposer community. Julian Orr at the Xerox Palo Alto Research Center has done a lot of work on how expertise gets propagated among copier technicians. One of the major ways is not through any of the technical documentation and manuals at all, but through stories that they tell informally at conferences, over drinks, and so forth. In some sense your teacher anecdotes of classroom successes are of the same kind. Maybe one of the things that this shows is that we need a way of rapidly disseminating these positive stories to give teachers a sense of the ways in which the kids and they can be empowered.

Judah Schwartz. Indeed. In a somewhat Edwardian fashion, the publisher of the Geometric Supposer has established something called the Geometric Supposer Society. You buy in, as it were, by sending in an anecdote which then gets included in the newsletter. And so this strategy is in fact used.

Uri Treisman. I'm not sure that's enough, and I'm worried about the "mom and apple pie" romanticism that says teachers must be involved, and that magic will happen. In fact, when you look at the teachers who are involved in such projects you find that they are a very special breed. Many of them are actually hired away from their classrooms by universities to do work on such projects.

I think it's really important to develop a careful understanding of the settings in which teachers learn to use these materials, and what works in them. We need accurate estimates of what the real cost is for teachers to learn about this stuff. Generally, teachers can take anything and turn it into a resource for what they want to do, and that sword cuts both ways.

Marcia Linn. That is the point. How do we empower teachers to create an experimenting society in the classroom? It is very difficult. What kind of support is really needed to create the kind of experimenting society where teachers really think they can try out a curriculum, listen to what students say, make some adjustments, and try it again? I think the anecdotes Judah described will help.

But I think that a more collaborative group where teachers could come together and talk to people like Judah would be better. Teachers could say, "Look, this is working, this isn't working," and you could say, "Have you considered this?," or "You are using cooperative groups, but you are encouraging cooperation for an activity where cooperation is very ineffective. Cooperative groups generate ideas well but are less successful at synthesizing ideas." This kind of discussion would move the process along. Materials such as these require changing the entire way one thinks about how teaching takes place, and teachers need support in doing that kind of thinking.

Andy diSessa. I wanted to ask a variant of one of the questions we've been discussing. Let me start by saying that I've been trying to think of myself lately as a designer of mediated activities, rather than a designer of artifacts. That change of focus may seem small, but it makes a big difference. Along those lines, I think we ought to be doing more thinking about the structure of activities as kids engage in them. In particular, when you get your good stories, what is the character of the activities that are taking place at that point?

I would like to ask you a question about that, but in order to frame it, I need to explain my mini-theory of the activities that I think you want kids to get engaged in. You want them to engage in a certain kind of mathematical game that we all understand, but it's not clear that a kid has ever participated in something like that. This games goes as follows: In stage number one, you have some domain or artifact, and you kick it a little bit, in ways that you understand you can kick these sorts of things. In stage number two, you notice that there's something interesting in the way it reacted to your kicking. Well, where does that notion of "interesting" come from, and how do you select what's interesting out of what happens? In stage three, and I think maybe this is a critical one, you have to explain the interesting thing. And that's where you invent the mathematics or you discover the mathematics. That's where you ask the question, does this kid have a notion for what is an adequate explanation for this or what kinds of moves to make to explain it. Are you collecting stories that look at kids' versions of "How do I kick this thing," kids' versions of "Oh, that's interesting," and kids' versions of "This is an adequate explanation"?

Judah Schwartz. First, I accept your shift from "artifact design" to "design of mediated activities" as a friendly amendment—but with the following reservation. In terms of school culture, I would want to say mediated activities and artifacts, because if you don't focus on artifacts the necessity for artifacts will tend to get lost. I think it's important that there be artifacts, and activities in the world of education do not always imply new artifacts. So I wanted to make sure that I continue to say artifacts, but I completely accept the notion that it's really mediated activities.

Now, about stories. We are indeed collecting stories of the sort you suggest.

We are doing that in algebra, at a very early stage. It is our hope that we can do the same kind of things that we've done in geometry, where we are now capable of building at every level moderately sensible islands of theory about what works, that is, what works in the approach to geometry, what works in the change of classroom behaviors, and what works in the kind of organizational context for the instruction.

Marcia Linn. I'm curious as to why you think the artifacts are necessary. Do you see them as a catalyst for change, or as an opportunity to justify change?

Judah Schwartz. It's not a philosophical issue, it's a tactical issue. The very existence of the artifact serves as a stimulus for things to happen, and to happen differently.

Jere Confrey. I think it is really important to spend some time talking and thinking about design, because I think there is an interesting relationship between the mediated activity and the artifact you're designing. You're not just creating an artifact, you're creating an artifact that's designed to get at particular things. There are certain things that are in essence hardwired into the design of your software, because those are the conceptions that you are instantiating in that piece of software—for example, your conception of what a function is and how you want the software to handle it. Then you want to leave certain things open as possibilities for a wide range of activities when students use the software. So both aspects of design are important for me. My software design needs to be informed by my working very closely with students and trying to imagine what their conception of the thing is that we're working with. It's also really important for me to articulate what my conception of it is, as I design my software. We need to get those issues out in the open as designers and say more than the fact that we're creating artifacts.

Marcia Linn. Let's turn to the last topic we wanted to address, which is research questions. You've said that the classroom environments you want to establish are radically different from the classical model of classrooms: patterns of authority will be different, discourse patterns will be different, even the patterns of engagement with mathematics will be different. Supposers will be in a wide variety of classrooms, ranging from those where there are rigid constraints on student behavior to those that have religious nonconstraints on student behavior. That provides a really interesting opportunity to look at the relationships between the kinds of constraints imposed on student interactions and behavior and what develops in those classrooms.

One set of questions centers around the conversations themselves. Which students get involved in the exciting exchanges of information? Which students feel empowered, and which students feel reluctant to enter into the conversations? What are the mediating events that lead to a change in the conversation?

A second set of questions deals with benefits. Who benefits, given that these conversations take place? Often, modeling effective discourse behavior is helpful for other members of the classroom community. Observers of the conversations may profit. Note Roy Pea's comment on work practices.

A third set of questions deals with the establishment of the classroom environment. How do we get teachers to model the kinds of discourse behaviors that we'd like to see, for example? It seems to me that the proposed kind of classroom environments cannot be fully assessed unless these questions are addressed seriously.

Roy Pea. We should note that the Supposers are not only Trojan mice for curriculum, but they're also Trojan mice for researchers. For example, the question, "How are we going to assess what students are learning in contexts like these?" takes on a new character in these environments and will force significant changes in assessment methods. We may need measures that look at the kinds of conjectures that students generate. Judah, have you worked on that problem?

Judah Schwartz. One thing that environments like the Supposers allow you to do is to move relatively easily from instances to classes of instances. Certainly that's clear in geometry. It's also the case in algebra; when you use the graphical transformation functions, you necessarily build a family of functions. Such families will have one or several invariants. So you can then pose questions not about tokens but about types. You can say "Build me any function that has the following properties." When you ask a question of that form, you're engaging the student in what is fundamentally a design problem, which has a nonunique set of right answers. It seems to me that's a different kind of assessment than we ordinarily perform. These environments lend themselves to the posing of that kind of question, and I think it's a much richer kind of question.

Marcia Linn. Following that, you said initially that you wanted to change students' habits of mind. The sense that I got was that you wanted to change them very broadly. It isn't the case that we want them to be poor representations of $100 graphing calculators, but that we want them to be able to think about problems they might encounter in their everyday life. I was curious as to how you see that generalization gradient getting established.

Judah Schwartz. That's too hard for me. . . .

Marcia Linn. Well, it really is one of the fundamental questions. . . .

Judah Schwartz. I didn't say it wasn't fundamental! Absolutely. I would love to think that new habits of mind acquired in this kind of arena seep out, and this person grows up to be someone who, say, votes intelligently. I would like to think that, but. . .

Jim Kaput. We should recognize that there's a difference between the curriculum change you're trying to make now and the kinds of curricular moves that might be possible a decade or two from now. You are talking about change in a system where you assume that kids and teachers are as we have them now. But another angle is to assume that, not long from now, in grades K through 6, kids will spend a tremendous amount of time building functions, modeling situations numerically, finding numerical patterns, drawing pictures, and playing with other models of situations. By middle school they could be pretty fluid at manipulating the models they've built. I'm wondering, given that scenario, what do you think about the role of the kinds of stuff you're advocating here, with respect to graphical manipulations of functions and so on? Where does that fit?

Judah Schwartz. I think it would be wonderful if we could postulate that all of the things that should have been done intellectually for and by kids have been done, and that they arrive at the secondary doorstep being intellectually nimble and agile and facile with these things and can use them in rich ways. I would love to believe that. It would be great if that worked so well that the universal notion of function in 6th graders was no longer that of a mapping from one set of numbers to another set of numbers given by an algebraic rule, but something much more general than that. Then I would say this tool has outlived its usefulness—because we could make a much richer tool. But for the moment I think it's a useful tool, because it can provoke all kinds of things that ought to be provoked.

Uri Treisman. There's another research issue I'd like to pursue. Tools like the Supposers help us to create settings in which students can develop intimacy with basic objects to the degree that in fact they can have intuitions about what they're doing. How many hundreds of hours do you need to play with numbers to have number sense? How many hundreds of hours do you need to play with geometric objects, or functions, to develop good intuitions about them? How many hundreds of hours of playing around do you really need before what you know is automatic enough and ingrained enough so that you have fodder for creative work and problem solving and design? Because these tools are appealing to children, we may have the opportunity to explore such questions.

Ron Wenger. In the short run a large part of the impact of artifacts on the curriculum will be the extent to which they influence the design of traditional textbooks. When we look at current practice from the point of view that the artifact provides us, we see enormous gaps in the existing texts. One of the test cases could be your Geometric Supposer case. Have you seen textbooks or interesting tasks being designed into textbooks, that don't assume the kids have access to software tools, that are qualitatively different from the contents of textbooks before people had that tool?

Judah Schwartz. Two-thirds of a yes. There are a lot of manifest influences of the Supposer on new geometry texts—statements like "If you have the Geometric Supposer here's a problem to try," with publishers all over the place. Those aren't always driven by altruistic motives. When a state like Texas says "You gotta link to software," the response is "Yes, sir, we'll link to software," and they put in some due obeisance to software, the Supposer among other things. Those are the bad cases. The good cases are where textbooks really do that; they say, "If you have the Supposer try the following." It's much harder to detect the presence of nonspecific references, that is to say, a problem that you think would not have appeared before but now does appear. How would I know such a problem if I saw it? An open conjecture is not likely to have been a problem in many geometry books earlier, and if it appears now, one can speculate that maybe the Supposer had an influence on that. But it's very hard to detect.

Ron Wenger. I don't think that will be nearly as ambiguous in algebra. I think that there are profoundly different tasks that we could put in books now, which don't assume a kid has access to these kinds of artifacts, which are qualitatively different.

Judah Schwartz. That is absolutely true. When you've done some mucking around with graphs, you can begin to ask questions like, "Tell me all the things you can't change about a cubic, no matter how you scale it. And explain why you believe what you are saying." That's an essay question, unlike the kinds of questions that are asked now. And it could be asked even in an environment in which graphing software isn't available. I think you're right, that a very good test of duration and depth of the impact of software tools is their impact on printed materials.

2 Some Thoughts on Writing for the Putnam

Bruce Reznick*
University of Illinois, Urbana

WHAT IS THE PUTNAM?

This chapter describes the process of composing problems for the William Lowell Putnam Mathematical Competition. Inevitably, this leads to the more general issue of mathematical problem writing. I shall be anecdotal and probably idiosyncratic, and do not purport that my opinions are definitive or comprehensive. The reader should not look for "How to Pose It," but rather sit back and enjoy the heart-warming tale of a boy and his problems.

The following is a quotation from the official brochure (Putnam, 1992). The William Lowell Putnam Mathematical Competition "began in 1939 and is designed to stimulate a healthful rivalry in mathematical studies in the colleges and universities of the United States and Canada. It exists because Mr. William Lowell Putnam had a profound conviction in the value of organized team competition in regular college studies." The Putnam, as it is universally called, is administered by the Mathematical Association of America. It is offered annually (since 1962, on the first Saturday in December) to students who have not yet received a college degree. In 1991, 2,375 students at 383 colleges and universities took the exam. (For more on the history of the Putnam, see Alexanderson, Klosinski, & Larson, 1985; Gleason, Greenwood, & Kelley, 1980.) The problems and solutions for Competitions through 1984 are in Alexanderson et al. (1985) and Gleason et al. (1980). The problems, solutions, and winners' names are also published in the *American Mathematical Monthly,* usually about a year after the exam.

---

*Author supported in part by the National Science Foundation.

The Putnam consists of two independent 3-hour sessions, each consisting of six problems arranged roughly in order of increasing difficulty. The exam is administered by proctors who cannot comment on its content. Contestants work alone and without notes, books, calculators, or other external resources. They are ranked by their scores, except that the top five are officially reported en bloc. Teams are preselected by their coach, and team rankings are determined by the sum of the ranks (not the sum of the scores).

> The examination will be constructed to test originality as well as technical competence. It is expected that the contestant will be familiar with the formal theories embodied in undergraduate mathematics. It is assumed that such training, designed for mathematics and physical science majors, will include somewhat more sophisticated mathematical concepts than is the case in minimal courses.
>
> Questions will be included that cut across the bounds of various disciplines and self contained questions which do not fit into any of the usual categories may be included. It will be assumed that the contestant has acquired a familiarity with the body of mathematical lore commonly discussed in mathematics clubs or in courses with such titles as "survey of the foundations of mathematics." It is also expected that self contained questions involving elementary concepts from group theory, set theory, graph theory, lattice theory, number theory, and cardinal arithmetic will not be entirely foreign to the contestant's experience. (from official brochure)

Between 1969 and 1985, I participated in all but three Putnams. I competed four times, the last two wearing the silks of the Caltech team, which placed first. As a graduate student and faculty member, I coached or assisted at Stanford, Duke, Berkeley, and Urbana. I was a grader in 1982 and a member of the Problems Subcommittee for the 1983, 1984, and 1985 Competitions. (I had been living quietly in Putnam retirement when Alan Schoenfeld invited me to this conference.)

WHO WRITES THE PUTNAM?

The Problems Subcommittee of the MAA Putnam Committee consists of three question writers, who serve staggered 3-year terms. The most senior member chairs the Subcommittee, the other three members of the full Committee are "permanent." During my service, the Problems Subcommittee consisted of successive blocks of three consecutive people from the following list: Doug Hensley, Mel Hochster, myself, Richard Stanley, and Harold Stark. Outgoing members are invited to suggest possible replacements, but these are not acted on immediately. This is an old-boy network, but in practice one often suggests the names of strangers whose problems one has admired. (It turned out that the 1985 group inadvertently consisted of three Caltech alumni, and, as chair, I was relieved when the Harvard team won.)

The rest of the Putnam Committee consisted of three permanent members, two of whom (Jerry Alexanderson and Leonard Klosinski) arranged the massive logistics of the Competition and a liaison with the Problems Subcommittee (Abe Hillman, then Loren Larson). They do an incredible amount of work, which is not germane to this chapter, but should not go unappreciated.

HOW WAS THE 1985 PUTNAM WRITTEN?

The following is an overview of how the 1985 Putnam was written. In November 1984, I wrote a welcoming instructional letter to Richard and Harold, describing our timetable and goals. In early December, each of us circulated about a dozen problems to the other two. After a decent interval, we circulated solutions to our own problems and comments about the others and added some more into the pot. A few more rounds of correspondence ensued. In March, we met for a weekend with the rest of the Committee to construct the Competition. After a small flurry of additional correspondence, the material was handed over to the logistics team.

The greeting letter of the Chair of the Problems Subcommittee is a quilt to which each chair adds (or rips up) patches. The following excerpts, lightly rewritten for style, are thus simultaneously traditional and fully my own responsibility.

> It used to be said that a Broadway musical was a success if the audience left the theater whistling the tunes. I want to see contestants leave the Putnam whistling the problems. They should be vivid and striking enough to be shared with roommates and teachers.
>
> Security should be a major concern . . . the problems should be handwritten or typed by ourselves, and our files should either be unmarked or kept home. Putnam problems ought to be pretty enough that you want to tell your friends about them. Do your best to resist this temptation.
>
> Lemmas in research papers are fair game, but material from well-known textbooks or problem collections are not. (If you submit a problem from a known source, please include this information.) Seminar material is OK unless undergraduates were present, and anything you have taught to an undergraduate honors class ought to dry out for a few years before gaining eligibility. In general, problems from other people are not reliably secure unless your source can vouch for their originality.
>
> As for the problems themselves, my feeling is that any problem solved by only one or two contestants is a failure, no matter how beautiful it might be. In the last couple of years, we have sought to turn A-1 and B-1 into "hello, welcome to the exam" problems, and their relative tractability has been appreciated. It is better to require one major insight than several minor ones (partial credit is undesirable). It is better to write a streamlined problem without many cases, so that we test perceptiveness, rather than stamina. Proofs by contradiction are, in general, unsuitable, both because they are ugly, and because they are harder to grade.

Although concern for the graders is not our primary consideration, we should keep them in mind. There is no reason to exclude a problem such as 1983 B-2, merely because there are many different legitimate proofs. On the other hand, we must be at pains to write unambiguous questions even at the expense of simplicity in the phrasing. Answers in a particular numerical form are often desirable so that students won't puzzle over the phrase "simplest terms"; this is one reason that the current year stands in for "n". Answers that turn out to be 0, 1 π, $\sqrt{2}$, etc., should be avoided to eliminate the lucky guess, and we should not present problems in which the solution is easy to guess but hard to prove. (The reverse is preferable.) I confess to a predilection for "garden variety" mathematical objects, such as powers of 2, binomial coefficients, pentagons, sines, cosines, and so on. I dislike problems with an elaborate notation, whose unraveling is a major portion of the solution.

WHAT DOES THE PUTNAM MEAN?

The previous three sections have given a theoretical description of how the Putnam works, and what it is intended to accomplish. I turn now to the Putnam in reality.

There is some evidence that the Putnam achieves its intended goals. Many schools run training sessions for contestants, in which interesting mathematics and useful techniques of problem solving are presented. A successful individual performance on the Putnam leads to fame and glory and an increased probability of a fellowship. (However, the results are announced in March, very late for seniors applying to graduate school.) There is also money: I was entertained for years by the William Lowell Putnam Stereo System. More importantly, a contestant can properly be satisfied in solving any Putnam problem, though this is tempered by the (larger number of) problems one does not solve. Putnam problems have occasionally led to research, and a problem may stick in a contestant's mind for years. The ultimate source of Reznick (1986) was 1971 A-1.

The phrase "Putnam problem" has achieved a certain cachet among those mathematicians of the problem-solving temperament and is applied to suitably attractive problems, which never appeared on the exam. One motivation for my joining the Problems Subcommittee was the aesthetic challenge of presenting to the mathematical community a worthy set of problems. In fact, the opportunity to maintain this "brand name of quality" was more enticing to me than the mere continuation of an undergraduate competition. Of course, the primary audience for the Putnam must always be the students, not one's colleagues.

At the same time, the Putnam causes a few negative effects, mainly because of its difficulty. Math contests are supposed to be hard, and the Putnam is the hardest one of all. In 1972, I scored less than 50% and finished seventh. In most years, the median Putnam paper has fewer than two largely correct solutions. For this reason, the first problem in each session is designed to require an "insightlet," though not a totally trivial one. We on the Committee tried to keep in mind that median Putnam contestants, willing to devote one of the last Saturdays before final exams to a math test, are likely to receive an advanced degree in the sciences. It is counterproductive on many levels to leave them feeling like total idiots.

Success on the Competition requires mathematical ability and problem-solving experience, but these are not sufficient; a "Putnam" temperament is also necessary. A contestant must be able to work quickly, independently, and without references and be willing to consider problems out of context. I have been saddened by reports of students who were discouraged in their academic careers by a poor performance on the Putnam. Fortunately for the mathematical community, there are many excellent, influential, and successful mathematicians who also did badly on the Putnam. As a result, the absence of Putnam kudos has a negligible effect on one's career. (At the same time, I confess to enjoying the squirmy defensiveness that the term "Putnam" evokes in some otherwise arrogant colleagues of the "wrong" temperament. They loudly deny an importance to the Competition that nobody else asserts.)

Among those who do very well on Putnam problems, there is little hard evidence that doing extremely well is significant. The best papers usually average about twice as many correct solutions as thirtieth. My impression is that the likely future mathematical outputs of the writers are comparable.

In sum, the Putnam plays a valuable, but ultimately inessential, role in undergraduate mathematics. This is a test; this is only a test.

WHAT MAKES A GOOD PUTNAM PROBLEM?

Other considerations besides pure problem aesthetics afflict the Putnam writers. We wish to have a balance of questions in various subject areas and solving styles. We need an "easy" question for A-1 and B-1. As college math teachers, we are often astonished at how poorly we know what it is that our own students do and do not understand. This is magnified on the Putnam, in which contestants come from hundreds of different programs. The Committee tries to be sensible. It's unreasonable to have the trace of a matrix in one of the easier problems, but we used it in 1985 B-6, on the grounds that a contestant who had not heard of a trace would probably be unable to do the problem anyway.

We want to test, if possible, abstract problem-solving ability, rather than classroom knowledge; maturity "yes," facts "no." We try to avoid the traditional corpus of problem-solving courses to minimize the reward in "studying" for the Putnam. This leads to a tradeoff between familiarity and quality. We occasionally receive a complaint that a problem is not new or has even appeared in material used to train Putnam competitors at a particular school. This is unfortunate, but probably inevitable. It would be easy to write an exam with twelve highly convoluted, certifiably original, and thoroughly uninteresting problems.

Otherwise acceptable questions have been rejected on the grounds that they "sound" familiar or "must have appeared somewhere," even when no member of the Committee can cite a reference. Here's an example: A projectile is to be fired up a hill which makes an angle α with the horizontal. At what angle should the projectile be launched in order to maximize the distance it travels?

One April, the day after the Committee completed its deliberations, someone discovered that our A-1 was a problem posed in the most recent *Two-Year College Mathematics Journal*. Fortunately, we had bequeathed an easy problem to the following year's exam, and a few phone calls resolved the crisis. I do not know what we would have done if this had happened in November.

What follows are some representative comments evoked by the first round of proposed problems in the Subcommittee:

> "Seems routine, too easy."
>
> "I found the computations too messy, and it was easy to head off in the wrong direction."
>
> "Can't use, it's been around for years."
>
> "Hard (or did I miss something). A good problem."
>
> "The trouble with this one is that an intuitive guess . . . is correct. OK, not inspired."
>
> "I was stumped, but it's a nice hard problem."
>
> "I couldn't do this one either."

Every Putnam I helped write contained at least one problem I could not do on first sight. More comments follow:

> "This looks very messy. I saw nothing that motivated me to take up pencil and scribble."
>
> "I wasn't lit up by waves of excitement."
>
> "We must use this one, I love it."
>
> "Yes, yes, yes."

(The last block suggests the sensuality of a good problem to the discriminating solver. For more on this subject, see Reznick, 1984.)

HOW ARE PUTNAM PROBLEMS POLISHED?

The Committee acts by consensus; I do not recall voting once on a problem. Most problems have one primary author, although the full group polishes the final version. Often, though, this version is a special case of the original problem. We tried to have at least one problem from calculus, geometry, and number theory on the exam. I was always amazed at the ability of the Committee to find unfamiliar problems in such fully excavated fields. It is hard to write serious algebra problems that are not basically manipulative, rather than conceptual, because we can assume so little knowledge. In analysis, any integrals must be innocent of measure theory.

We do not accept a problem until we have seen a solution written out in full; sometimes we produce more than one solution. It is not unusual for contestants to find new (and better) solutions. One problem (1983 B-2) evolved from the remarkable fact, familiar to our silicon friends, that a nonnegative integer n has a unique binary representation. Let $f(n)$ denote the number of ways that n can be written in the form $\Sigma a_i 2^i$, if the a_i's are allowed to take the values 0, 1, 2, or 3. (It turns out that $f(n) = [n/2] + 1$. Here, and below, $[x]$ represents the largest integer $\leq x$. This problem has at least three different solutions: by generating functions—$\Sigma f(n) x^n = \{(1 - x)(1 - x^2)\}^{-1}$, by induction—use the recurrence $f(2n) = f(2n + 1) = f(n) + f(n - 1)$, or by direct manipulation—write $a_i = 2b_i + c_i$, where b_i and c_i are 0 or 1, this gives a bijection onto sums $n = 2k + m$, where $k = \Sigma b_i 2^i$ and $m = \Sigma c_i 2^i$. Elsewhere, I (Reznick, 1990) have explored this topic more extensively.

Several times, there was true collaboration. Doug Hensley called me to say that he wanted a problem in which an algorithm terminated, because a certain non-negative integral parameter decremented by 1 after each iteration. This reminded me of a situation I was playing with, in which n was replaced by $n + [\sqrt{n}]$; the relevant parameter is $(n - [\sqrt{n}]^2) \pmod{[\sqrt{n}]}$, which decrements by 1, usually after two iterations. It follows that one eventually reaches a perfect square (see 1983 B-4).

Another time, Mel Hochster had been playing with a problem using the tips of the hands of an accurate clock (as we ultimately phrased it—we received complaints from students who were only familiar with digital clocks!). This was a nice "trapdoor" situation. A reasonably competent student could parameterize the positions of the tips, and after a half-hour of calculation, solve the question. The intended insight was to make a rotating set of coordinates in which the minute hand is fixed, so only the hour hand is rotating. We would have a problem, if only we could find the right question. It occurred to me to look at the distance between the tips when that distance was changing most rapidly. Solving it the long way, I uninspiredly computed an answer that shouted out, "You idiot, use the Pythagorean theorem!" In fact, the derivative of the difference vector from one tip to the other has constant magnitude and is normal to the hour vector, and the distance is changing most rapidly when the difference vector and its derivative are parallel (see 1983 A-2).

The length of service as a Putnam writer seems optimal. In my first year, I was bursting with problems I had saved for the Putnam and discovered that some were unsuitable. In the second year, I tried to rework the leftovers and develop some techniques for consciously writing other problems (these will be discussed below). (A few Putnam rejects have appeared in the *Monthly* problem section, where the lack of time constraints relieve concern over messy or evasive algebra.) By the third year, I felt drained of inspiration; in fact, my impression all three years was that the Chair placed the fewest problems on the exam. Service on the Subcommittee was also beginning to have an adverse effect on my research. Ordinarily, a mathematician tries to nurture a neat idea in hopes that it will grow into a theorem or a paper. I found that I was trying to prune my ideas so that they would fit on the exam. Bonsai mathematics may be hazardous to your professional health!

DID ARCHIMEDES USE δ'S AND ϵ'S?

The story of one of my favorite problems (1984 B-6) serves as an object lesson in theft. In the course of researching the history of the Stern sequence and Minkowski's ?-function, I had run across a beautiful example of Georges de Rham. Let P_0 be a polygon with n sides, trisect each side, and snip off the corners, creating P_1, a polygon with $2n$ sides. Iterate. The boundary of the limiting figure, P_∞, has many interesting counterintuitive geometric properties (see the last paragraph of the discussion of this problem in Alexanderson et al., 1985). For example, it is a smooth convex curve which is flat almost everywhere. I was rather pleased with myself for having noticed a property of P_∞ itself. Suppose one corner, snipped from P_i, has area A. Then each of the two new adjacent corners snipped from P_{i+1} has one-third the altitude and one-third the base of the original corner, and so has area A/9. Further, if P_0 is a triangle, then P_1 is a hexagon whose area is two-thirds the area of P_0. This information can be combined with the formula for the sum of a geometric series with $r = 2/9$ to show that the area of P_∞ is four-sevenths the area of P_0. A sneaky new problem that requires only precalculus—Putnam heaven! I had mentioned this result in a seminar a few years before, but no undergraduates attended, and I was confident of security.

The Tuesday before the competition, I attended a Pi Mu Epsilon lecture on the approaches of Archimedes to calculus, given by Igor Kluvanik, an Australian mathematician visiting Urbana. To my horror, I learned that Archimedes had stolen my method in order to compute the area under a parabola! At least one colleague noticed that I lost my color, and I told her that I could explain the circumstances in about a week. Fortunately, our students did not do unusually well on that problem.

By the way, we stated this problem so that P_0 was an equilateral triangle of side 1 and asked for the area in the form $a\sqrt{c}$, where a is rational, and c is integral. The majority of solvers assumed, incorrectly, that P_∞ had to be the circle inscribed in the triangle and derived an answer involving π. (When I mentioned this problem at a colloquium, a famous mathematical physicist in the audience audibly made the same guess.) The reader might find it amusing to consider the following variation, in which the resulting figure is not a circle, but is piecewise algebraic: Suppose that, rather than a trisection, each side is split in ratio 1:2:1 before the corners are snipped off. Describe P_∞.

WHAT MAKES A GOOD PUTNAM PROBLEM?

The instructions sent to composers do not include a description of the characteristics of a good Putnam problem. The aesthetic seems to be fairly universal among dedicated problem solvers and can be applied more generally to describing good mathematical problems. (Perhaps this reluctance to be specific also reflects the mathematician's cowboy taciturnity on such woolly subjects.)

I will hazard some definitions. A mathematical problem is simply a mathematical situation in which some information is implied by other information. The principal characteristics of a good problem are simplicity, surprise, and inevitability. By *inevitability*, I mean two things: once you've solved the problem, you cannot look at it, without also seeing its solution; once you see the problem, you feel you *must* solve it. (A tacit rite of passage for the mathematician is the first sleepless night caused by an unsolved problem.)

As an undergraduate competitor, I told my housemates that doing well on the Putnam reflected one's ability to do very quickly, other people's tricky, solved problems. I'll stand by that today. Three of the most important preliminary questions a problem solver must face are: (a) Is there a solution? (b) What do I need to know to find the solution? and (c) What does the solution look like? These questions are all answered in advance for the Putnam competitor. You know that there *is* a solution, which is probably short and clever and does not require a great deal of knowledge. You know that you will recognize the solution when you see it. Tables and computer data and other references are irrelevant, and inaccessible in any case. You cannot collaborate, or even describe the problem to someone else in hopes of understanding it in the retelling. It is for these reasons that I am extremely unhappy when I hear that some problem-solving courses use the Putnam as a final exam. Chocolate decadence cake à la mode is a delicious dessert, but makes an unfilling main course.

This discussion begs the larger question of the role of problems within mathematics. Simplicity and surprise may be enjoyable, but they do not accurately characterize much of the mathematician's world, in which correct insights are often wrested after much reflection from a rich contextual matrix and are as snappy as a tension headache.

OKAY, SO HOW DO YOU SIT DOWN AND CREATE A PUTNAM PROBLEM?

Okay, so how do you sit down and create a Putnam problem? One way is to keep your eyes open for anything in your own work that looks like a Putnam problem.

You can make a votive offering to Ineedalemma, the tutelary goddess of mathematical inspiration.

You can also be somewhat more systematic. I was fortunate to have a father who addressed very similar questions in his own work: comedy writing. He wrote a book about writing jokes, from which I take the following quotation:

> Very few writers can pound out a huge batch of jokes week after week relying solely on sheer inspiration. They need the help of some mechanical process. When a writer "has to be funny by Tuesday" he's not going to wait for hot flashes of genius, especially if he doesn't happen to be feeling too "geniusy." . . . Switching is the gag writer's alchemy by which he takes the essence of old jokes from old settings and dresses them up in new clothes so they appear fresh. (S. Reznick, 1954, p. 15)

(Putnam punks such as myself are often inveterate punsters. Punning requires the rapid formal combinatorial manipulation of strings of symbols, without much concern for content. This skill is also very helpful on the Putnam. More serious connections between humor and creativity are discussed in Koestler, 1964.)

The details of switching problems and switching jokes are substantially different, but the principle is the same. The following is one practical illustration of problem switching. I heard a seminar speaker refer to a beautiful result of Mills: There is a positive number α with the property that $p_n = [\alpha^{3^n}]$ is a prime for every $n \geq 1$. The construction is recursive, based on the observation that α must lie between $(p_n)^{3^{-n}}$ and $(p_n + 1)^{3^{-n}}$, and there is always a prime between any two consecutive cubes. It occurred to me that something similar might be wrought out of the simpler expression $[\alpha^n]$. One of the most familiar properties of α^n is that it is always even if α is an even integer (and $n \geq 1$) and always odd if α is odd. The most counterintuitive behavior for $[\alpha^n]$ would thus be for it to alternate between even and odd. If you start with $[\alpha] \geq 3$, then Mills's interval argument will work, and this is how 1983 A-5 came to life.

Although this problem was solved by fewer contestants than we had hoped, perhaps some contestants later realized that the alternation of even and odd is basically irrelevant to the problem, and that any pattern of parities (mod 2) can be achieved using the same proof, as well as any pattern (mod m). Later on, William Waterhouse found an explicit algebraic integer α with the property that $[\alpha^n]$ alternates in sign and submitted this version of A-5 to the *Monthly* Problems Section. I refereed it, and we received dozens of correct solutions; see Waterhouse (1985, 1989).

Another way to create Putnam problems is via Fowler's method. Gene Fowler once explained that it's very easy to write. All you have to do is sit at a typewriter and stare at a sheet of blank paper until blood begins to appear on your forehead. I often applied this technique at less exciting seminars and colloquia, when my neighbors thought I was doodling. I'd take a combination of simple mathematical objects and stare at them until I could see a Putnam problem. Sometimes it worked. For example, 1984 A-4 asks for the maximal possible area of a pentagon inscribed in a unit circle with the property that two of its chords intersect at right angles.

Contrary to popular opinion, it's unhelpful to read through old problem books very much for inspiration, because subconscious plagiarism is a great danger, and our larger audience is very alert. (It might be more useful to look through the back of books, because switches based on solutions are less transparent.)

So how do you sit down and create a Putnam problem? Let's apply Pólya's rules and generalize the question. How do you sit down and create? This is a very difficult and personal question. (It might not even have an answer; our romantic culture tends to identify the results of algorithmic thinking as mechanical, rather than creative.)

In the end, you can do everything you can, rely on the rest of the Committee for inspiration, and visualize two thousand fresh minds on the first Saturday in December, eager to be challenged.

REFERENCES

Alexanderson, G. L., Klosinski, L. F., & Larson, L. C. (1985). *The William Lowell Putnam Mathematical Competition—Problems and solutions 1965–1984*. Washington, D.C.: Mathematical Association of America.

Gleason, A. M., Greenwood, R. E., & Kelley, L. M. (1980). *The William Lowell Putnam Mathematical Competition—Problems and solutions 1938–1964*. Washington, D.C.: Mathematical Association of America.

Koestler, A. (1964). *The act of creation*. New York: Dell.

Putnam Brochure. (1992). *The fifty-third annual William Lowell Putnam Mathematical Competition*. Washington, D.C.: Mathematical Association of America.

Reznick, B. (1984). Review of "A Problem Seminar," by D. J. Newman. *Bulletin of the American Mathematical Society, 11*, 223–227.

Reznick, B. (1986). Lattice point simplices. *Discrete Mathematics, 60*, 219–242.

Reznick, B. (1990). Some binary partition functions. In B. C. Berndt, H. G. Diamond, H. Halberstam, & A. Hildebrand (Eds.), *Analytic number theory: Proceedings of a conference in honor of Paul T. Bateman* (pp. 451–477). Boston: Birkhauser.

Reznick, S. (1954). *How to write jokes*. New York: Townley.

Waterhouse, W. C. (1985). Problem E3117. *American Mathematical Monthly, 92*, 735–736.

Waterhouse, W. C. (1987). Even odder than we thought. *American Mathematical Monthly, 94*, 691–692.

Comments on Bruce Reznick's Chapter

Loren C. Larson
St. Olaf College, Northfield, MN

I have served as the MAA liaison to the Questions Committee since 1983 and have helped prepare problems and solutions for publication. This means that I have had an insider's view of the Putnam exam, so I can attest to the accuracy of Bruce's delightful and engaging account. His thoughts reflect the combination of playfulness, good humor, and seriousness that goes into creating this exam. I like his final injunction to himself as a creator of Putnam problems: "In the end, you do everything you can, . . . , and visualize two thousand fresh minds . . . eager to be challenged." This really cuts to the core of our larger educational task.

I believe that the spirit of the Putnam, as presented so exquisitely in Bruce's paper, could (and should) be communicated to a larger population of students. Therefore, my comments amount to a response to Bruce's summary statement: "In sum, the Putnam plays a valuable, but ultimately inessential, role in undergraduate mathematics." I believe there are ways that the Putnam could be made more valuable and ultimately less inessential. Along these lines, I consider three points: (a) its value as a supplement to coursework, (b) its value as a standard of excellence, and (c) its value as a source of enrichment and stimulation.

Halmos suggests that problem solving is the heart of mathematics (Halmos, 1980), and his point does not need elaboration in this group. Certainly we aim to do more than teach students what they do not know. Having students work on Putnam and Putnam-like problems is one way of meeting some of these more intangible goals. Most Putnam problems are accessible and instructive. Each requires an idea and a clear, coherent argument. Skills developed in doing Putnam problems are not lost in advanced studies.

Most Putnam problems come from classical subjects common to every undergraduate mathematics student. Nearly the entire test is accessible to students who have had the lower division courses: calculus, linear algebra, beginning analysis, and the rudiments of number theory. The objects of interest are the basic building blocks of mathematics: numbers, lines and circles, two- and three-dimensional space, functions, matrices; and the ideas involve limits, convergence, rates of change, area, and linear transformations. These topics are part of the common background of all students, regardless of what directions they choose to pursue in their advanced mathematics coursework.

Working and thinking about Putnam problems should not be an end in itself, but a means to an end, namely, achieving a deeper understanding of fundamental concepts in undergraduate mathematics.

It has been said that the Putnam exam presents a narrow view of mathematics. Obviously, problem solving is broader than Putnam problem solving. Mathematicians of today and tomorrow are more likely to work in teams with today's tools: computers and library resources. Moreover, some mathematicians are not interested in problem solving at all and are mainly concerned with theoretical issues and interconnections between ideas.

One cannot quarrel with these observations, but they really do not affect the Putnam as a supplement to coursework, especially if problem-solving seminars devoted to Putnam problems focus on having students understand the concepts involved rather than on getting answers. An understanding of the concepts included on the Putnam are a prerequisite for advanced courses in mathematics, and the Putnam can deepen this understanding; in that sense, it is a useful supplement.

A STANDARD OF EXCELLENCE

The Putnam exam was set up to test "originality as well as technical competence." The Putnam exam is a way for the professional community to define, in an implicit way, what it regards as important, using the content of lower division mathematics and assuming the mathematical maturity of upper division students. The Putnam aims to acknowledge and reward abstract problem-solving ability (know-how) rather than just classroom knowledge. The objective is to reward those intangible qualities that are associated with creative problem solving: insight, understanding, ingenuity, and so forth. What other national exam attempts to honor those who excel in these ways?

The situation is somewhat similar to that in music education, where aspiring pianists, for example, enter competitions for the sake of improvement. Many of the players, perhaps nearly all, will never make the concert stage, but the competition is valuable nevertheless.

The Putnam makes no pretense at covering all kinds of mathematical talent. It does not purport to measure deep theoretical understanding of a subject or the qualities of mind that are important in mathematical research: perseverance, initiative, powers of concentration over extended periods of time, the ability to ask worthwhile questions that open up new ideas and new vistas, or the ability to frame truly worthwhile abstractions and generalizations of particular and specific results. Nonetheless, it has its virtues. An important component of research is the small problem, such as a technical lemma, which requires one (or several) good ideas. Putnam problems simulate this well, under reasonable time and grading restrictions. The depth of thinking and facility with technique required in doing Putnam problems represent the ideal that our undergraduates should strive toward. The Putnam encourages the search for elegant solutions and clear writing, both worthy aspirations.

As Bruce notes, the Questions Committee must perform a number of delicate balancing acts. It should draw from topics in the standard curriculum, but not reward subject matter study per se. It should feature problem-solving skills and problem aesthetics, but avoid arcana or trick problems. (Most problems are not tricks, but are nuts to be cracked (see below.) And it must be challenging, even for the best students, but not *too* hard.

My view is that the exam committees have succeeded rather well in this balancing act. By and large, the Putnam has been successful in identifying and rewarding mathematical talent among those who take the test. The record shows that most of the top finishers do continue on to distinguish themselves; some at very high levels. (Obviously, the Putnam does not recognize all types of talent. For one thing, it only tests one dimension of excellence, discussed above, and for another, only a small percentage of mathematicians have taken the exam.)

A SOURCE OF ENRICHMENT AND STIMULATION

What is it that attracts us to mathematics? What is it that keeps us interested in a problem for days, weeks, years? What is it that draws us back, time and time again, and renews us with energy? To many, the answer would be the joy of good problems and a desire for complete understanding.

"This is a test. This is only a test." But what a test! It is a treasure! A gift! It is a collection of excellent problems, which have the power to produce pleasure, like music. The Putnam puts a premium on problem aesthetics. It aims to feature new and original problems that are interesting and beautiful and inviting. I like Bruce's way of putting it: "Students should leave whistling the problems."

I know undergraduates whose most memorable and satisfying mathematical experience was that of working a single Putnam problem. The pleasure of intellectual achievement is one of the most important values we can pass on to our students. The Putnam offers us that opportunity.

The Putnam was originally set up to take advantage of the natural instinct some students feel for competition. Competition can give a focus and a sense of excitement. But not all students react positively to competition; some simply do not feel the need to assert themselves in this way. At the present time, less than 1% of eligible students actually take the Putnam. Large numbers of students do not even know about it, but even for those who do, most stay away from it. Of course the Putnam is not for everyone, but with proper coaching and a realistic point of view, more could enjoy it, either by direct participation or in other ways, such as in a class presentation, a problem seminar, a special assignment, a colloquia, an independent study, or in personal reading.

Contestants should know, when they go into the exam, that it is difficult. This past year (1989), the median score was 0 (out of a possible 120), and this is not unprecedented. As a rule of thumb, the number of survivors is cut in half with each additional problem solved. In most years, fewer than 500 students solve three or more problems (of 12).

The upshot of this is that unless students have had considerable experience and training, they probably will not be among the leaders. The same could be said for running a marathon. Very few who run a marathon have any hope whatsoever of finishing among the leaders, even in their age group. The point is, runners set their own goals and participate to the best of their abilities. Perhaps we should think of taking the Putnam in the same way. For many students, the goal might be to get the "easy" problem in both the morning and afternoon session. The important thing is that students can set their own expectations and complete on their own terms (against themselves). The exam can be taken on a number of levels. Every finisher is a winner.

Can students who fail to work a single problem feel good about the exam? My answer is a bit idealistic, and there may be differences of opinion, but I say it depends on expectations. A coach needs to reassure them, in advance, that there are benefits to working hard on a problem for three hours, even if that problem is not solved. In the process of thinking about a problem, students may generate a number of interesting questions, they will be making connections, and they will be primed for learning and won't likely forget the solution when they see it. If students are taught to experience the joy of the problems, that is their success. And this is the case for every student; for to miss the joy is to miss everything.

AN EXAMPLE

I have found that many people have misconceptions about the nature of Putnam problems. So in this section, I'd like to consider a problem, Problem B-5, from the 1989 exam. This problem is not a particularly clever problem, nor does it require a particularly clever idea, but it is typical in that it goes slightly beyond the routine textbook setting and gives some opportunity for creative thinking.

> **B-5.** (1989). Label the vertices of a trapezoid T (quadrilateral with two parallel sides) inscribed in the unit circle as A, B, C, D so that AB is parallel to CD and A, B, C, D are in counterclockwise order. Let s_1, s_2, and d denote the lengths

of the line segments AB, CD, and OE, where E is the point of intersection of the diagonals of T, and O is the center of the circle. Determine the least upper bound of $\frac{s_1 - s_2}{d}$ over all such T for which $d \neq 0$, and describe all cases, if any, in which it is attained.

This problem should look familiar to anybody who has studied calculus. It is a variation on a common theme: "Among all figures that can be inscribed in another figure, find the one (or ones) that maximize (or minimize) some property." In this case, we consider the set of trapezoids that are not rectangles ($d \neq 0$) and among those that can be inscribed in the unit circle, we wish to describe those (if any) that maximize the quantity $(s_1 - s_2)/d$.

Even though this problem has a familiar contextual setting, it suffers from not having immediate appeal. The quantity $(s_1 - s_2)/d$ seems artificial and contrived. It isn't a natural question. Why should we be interested in this parameter? Does it have a geometrical significance? Is it related in some way to area, or slope, or perimeter?

As in many problems, the problem (treasure) starts to take on intrigue when one begins to sketch out a few cases. We may assume that the bases of the trapezoids are horizontal. There are two quantities of interest: (a) the difference between the length of the bases, $s_1 - s_2$, and (b) the distance d between the origin and the point at which the diagonal crosses the y-axis. We need to investigate what happens to these as the bases AB and CD slide up and down the y-axis. As CD moves up, $s_1 - s_2$ increases, but so does d. If we let it reach the top, and then push AB up to the diameter, we have the degenerate case in which the trapezoid is really a right triangle. In this case, $(s_1 - s_2)/d = 2$. Is this a maximum?

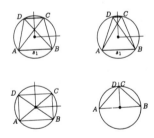

As CD moves down to make a rectangle, $s_1 - s_2$ decreases to 0, but so does d. What is the limit of this ratio as the trapezoid approaches a rectangle (that is, as d goes to 0)? Perhaps there is no least upper bound.

After considering a few special cases, we discover the following.

1. The trapezoids are isosceles, and the intersection point E lies on the y-axis.
2. We may assume that one base is above the x-axis and the other is on or below. This is because reflecting one of the bases across the horizontal

axis does not change the value of $s_1 - s_2$, but the larger trapezoid corresponds to the smaller d value.

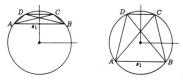

3. The trapezoids form a two-parameter family. In some way or other, locating the two bases of the trapezoid will require two pieces of information.

There are many ways to solve this problem, depending to a large extent on how the parameters are identified. We will sketch five solutions.

Solution 1

Probably the most immediate and (seemingly) natural parameterization for most students is the following.

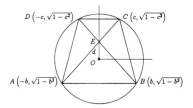

Here, the x-coordinates of B and C are real numbers b and c, $0 \leq c < b \leq 1$. With this beginning, one finds that

$$\frac{s_1 - s_2}{d} = \frac{2(b^2 - c^2)}{b\sqrt{1-c^2} - c\sqrt{1-b^2}}$$

$$= 2\left(b\sqrt{1-c^2} + c\sqrt{1-b^2}\right). \quad (1)$$

Variation 1. From (1), we can take the two partial derivatives, set them to 0, and solve. This is not a particularly elegant way to do the problem, but nonetheless, one can carry it out, and we find that the least upper bound is 2; it is obtained precisely when $b^2 + c^2 = 1$.

Variation 2. A student who has mastered the art of "maximization without calculus," could proceed in (1) by applying the Cauchy-Schwarz Inequality,

$$2\left(b\sqrt{1-c^2} + c\sqrt{1-b^2}\right) \leq$$

$$2\left(\sqrt{b^2 + (1-b^2)}\sqrt{(1-c^2) + c^2}\right) = 2,$$

with equality if and only if $bc - \sqrt{1-b^2}\sqrt{1-c^2} = 0$, $b \neq c$, which is equivalent to $b^2 + c^2 = 1$, $b \neq c$.

Variation 3. If we switch to polar coordinates, and write $b = \cos\varphi$ and $c = \cos\theta$, we find that

$$\frac{s_1 - s_2}{d} = 2\left(\cos\varphi\sin\theta + \cos\theta(-\sin\varphi)\right) = 2\sin(\theta - \varphi) \leq 2,$$

with equality if and only if $\theta - \varphi = \pi/2$; that is, if and only if $\angle COB$ is a right angle. This gives a different characterization of the maximal trapezoids.

Solution 2

Here, the idea is that the quantity $(s_1 - s_2)/d$ is independent of the size of the circle, so consider the following figure, where C and D have coordinates $(-1,0)$ and $(1,0)$ respectively, and A and B have coordinates (a,b) and $(-a,b)$, $a > 1$, $b > 0$.

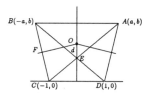

From the coordinates of F, the midpoint of BC, we find the coordinates of O (by writing the equation of the line through F perpendicular to BC) and the coordinates of E (from line AC). This leads to an expression for d, and we find that

$$\frac{s_1 - s_2}{d} = \frac{AB - CD}{OE} = \frac{2a - 2}{\dfrac{b}{2} + \dfrac{a^2 - 1}{2b} - \dfrac{b}{a+1}}.$$

Now, instead of maximizing this expression, it is equivalent to minimize the reciprocal. For this, we find that

$$\frac{b}{4a - 4} + \frac{a+1}{4b} - \frac{b}{2(a+1)(a-1)} = \frac{b(a+1) - 2b}{4(a+1)(a-1)} + \frac{a+1}{4b}$$

$$= \frac{1}{4}\left(\frac{b}{a+1} + \frac{a+1}{b}\right)$$

$$\geq \frac{1}{4}(2),$$

the last step by the arithmetic mean–geometric mean inequality, with equality when and only when $b = a + 1$. Geometrically, this condition means that the diagonals meet at right angles (because then, line AC has equation $y = x + 1$, and *BD* has equation $x + y = 1$, and the slopes, 1 and -1 respectively, are negative reciprocals). The case when *O* and *E* coincide, precisely when $a \neq \pm 1$, $b \neq 0$, must be excluded. This is a third characterization of the maximal trapezoids.

Solution 3

One of the nicest parameterizations for this problem is based on fixing *d* and then varying the angle θ between the diagonal and the *y*-axis. The trick is to figure out how to express $s_1 - s_2$ in terms of *d* and θ. By using either the Law of Cosines or the Law of Sines, one finds that $s_1 - s_2 = 2d \sin 2\theta$.

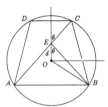

From this, it follows that the upper bound of 2 is attained for those trapezoids that are not squares precisely when $\theta = \pi/4$.

Solution 4

Perhaps the most elementary solution is based on the following diagram. Here, *H* is the foot of the perpendicular line from the origin to the diagonal *BD*, and *F* is the projection of *H* onto the *y*-axis, $a = DE$ and $b = EB$.

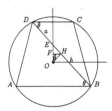

From the figure, $EH = (a + b)/2 - a = (b - a)/2$, and therefore

$$s_1 - s_2 = 2b \cos \theta - 2a \cos \theta = 4 \left(\frac{b - a}{2} \right) \cos \theta$$

$$= 4 \, EH \cos \theta = 4FH.$$

As θ varies, for a fixed value of *d*, *H* traces a circle with diameter *OE*. Thus, $s_1 - s_2$ is maximized when $FH = d/2$. It follows that the maximum value of $(s_1 - s_2)/d$ is 2, and it occurs precisely when $\theta = \pi/4$. This is a fourth characterization of the maximal trapezoids.

Solution 5

Using the same notation as in the previous solutions, let A and C be opposite vertices of the trapezoid on the intersection of the diagonal line $y = mx + d$ and $x^2 + y^2 = 1$. Then,

$$x^2 + (mx + d)^2 = 1,$$

$$(m^2 + 1)x^2 + 2dmx + (d^2 - 1) = 0,$$

$$x^2 + \left(\frac{2dm}{m^2 + 1}\right)x + \frac{d^2 - 1}{m^2 + 1} = 0.$$

Now (consider the x coordinates), $s_1 - s_2$ is twice the sum of the roots of this equation, and therefore

$$s_1 - s_2 = 2\left(\frac{2m}{m^2 + 1}\right)d \leq 2d,$$

with equality if and only if $m = 1$. Thus, for $d \neq 0$ the maximum value of $(s_1 - s_2)/d$ is 2, and it is attained precisely when the diagonals intersect at right angles, and the trapezoid is not a square.

About 800 students attempted this problem (of the approximately 2,100 students), and about 200 solved it correctly. The problem tests a standard optimization routine that is part of the mainstream (calculus), yet it leaves room for considerable creativity. The lessons learned in seeing different approaches, some much more elegant than others, adds interest to the problem. For those who do computer calculus, it should be pointed out that a computer solution would involve the same degree of creativity; namely, one must identify the relevant parameters and express the function of interest in terms of those parameters.

The problem might have been made superficially more difficult had it been posed for an arbitrary convex quadrilateral inscribed in a circle.

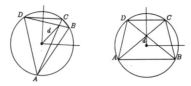

Here, $s_1 - s_2$ is unchanged as AB slides about the circle, and d is minimized when the figure is a trapezoid, and we're back to problem B-5 as stated.

REFERENCES

Halmos, P. R. (1980). The heart of mathematics. *The American Mathematical Monthly*, 87, 519–524.

A Discussion of Bruce Reznick's Chapter

Ingram Olkin
Alan H. Schoenfeld

> **Ingram Olkin**, *Stanford University*
> **Bruce Reznick**, *University of Illinois*
> **Alan H. Schoenfeld**, *University of California, Berkeley*
> **Ed Dubinsky**, *Purdue University*
> **Lester Senechal**, *Mount Holyoak College*
> **Gaea Leinhardt**, *University of Pittsburgh*
> **John Addison**, *University of California, Berkeley*
> **Robert Davis**, *Rutgers University*

Ingram Olkin. One picks up tricks of the trade over time, and one that you have to develop is what to do when you are scheduled to be a discussant at a paper that you haven't seen, where you have no idea of format or what is being planned. I learned how to deal with this problem at a meeting of the Royal Statistical Society. The procedure is for a speaker to give a scientific lecture, after which there are a number of discussants. The first and second discussants play a special role in that the first discussant proposes a vote of thanks, and the second discussant endorses the vote. The beginning of such a discussion is something such as "I am pleased to offer a vote of thanks." This is followed with a large *but* or *however,* after which he or she feels perfectly comfortable tearing the paper apart.

Once when I attended a meeting, Frank Anscombe of Yale University had just arrived from the United States. The chair said, "We have an old friend from the United States who used to be here, and I wonder if he would like to comment on the paper," which happened to be in his area. Anscombe was nonplussed and said, "Many years ago when I used to frequent these halls, it was not unusual for the discussant to speak on a topic totally alien from that of the speaker." He then launched into a presentation of his own work.

So I now feel free to deviate somewhat from the main topic. I was the third member of that coffee klatsch at which Alan Schoenfeld, Jim Greeno, and I talked about setting up a conference such as this. The motivation stemmed from the fact that there are three main constituencies concerned with mathematics education: the research mathematician, the mathematics educator, and the cognitive psychologist. These groups are somewhat isolated from one another. They generally distrust each other and often don't believe that the other constituencies have anything to contribute. Indeed, it is rare for a meeting of cognitive psychologists to invite a research mathematician to air his or her views. It may be even more rare for the reverse. Thus, what this conference can accomplish is to provide an opportunity for the various groups to come together. If we do nothing else but listen to one another, the conference will be a success in that it cannot but help alter some previously held conceptions or misconceptions.

A repeated theme that I hear in many talks—not just today, but at other conferences—is an emphasis on case studies. A special curriculum, for example, will be given to a class, and observations of its effect will be made. As a statistician I need to know the larger population from which this sample is taken, and to which group generalizations will be made. If this is not done, then this experiment becomes an isolated observation. One procedure used in science is to randomize individuals into a control and experimental group. I urge that experiments be thought of in this light.

Now, back to the Putnam. I was very interested in the talk, and the jokes were first rate. A point that struck me was that the sum of ranks was used in arriving at a ranking of the competition. This raises an interesting statistical question, namely, how many questions should you have on a test to be able to distinguish between individuals. This problem, or variants thereof, are well known in a variety of contexts. For example, the world series with a fixed number of seven games is a poor discriminator in the sense that if the teams are close in ability, as is usually the case, there is a sizeable likelihood that the poorer team may win. As the number of games increases, this likelihood decreases. So, it would be interesting to estimate the proportion misclassified. This could be studied with external criteria.

I question how good the Putnam is in terms of predicting research ability. In our own case at Stanford, I don't think that we have a great track record in choosing candidates we expect to be successful. Of course, here the definition of success is at the heart of how well we do.

"The Putnam temperament" is an interesting phrase and perhaps points to the central issue. I have just read Richard Feynman's book *Surely You're Joking, Mr. Feynman*. He was a member of a mathematics club which was similar to the Putnam in that there was a mathematics contest every or every other Saturday. Feynman made the comment that what he noticed was that the problems took 50 seconds to solve, but they were given only 40 seconds. So the question was how can you see the answer in less time? There is a certain type of insight that is different—hard to describe—but is central to solving problems. To some degree insight involves always questioning how do you know. In Feynman's book he goes to an extreme to test whether ants walk in a straight line by putting sugar in the bathtub and observing their path. But the question—how do you know—is the essence of a proof. We need to understand the mechanism by which individuals raise such a question.

In mathematics we use the word "beautiful" to refer to problems, theories, or solutions, and the term is generally understood. But the inexperienced individual doesn't know what beautiful means. Other terms that you use in the paper are simplicity, surprise, inevitability, or want to know the answer. Again, there is general understanding of these terms by the mathematical community. I think that the research mathematician is consumed by the need to know the answer. One question that we should examine is whether the nonmathematical individual who is trying to solve a problem wants to know the answer. That is, is there a real need to know?

You mentioned that a problem solved by only one or two contestants is not a good problem. This reminded me of a trip I took with Pólya. I was driving him to San Luis Obispo to a mathematics association meeting, and I asked him how he decided whether a dissertation problem is a good one. He said that he first solves the problem and notes how long it took him. If he solves it within a certain period of time, then he knows that it is both feasible and a thesis topic. I don't know that I agree with this procedure, but it relates to one of your points in how to teach a student. One procedure is to ask a student to generate new problems from a given one. In order to do this the student will have to understand what makes the problem and solution work.

Bruce Reznick. I was interested that you brought up the ranks on the Putnam, because I once did a study which I didn't pursue. We got the complete list of the point scores and the rank, and I plotted the log of the rank against the point score. It was virtually linear for most of the entire range. Depending on the exam, each problem reduced your rank by a factor of 2.1 or 1.8 or something like that, but it was consistent from one to the other. I don't know what model of distribution of abilities that corresponds to. I asked a statistician once and I didn't get an answer, and I didn't pursue it.

Ingram Olkin. Well, it's interesting at least that there is a good relationship.

Bruce Reznick. The relationship held for the years I looked at it.

Ingram Olkin. It probably would change if there were more questions or fewer questions, so it may be that the number you're asking is about right.

Bruce Reznick. Another point about the Putnam temperament is the fact that many "Putnam punks" like myself are inveterate punsters. This is a clearly recognized trait. Both activities are really manifestations of the same thought process, a combinatoric manipulation of symbols without too much concern for content.

Alan Schoenfeld. Bruce Reznick and Loren Larson's comments on the Putnam exam provide a nice opportunity for a Type A *versus* Type B comparison of problem-solving goals and styles. On the one hand, the Putnam exam is, in many circles, viewed as the ultimate test of (a particular kind of) mathematical problem solving. And, as Bruce and Loren have noted, Putnam problems have a very particular flavor. On the other hand, over the past 15 years I have offered a number of courses in mathematical problem solving. The goals of my courses differ substantially from Putnam goals, and for a variety of reasons, my courses have made little use of problems taken from past Putnam exams. I thought it might be of interest to pursue a comparison of (a) the goals for each type of experience, and (b) the means we use to get there, namely the problems (and thus the criteria for their selection). How much are we in the same game, how much not?

By way of introduction, I should point out that "problem solving" means many things to many people. A 1983 survey, for example (Schoenfeld, 1983), indicated that five rather different sets of courses all marched under the problem-solving banner:

1. Seminars to prepare students for competitions such as the Putnam.
2. Courses designed to "provide my students with an introduction to what it means to think mathematically."
3. Courses for future teachers of mathematics, with an emphasis on learning to solve problems, so that one could then teach students to do so.
4. Courses in mathematical modeling.
5. Remedial courses, in which slightly nonstandard problems were used as a means to help students "develop basic thinking skills."

Clearly, the problem sets used in some of these course types—for example, numbers 1 and 5—are virtually if not completely nonoverlapping. But what about types 1 and 2? As it happens, my courses are of type 2. How much do Putnam-type problems have to do with the problems chosen to help students "learn to think mathematically?"

Let me say a bit more about the history of my courses and my goals for them. I began developing problem-solving courses in the mid-1970s. When I did, I had to decide the level at which they would be offered. Should they be at the doctoral level, say, in preparation for qualifying exams? While such courses have been rewarding (and some, for example, those given at Stanford in the 1960s, are legendary), it seemed to me that for the most part doctoral students can and should fend for themselves. That is, if they really needed instruction in problem solving above and beyond their course work, there is some reason to be concerned about their potential as mathematicians. Dropping down a level, what about a seminar to prepare students for the Putnam? I decided against that too. As it happens, I'm not especially good at Putnam problems; I tend to be better at chewing over larger problems, more slowly. Odds are that I couldn't give much help, if any, to budding Bruce Reznicks and Loren Larsons; those folks have a talent that I don't have. But more importantly, my goals were to reach a larger audience, with a somewhat different purpose. By the mid-1970s it had become clear to me that the vast majority of our mathematics majors, by their junior or senior years, have no real idea of what it is to do mathematics. Broadly speaking, they have spent their time, through all of elementary and secondary school, through calculus, and most likely their undergraduate courses in differential equations and linear algebra, simply mastering techniques and working exercises. For most of them, doing mathematics has meant studying material and working tasks set by others, with little or no opportunity for invention or sustained investigations. The joy of confronting a novel situation and trying to make sense of it—the joy of banging your head against a mathematical wall, and then discovering that there might be ways of either going around or over that wall—is one that many mathematics majors don't experience until late in their careers. Indeed, some don't meet it until their second or third year of graduate school, after fording their way through difficult but still "this is the content you have to know" graduate courses in real analysis, complex analysis, and algebra. I wanted my students to have more of a sense of mathematical *engagement* than they have in standard courses.

My first course was offered at Berkeley in 1976, at the upper division level. It was interesting to note that very advanced students—many were taking graduate courses and doing well—could fail to solve very simple problems such as "prove, noting the conditions under which the relationship holds, that if a central angle and an inscribed angle in a circle subtend the same arc, then the measure of the central angle is twice that of the inscribed angle." Out of context, simple problems were hard, when students had forgotten the tricks used to solve them. Many of those tricks were applications of Pólya's heuristics—strategies I did (and do) believe that most mathematicians pick up in the course of their careers, but which these students hadn't yet learned. My early problem-solving courses focused on problems amenable to solutions by Pólya-type heuristics: draw a diagram, examine special cases or analogies, specialize, generalize, and so on.

Over the years the courses evolved to the point where they focused less on heuristics per se and more on introducing students to fundamental ideas: the importance of mathematical reasoning and proof (which many students did not take seriously), for example, and of sustained mathematical investigations (where my problems served as starting points for serious explorations, rather than tasks to be completed). Masochists with a thirst for detail will find the most extensive descriptions of early versions of the course in my book *Mathematical*

Problem Solving (1985). My chapter in this volume, "Reflections on Doing and Teaching Mathematics," describes some of the things I focus on today. The most recent versions of my course have been offered at the lower division level. Here's why. Students who are not going to be mathematics majors, or who don't need mathematical tools for use in the sciences, profit from the course because it offers them an opportunity to think—an opportunity generally not present in cookbook courses like the standard calculus course. It gives them a taste of what it is to do mathematics, a taste that they would otherwise be deprived of. The mathematics majors—and a substantial number of them enroll in the courses—are given the opportunity and leisure to play with ideas, to invent, and to pursue intriguing notions to see what they might contain. This often gives them a sense of what it is to do mathematics, long before they have the opportunity in standard courses. (It is clear from Loren's comments, and at least implicit in Bruce's, that this can be a property of Putnam seminars. The best students get to play with the mathematics, to whistle the problems, and maybe even to improvise on their melodies.)

As my problem-solving courses have evolved, so has my idea of what constitutes a good problem for them. My *problem aesthetic* consists of five criteria by which I judge the potential usefulness of a problem for my course. The basic idea is summarized in Table 1. An elaboration follows, although most of the items should be largely self-explanatory.

First, my preference is for "easy access" problems—problems that require little by way of formal background, or specialized knowledge or methods, at the entry level. The reasons should be obvious: I don't have to provide extensive background information, and students are not handicapped by differential backgrounds. As noted above, this does not mean that the problems we work on are necessarily trivial. Out of context, rather simple problems can prove to be surprisingly challenging. (A case in point: readers might want to make their way through the geometric construction problems in Chapter 1 of Pólya's *Mathematical Discovery* [1981].)

Second, "multiple access" problems have some nice properties. They allow me to point out to students that there's often more than one way to skin a mathematical cat, and that what counts is more than simply getting an answer,

TABLE 1
A Problem "Aesthetic" for Problem-Solving Courses

Good problems should:
1. Be accessible (not require a lot of machinery).
2. Be solvable (or at least approachable) a number of ways.
3. Illustrate important ideas (good mathematics).
4. *Not* have trick solutions.
5. Be extensible and generalizable (lead to rich mathematical explorations).

but seeing connections or mathematical ideas. This point is clearly an important part of both Bruce Reznick and Loren Larson's problem aesthetics: Bruce mentions a problem with three solutions, and Loren presents five solutions to his exemplary problem. (In class I noted the large number of extant proofs of the Pythagorean theorem, for example, and that any member of Berkeley's mathematics department would be happy to find a new proof. This is one of the first salvos in my attempt to get students beyond the "obtain the answer and move on" syndrome.) In addition, the fact that one might try *lots* of different things on a problem, only some of which will prove successful, serves as an opportunity to raise "logistical" (self-regulatory, or "control") issues: How do you decide what to pursue, and for how long, before considering other options? (For detail, see chapters 1, 4, and 9 of my *Mathematical Problem Solving* [1985].)

The third and fourth themes are closely related. On the positive side, I want the problems to serve as vehicles for introducing students to some real, honest, valuable mathematics. That means that either the topics themselves should be valuable (and we can make good progress on them, illustrating important ideas), or that the reasoning patterns involved in solving the problems are. This latter category splits two ways. On the one hand, there are generically useful mathematical modes of thought: various types of proofs, abstraction, representation, etc. On the other, some problems provide nice contexts for illustrating particular heuristic strategies and for showing how one might make progress on problems that, at first pass, seem intractable. In contrast, I try to avoid problems with trick solutions.

From the audience. What is a trick solution?

Bruce Reznick. I define a trick as a technique you haven't seen before.

Alan Schoenfeld. . . . and are not likely to use again.

Fifth and most important, I think of my problems as starting points, as introductions to potentially rich mathematical explorations. My goal is to give my students an opportunity to *do* mathematics, and the problems I choose are intended to facilitate that goal. I'll mention two such problems here; they've been discussed extensively elsewhere.

One problem, borrowed from Steve Brown and Marian Walter's *The Art of Problem Posing* (1990), is to ask students to pursue extensions of the Pythagorean theorem. (See my chapter "Reflections on Doing and Teaching Mathematics" for details.) The one-liner is that my students wound up conjecturing and proving some (relatively minor, but nontrivial) results that I didn't know. Another problem starts with my asking students to find a 3×3 magic square. Once they've found it, we begin with the interesting questions. Is their solution unique? What if we used integers other than 1, 2, 3, 4, 5, 6, 7, 8, 9? How about 31 through 39? How about 7, 14, . . . , 63? How about any arithmetic sequence? Those problems are trivial, and the students soon realize that if S is a magic square with the digits 1 through 9, then—using the obvious matrix notation—$(aS + b)$ is a magic square. So, you can generate infinitely many magic squares from the basic square. Here's a tougher question. Suppose you have a 3×3 magic square—a set of nine digits entered into a 3×3 matrix with the property that the sum of each row, column, and diagonal of the matrix is the same. Must this square be of the form $(aS + b)$, where S is a magic square with the digits 1 through 9? And so on. The point is that there's lots of honest-to-goodness mathematics involved here: The original problem merely serves as a jumping off point for mathematical investigations.

This last criterion contrasts in an interesting way with the criteria laid out by Bruce Reznick.

> Three of the most important preliminary questions a problem solver must face are these: (a) Is there a solution? (b) What do I need to know to find the solution? (c) What does the solution look like? These questions are all answered for the Putnam competitor, who knows that there is a solution, which is probably short and clever and does not require a great deal of knowledge.

At the same time, it should be noted that Bruce's comments apply to the problems in the context of the exam itself. As both Bruce and Loren point out, at least some Putnam problems can serve as the starting points for extensive investigations—and some of the problems haunted Bruce for years, with productive results. I think Bruce hits the nail on the head at least twice. First, there's his comment that the principal characteristics of a good (Putnam) problem are simplicity, surprise, and inevitability. These certainly resonate with the mathematician in me, and I expect with most mathematicians. There is an "Occam's razor" aesthetic among mathematicians, which looks for clean and piercing arguments; it's no accident that "slick" and "elegant" are words of high praise in the mathematical community. Simplicity counts, as does inevitability—that means there's something solid, and not frivolous, about the problem. As for surprise, and the fact that so many Putnam problems have a bit of all three, Bruce's second comment says it better than I could: "Chocolate decadence cake à la mode is a delicious dessert, but makes an unfilling main course."

Bruce Reznick. I want to raise one issue with the research mathematicians here. The question of a problem leading to another problem leads to some unpleasant questions. Are there people who solve problems who even care about them once they solve them? There are many people for whom a problem is no longer interesting once it's solved—it's almost like an earthworm going through the earth and perhaps enriching it a little bit, but once you're done you forget you were ever through this particular patch of soil. That particular attitude is found more than just among the problem solvers. I was at the Toronto AMS meeting about seven years ago, where Dieudonné was talking about Bourbaki. He was asked why there was no Galois theory in Bourbaki, and he said, "Because it doesn't lead to any interesting questions." This seemed to me to be such a sterile idea of what the beauty of mathematics was that, although I didn't like Bourbaki before, I decided at that point that I'd never buy the books, even if they were in paperback.

Ed Dubinsky. I'd like to make a comment. It seems to me, having a conference about mathematical thinking, and in particular with a great interest in mathematics education, that talking about the Putnam exam really raises a very important question, and it's one about which I feel I don't know very much at all. I think it's very important and needs research. The question is, is there one kind of mathematical knowledge, or are there two?

The issue of concern for most professional mathematicians is the continuation and preservation of the species: the education and production of first-rate research mathematicians, people who are from the beginning very talented in mathematics. A second issue, perhaps the fundamental issue of concern for mathematics educators, is that of mathematical literacy—the raising of the level of mathematical understanding among the general population that will be necessary in the future. Those are two fundamental issues.

I think what we really need to ask, in the context of this situation, is: What are the connections between the two? To what extent are those two questions the same? Where are they really very different questions? Although I have some experience in both these areas, I find myself totally at a loss. I need to know more about what mathematical knowledge is, and I think that issue should be part of our long-term agenda.

Lester Senechal. I'd like to comment on the value of the indicator. I think the Putnam is a much better indicator than you might suppose, but I think it's a good indicator for men only. I think the Putnam is a very male-oriented activity. I don't think this is a good way for women. I myself, throughout my education in high school, enjoyed taking examinations, and I supposed for a long time that this was true of everyone. It's gradually dawned on me that my women students, and I teach women, don't enjoy examinations. They're terribly threatened by them. I think there is something very threatening about the Putnam that applies to women, and I'm afraid this might apply to minority students as well. I think the Putnam has a very great value, but we also have to be aware of that side of it.

Gaea Leinhardt. I want to return to a theme raised by Ed Dubinsky, which has to do with the general literacy goals we have for mathematics instruction. One thing I've been thinking about is the "pool problem," and it relates directly to this conference. What I mean by the pool problem is, you need to expand the pool of mathematically competent citizenry.

The United States needs to get its base level above fourth grade in mathematics. I don't care what reforms one is talking about. At the moment, elementary school teachers have about a fourth-grade competency. That is, they are fairly competent at fourth-grade manipulations, but they have real serious trouble with anything beyond that. That suggests that we need a tremendous enriching of the mathematics knowledge base, from that level up. I think all of those enrichment activities will automatically have an effect on the more selective group. You can't help but increase the number of people who get much further, on the assumption that part of our problem is caused by the fact that some people—often minorities and women—have been eliminated too early and inaccurately. I think this is a very critical issue and it requires, I think, the kind of activities that you just described taking place at much lower educational levels.

The language that we heard in response to Judah Schwartz's presentation suggested some notions about what children ought to be thinking. I think we ought to concern ourselves, in exactly the same way, with the way teachers are thinking. We need to expose teachers to a comfortable and safe mode of exploratory inquiry. Unless they themselves feel comfortable with inquiry, they won't feel comfortable allowing their students the freedom to engage in mathematical inquiry.

Alan Schoenfeld. The pool problem is quite serious. Here's one indication. Two recent NRC (National Research Council) publications, *A Challenge of Numbers* (1990) and *Renewing U.S. Mathematics* (1990), show just how bad the problems are. Here's one set of figures. Twenty years ago, entering freshman declaring themselves as math majors were 4.6% of the population. Those who stuck it out to be math majors were less than 2%. Today the entering percentage of those declaring themselves to be math majors is 0.8%.

John Addison. I think those figures are a little misleading, because although I don't have data on this, I would guess that there is almost a complete switch in the population of incoming math majors versus the ones who go out. An enormous number of people come in thinking they're going to major in math based on high school math and find out, of course, that math is not like what they learned in high school. Indeed, one may end up almost rerecruiting the math major. I'm not sure how significant the number coming in is for us. I think the more significant question is how do we recruit math majors in the university.

The question that I really wanted to raise was this. I wondered what Bruce or anybody else thought the consequences of the Putnam exam are for mathematics. You read to us what the goals and the original intention were. What are the consequences of the fact that we have the Putnam exam? Are there consequences for mathematics? Take the question at any level you want.

Bruce Reznick. I don't think there are major consequences. I think there is a certain enriching problem pool. There are some departments that are probably inspired to spend more time talking about problem solving. There are some students who are inspired to devote a lot of energy to problem solving. Pathology is too strong a word to use for heavy Putnam involvement. I don't know what the right word is, but it characterizes someone who is willing to devote a lot of time to mathematical problems on their own. That is basically a prerequisite to doing a competition like this: You really need to have done a lot of problems and to enjoy doing these things on your own. Such behavior is at the very least a social aberration, and I suspect it's an aberration that is much less well tolerated among girls than among boys. I don't know that for sure. Speaking narrowly, I don't think that the content of the exam is gender related. The fact that it is a 6-hour exam on one day and you can't work with anybody else might prove to be more significant. I just don't know.

Gaea Leinhardt. I don't think it's gender related in the way you suggest. I watched six 11- and 12-year-old girls spend almost the whole summer this past year writing 200–300-page novels. These were completely solitary activities. The girls happen to be very verbal, and I don't want to stretch on that point at all. I think that you will find a collection of females at early adolescence that are solitary creatures as well. I don't think that's the discriminating factor.

Bruce Reznick. And at 18 or 19?

Gaea Leinhardt. There too.

John Addison. As Bruce notes, you can see the Putnam glass as being half empty or half full. I happened to have one three-time Putnam winner as a student in a graduate course. Nothing has really ever happened in his career, and that's certainly interesting. Far more impressive to me, however, is the number of famous mathematicians who are on the list of Putnam winners. I think there really is a much bigger correlation than one would like to admit sometimes. I did a quick check. In the Berkeley math department in our algebra combinatorics section which isn't all that big we have five Putnam fellows.

From the audience. How big is the math department?

John Addison. Seventy-three, but probably only 10 or 15 that you would say are related enough to algebra to be in that category.

Ingram Olkin. I think your numbers show that it's just the other way around. I would expect many more to be Putnam fellows. That is, take the *Transactions of the American Mathematical Society* and look at the number of people who published in the last 10 years and see how many are Putnam prize winners. I suspect that number is not great relative to the total population.

John Addison. Well, there is this curious thing that maybe is unfortunate, that it seems to be turning out more people in algebra and combinatorics. However, I'd say 5 out of 10 is pretty impressive. It raises the interesting question, of course, about whether many of these Putnam people started competitions in high school. There is now a three-tiered series of examinations. It's interesting to ponder whether or not this long series of experiences that most of those people have had including the Putnam has actually assisted at all in their mathematical careers. Is this something that these people would have done equally well on anyway, or is it really making any kind of positive input.

Bob Davis. Lurking under a lot of this discussion is a lot of the substance of Judah's presentation and Andy's remarks in reaction. Ed Dubinsky pointed out earlier that there are two fundamental issues we face—the preparation of research mathematicians and general mathematical literacy. He wondered how much they have in common. An interesting question that has bothered me for some time is: Are there two things known as mathematics or in some sense one?

The schools that I looked at are fixated on things like 9th-grade algebra, where you spend a certain amount of days combining like terms. In those schools, and in most I think, mathematics is a very specific thing. You are always told what to do; you are told exactly how to do it; and you're in big trouble if you do it any other way. This is known as mathematics. The basic thing is following instructions, doing what you were told, and remembering what you were told. It's a quite different activity than some of us, myself included, have thought of as mathematics. Some of the Putnam problems provide examples of this more freeform mathematics. They're quite different from school math problems, in that you will not do any of these problems by doing what you were told to do or something closely resembling it.

There are a lot of positions that you could take on the relationship between the two kinds of problems. A very popular position says that you have to go through all the do-what-you-are-told stuff, and eventually you can get to the interesting stuff. I don't like that for a lot of reasons. One is looking at children. I find that children do interesting things at age 6, so I don't understand why you want them to follow this do-what-you-are-told stuff for so long. My personal suspicion is that you don't need most of the mimicry mathematics we teach, and that we can get rid of an awful lot of it. This is also a social issue. Looking at New Jersey schools, for example, I see that minorities are being badly shortchanged, because they are told if you learn how to do the do-what-you-are-told stuff there are going to be all kinds of jobs out there for you—and there aren't.

Alan Schoenfeld. Let me bring the discussion to a close by pointing to an authoritative reference that seconds Bob's evaluation of the current mathematics curriculum, and Ed's suggestion that there may be school math and "real" math.

Webster's New Universal Unabridged Dictionary offers the following definitions of the term *problem*.

Definition 1: A question that is perplexing or difficult.
Definition 2: In mathematics, anything required to be done.

REFERENCES

Brown, S., & Walter, M. (1990). *The art of problem posing* (2nd ed.). Hillsdale, NJ: Lawrence Erlbaum Associates.
National Research Council. (1990). *A challenge of numbers.* Washington, DC: National Academy Press.
National Research Council. (1990). *Renewing U.S. mathematics: A plan for the 1990s.* Washington, DC: National Academy Press.
Pólya, G. (1981). *Mathematical discovery* (combined paperback ed.). New York: Wiley.
Schoenfeld, A. H. (1983). *Problem solving in the mathematics curriculum: A report, recommendations, and an annotated bibliography.* Washington, DC: Mathematical Association of America.
Schoenfeld, A. H. (1985). *Mathematical problem solving.* Orlando, FL: Academic Press.

3 Reflections on Doing and Teaching Mathematics

Alan H. Schoenfeld
The University of California, Berkeley

This two-part chapter is concerned with issues of mathematical philosophy and pedagogy. Part I deals with issues of ontology and/or epistemology—or in more down-to-earth language, what it means to *do* mathematics. Part II, which is grounded in epistemological issues but focuses on issues of instruction, provides descriptions of selected aspects of my problem-solving courses. Those courses are designed to engage students in the practices of doing mathematics and, as a result, to have them develop a sense of discipline (i.e., a mathematical perspective) consistent with that held by mathematicians.

In a formal sense, Part I is neither necessary nor sufficient for Part II. One may do mathematics one way and teach it another, of course. Conversely, philosophy can inform pedagogy but not determine it. Yet for most people there is an extremely strong relationship between Parts I and II. Whether or not one is explicit about it, one's epistemological stance serves to shape the classroom environments one creates (Hoffman, 1989). In turn, our classrooms are the primary source of mathematical experiences (as they perceive them) for our students, the experiential base from which they abstract their sense of what mathematics is all about. Hence, getting our epistemology straight, or at least into the open for discussion, is a vitally important enterprise.

PART I: EPISTEMOLOGICAL ISSUES

The past few years have seen attempts on the part of philosophers and mathematicians to reconceptualize and redescribe the mathematical enterprise. The reconceptualization has its roots in the work of Pólya (1954), Lakatos (1977, 1978),

Benacerraf and Putnam (see, e.g., Benacerraf & Putnam, 1964), and more recently in the writings of Kitcher (1984). A main theme in that work is that the *doing* of mathematics is a (somewhat) empirical endeavor (see, e.g., Lakatos, 1978, pp. 30–34, "Mathematics is Quasi-empirical"). More pragmatically, mathematical authors such as Steen (1988) and Hoffman (1989) have tried to frame a popular notion of mathematics that accurately reflects the nature of contemporary mathematics and that also serves as a basis for a modern pedagogy of mathematics. The ideas expressed here are grounded in some philosophical reflections (discussed later), but they start off in a practical vein. Hoffman's ideas on the nature of mathematics and the need for reform in mathematics education are used as a starting point for discussion. The following are some of the main points made by Hoffman (1989):

A. The current system of mathematics education:
- misrepresents mathematics, presenting it as a dead and deadly discipline;
- is based on a false mastery model, in which isolated skills are taught in the hope they can then be used to solve prepackaged problems;
- dumps by the wayside, after 8th grade, roughly 50% of the kids each year—with much higher percentages for most minorities;
- is self-reproductive, in that the successes of the system are the ones who perpetuate it, and they have no models but the ones they've gone through;
- is hence in need of comprehensive overhaul.

B. We need a powerful shorthand description of what mathematics is to convey the flavor of the discipline and to guide our teaching of it.
- Mathematics is *the science of patterns*.

I agree with all of A, noting that there is widespread recognition of the problem and significant progress in working on it (see, e.g., California Department of Education, 1992; National Council of Teachers of Mathematics, 1989, 1991; National Research Council, 1989, 1990). Let me now turn to B.

I agree with Hoffman's arguments, as far as he takes them (see below for detail). Rather than as a descriptive end, however, I see the delineation of mathematics as "the science of patterns" as a point of departure. Describing mathematics that way raises some interesting issues and takes us into territory that Hoffman and others who have used the term (e.g., Steen, 1988) may not have anticipated. I'd like to venture into that territory.

Just What is Mathematics Anyway?

When mathematicians talk about mathematics, they usually mean the *products* of mathematics.

Hoffman (1989) begins with two questions, one metaphysical (What is mathematics?) and one epistemological (What does it mean to "know" mathematics?). The second question is a misdirection, albeit a subtle one. The question we should be asking instead is, "What does it mean to *do* mathematics, or to act mathematically?" The answer to this question comes from a liberal (well, perhaps radical) interpretation of Hoffman and Steen's answer to the first question, that mathematics is the science of patterns.

Steen's (1988) article on the mathematical enterprise is essential reading for everyone in the mathematics and mathematics education communities, including—perhaps most importantly—students. It describes the scope and depth of modern mathematics and its power in an increasingly mathematical world. Here are a few samples:

- Number Theory. "Fifty years ago G. H. Hardy could boast of number theory as the most pure and least useful part of mathematics; today number theory is studied as an essential prerequisite to many applications of coding, including data transmission from remote satellites" (p. 611).
- Applications. "The 1979 Novel prize in medicine was awarded to Allan Cormack for his application of the Radon transform, a well-known technique from classical analysis, to the development of tomography and computer assisted tomography (CAT) scanners. . . . Structural biologists have become genetic engineers, capturing the geometry of complex macromolecules in supercomputers and then simulating interaction with other molecules" (p. 614). Stochastic differential equations are now used to model chemical processes, stock market behavior, and population genetics.
- Core mathematics. The past 15 years have brought the solutions to some major unsolved problems, for example, the four-color problem and the Bieberbach and Mordell conjectures, and opened up new areas, such as prime factorization both of integers and polynomials.

Steen's main point is that old conceptions of mathematics were never quite accurate, and that they now fall short of the mark. Mathematics, classically defined as "the science of number and space" (Steen, 1988, p. 611) now includes the study of regularities of all sorts—not only in patterns of twin primes, but in the patterns emerging from CAT scans as well. Hence, he proposes, with Hoffman, that mathematics is the science of patterns.

I am about to stretch the implications of this description. To set up the coming contrast, let me delineate what I think will be the standard interpretation of the phrase—a view that may be narrower than Steen's and Hoffman's. From the typical mathematician's point of view, *mathematics* is the "stuff" characterized above (number theory, applications, core math, etc.); *learning mathematics* is finding out about that stuff (often by being told, but sometimes by being presented with the opportunity to develop it on one's own); and *doing mathematics* is reaching the stage at which one is producing more of that stuff by oneself or in collaboration with others. Starting from "a science of patterns," a quite different view can be pursued.

A Brief Rhapsody on the *Science* of Patterns

The patterns part of the phrase requires no elaboration. Mathematics consists of observing and codifying—in general via abstract symbolic representations—regularities in the worlds of symbols and objects. (Work in these two spheres comprises pure and applied mathematics respectively.)

The science part is more interesting. To begin, a general (and positive) entailment of the term is that science is about making sense of things—finding out what makes them tick. From my point of view (see, e.g., Schoenfeld, 1987, 1990), that's precisely what mathematics is all about—a particular kind of sensemaking, in which one's main tool kit consists of a set of symbolic tools, and there are well-established styles of reasoning for seeing how things fit together. Furthermore, "doing science" is generally recognized to be a social rather than a mere individual and solitary act. There is a scientific community that shares and builds ideas. So there is of necessity a premium on being able to communicate scientific results as well as on getting answers. It's that way in the mathematical community as well. To remind us that these are not the general perceptions regarding mathematics, let me briefly recall one shopworn example and introduce a fresher one. Consider these two problems:

1. An army bus holds 36 soldiers. If 1,128 soldiers are being bussed to their training site, how many buses are needed?
2. Imagine you are talking to a student in your class on the telephone and want the student to draw some figures. [They might be part of a homework assignment, for example.] The other student cannot see the figures. Write a set of directions so that the other student can draw the figures exactly as shown below.

Problem 1 comes from the Third National Assessment of Educational Progress (Carpenter, Lindquist, Matthews, & Silver, 1983). Seventy percent of the students who took the exam did the relevant computation correctly—and 29% of the students (41% of those who did the right calculation) went on to say that the number of buses needed is "31 remainder 12." Problem 2 comes from the 1987–88 California Assessment Program's statewide assessment of 12th graders' mathematical skills (California Department of Education, 1989). Only 15% of the high school seniors who worked the problem were able to describe the figures with any degree of clarity.

These are negative examples, which show in striking ways just how mathematics isn't learned—and hence point to what's missing in mathematics instruction. In the first case, you can't write down "31 remainder 12" if you are thinking about real buses. It's clear that for the students who wrote that answer, the problem wasn't about real objects at all. Many if not most students see mathematics word problems simply as cover stories that give rise to computations. Their learned behavior is that one does the computations and writes the answers down, period—never mind if the answer doesn't make sense outside that context. That's about as far from mathematics as sense-making as you can get. In the second case, the reason so few of the students could communicate about mathematics is very simple: They'd had little or no practice at doing so. When mathematics is taught as received knowledge rather than as something that (a) should fit together meaningfully, and (b) should be shared, students neither try to use it for sense-making nor develop a means of communicating with it.

These two examples represent just the tip of the iceberg, of course. Elsewhere (Schoenfeld, 1992, p. 359) I have written about student beliefs such as the following:

- Mathematics problems have one and only one right answer.
- There is only one correct way to solve any mathematics problem—usually the rule the teacher has most recently demonstrated to the class.
- Ordinary students cannot expect to understand mathematics; they expect simply to memorize it and apply what they have learned mechanically and without understanding.
- Mathematics is a solitary activity, done by individuals in isolation.
- Students who have understood the mathematics they have studied will be able to solve any assigned problem in 5 minutes or less.
- The mathematics learned in school has little or nothing to do with the real world (cf. the bussing problem).
- Formal proof is irrelevant to the processes of discovery or invention.

The roots of such beliefs reside, alas, in the students' classroom experience. But enough negativity; let me return to the theme of mathematics as the science of patterns.

Note that *hands on* and *empirical* (meaning "grounded in the results of data-gathering") are terms that at least *sound* natural with regard to science. Here is the official word, from Webster's *New Universal Unabridged Dictionary* (1979): "Science . . . systematized knowledge derived from observation, study, and experimentation carried on in order to determine the nature or principles of what is being studied."

In fact, that's precisely what I think mathematics is all about. The *result* of mathematical thinking may be a pristine gem, presented in elegant clarity as a polished product (e.g., as a published paper). Yet the path that leads to that product is most often anything but pristine, anything but a straightforward chain of logic from premises to conclusions.

Here is a generic description of the genesis of a mathematical result (e.g., a theorem) that takes a mathematician, say M, a few months to derive. Somewhere near the beginning of the process, M has the intuition that the result ought to be true, and thinks she knows why. So, she begins to sketch out a proof. Part of the argument goes fine, but then there is a place at which things bog down; she can't get an intermediate result that seems necessary. M tries three or four different ways of getting around the difficulty, without success. So, she begins to think the result might not be true. If not, there ought to be a counterexample—at the point in which she has run into trouble, of course. She tries to construct one, but it does not work. Nor does a second, a third, and so on, and then M sees that all the counterexamples fail for the same reason. That reason is the idea that has been missing from the proof, and M now gets past the roadblock. Of course, she encounters others as she continues working on the theorem. M is fortunate this time: The empirical data (attempts at counterexamples, etc.) work in her favor, and they result in her finding the ideas that allow for her proof. Other times she is less fortunate: Promising potential theorems turn out not to be true, and that's the end of the story.

In sort, mathematics is a "hands-on," data-based enterprise for those who engage in it. Doing mathematics is doing science, as defined above. It has a significant empirical component, one of data and discovery. What makes it mathematics rather than chemistry or physics or biology is the unique character of the objects being studied and the tools of the trade.

Another characteristic of the scientific enterprise is that it is, in large measure, a social enterprise. Many of the problems considered central are too big for people to solve in isolation. In consequence an increasingly large percentage of mathematical and scientific work is collaborative. Such collaborative work both requires and fosters shared perspectives, among collaborators in particular and across the field at large. When we say someone is a member of the scientific community, that phrase has significant entailments. It means that the person has the appropriate knowledge base, of course. But it also means that a person has picked up not only the tools but the perspectives of his or her discipline—a particular way of seeing the world, a style of thinking about it. (The stereotypes about doctors and lawyers, for example, do have a basis in reality; members of those groups tend to have particular ways of seeing the world. So do mathematicians, who develop their world views in the same way as do the others—by interacting with those who are already members of the community.)

I hope you are with me so far, because I am about to up the ante. The issue is the character of mathematical *knowing:* whether mathematicians can always be absolutely confident of the truth of certain complex mathematical results, or whether, in some cases, what is accepted as mathematical truth is in fact the best collective judgment of the community of mathematicians, which may turn out to be in error. I will argue the latter and will argue that taking this perspective has implications for classroom practice.

The notion of "fallible truths" and the role of the scientific community in defining those truths is more familiar in the case of the physical sciences. Popper (1959) and Kuhn (1962) highlighted the idea, and it has received a fair amount of recent discussion. The notion that absolute truth is unattainable in science is at least implicit in the language of science, in the use of the term *theory* for "tentative explanation." Again, thanks to Webster, a theory is "a formulation of apparent relationships or underlying principles of certain observed phenomena which has been verified to some degree." Basic science consists in large part of theory development and refinement, the construction of explanatory frameworks that account for data as well as possible. In that context, laws have a funny meaning. Scientists understand that the laws of science are not statements of absolute truth but merely theories that appear to have exceptionally solid grounding. New data, or different and more encompassing explanations, can result in the old laws losing credence and new versions taking their place (e.g., relativity supplants Newtonian mechanics, which supplanted the Aristotelian view).

Now the stereotype is that it's different in mathematics: It appears that you start with definitions or axioms and all the rest follows inexorably. However, as Lakatos (1977) shows in *Proofs and Refutations,* that isn't the way things really happen. The "natural" definition of polyhedron was accepted by the mathematical community for quite some time and was used to prove Euler's formula—until mathematicians found solids that met the definition but failed to satisfy the formula. How did the community deal with the issue? Ultimately, by changing the definition. That is, the grounds for the theory—the definitions underlying the system—were changed in response to the data. That sure looks like theory change to me: New formulations replace old ones, with base assumptions (definitions and axioms) evolving as the data come in. (In his 1978 work, *Mathematics, Science, and Epistemology,* Lakatos uses the term *quasi-empirical* to describe mathematics; he notes that mathematical theories cannot be true; they are at best "well-corroborated, but always conjectural" [p. 28].) What do we have then, regarding the nature of truth in mathematics (as in science)? To state things in the most provocative form: With regard to some very complex issues, truth in mathematics is that for which the vast majority of the community believes it has compelling arguments. And such truth may be fallible.

Serious mistakes are relatively rare, of course. For topics such as simple arithmetic or elementary real analysis, to pick two, there's no room for doubt. Once you make the definitions, the results follow—and the chain of logic that leads to the conclusions is sufficiently accessible so that anyone trained in the mathematics (i.e., who knows the rules of the game and plays by them) can confirm them. But, for complex results (e.g., a false proof of the Jordan Curve Theorem was widely known and accepted for a decade, and there was great controversy over the proofs of the four-color theorem and the Bieberbach conjecture), there is a social dimension to what is accepted as mathematical "truth." Once one accepts this notion, discussions of some traditional epistemological/ontological questions—questions of what it means to know mathematics and of mathematical authority (where does mathematical certainty reside?)—take on an interesting character. These are pursued in Part II.

Here is a distillation of my story so far: Mathematics is an inherently social activity, in which a community of trained practitioners (mathematical scientists) engages in the science of patterns—systematic attempts, based on observation, study, and experimentation, to determine the nature or principles of regularities in systems defined axiomatically or theoretically ("pure math") or models of systems abstracted from real-world objects ("applied math"). The tools of mathematics are abstraction, symbolic representation, and symbolic manipulation. However, being trained in the use of these tools no more means that one thinks mathematically than knowing how to use shop tools makes one a craftsperson. Learning to think mathematically means (a) developing a mathematical point of view—valuing the processes of mathematization and abstraction and having the predilection to apply them, and (b) developing competence with the tools of the trade and using those tools in the service of the goal of understanding structure—mathematical sense-making. Finally, some mathematical truths (results accepted as true by the community) are in fact "provisional truths," reflecting the field's best but possibly incorrect understanding.

The Bottom Line

Why raise all this fuss about the nature of mathematics? Because people develop their understanding of the nature of the mathematical enterprise from their experience with mathematics, and that experience (at least the part that is typically labeled as being "mathematics") takes place predominantly in our mathematics classrooms. The nature of that experience at present was described by Hoffman and summarized at the beginning of this chapter; the consequences of that experience are illustrated by the list of student beliefs summarized earlier in this section. When mathematics is taught as dry, disembodied, knowledge to be received, it is learned (and forgotten or not used) in that way. However, there is an optimistic counterpoint to the observation that one's experience with mathematics determines one's view of the discipline, and it has its own imperative: The activities in our mathematics classrooms can and must reflect and foster the understandings that we want students to develop with and about mathematics.

That is, if we believe that doing mathematics is an act of sense-making; if we believe that mathematics is often a hands-on, empirical activity; if we believe that mathematical communication is important; if we believe that the mathematical community grapples with serious mathematical problems collaboratively, making tentative explanations of these phenomena, and then cycling back through those explanations (including definitions and postulates); if we believe that learning mathematics is empowering and that there is a mathematical way of thinking that has value and power, then our classroom practices must reflect these beliefs. Hence, we must work to construct learning environments in which students actively engage in the science of mathematical sense-making, as characterized earlier. Part II describes aspects of my attempts in that direction.

PART II: PEDAGOGICAL ISSUES

Elsewhere (see, e.g., Schoenfeld, 1985) I have characterized the mathematical content of my problem-solving courses. Here, in an extension of the themes explored in a number of recent (and one not-so-recent) papers (Balacheff, 1987; Collins, Brown, & Newman, 1989; Fawcett, 1938; Lampert, 1990; Lave, Smith, & Butler, 1988; Lave & Wenger, 1989; Schoenfeld, 1987, 1989b, 1992) I focus on the epistemological and social content and means. The content of my problem-solving courses is *epistemological* in that the courses reflect my epistemological goals: By virtue of participation in them, my students will develop a particular sense of the mathematical enterprise. The means are *social,* for the approach is grounded in the assumption that people develop their values and beliefs largely as a result of social interactions. I work to make my problem-solving courses serve as microcosms of selected aspects of mathematical practice and culture—so that by participating in that culture, students may come to understand the mathematical enterprise in a particular way.

What follows are two illustrations of goals, practices, and results. Those illustrations might be called protoethnographic. Though they might appear anecdotal, I believe they contain the substance from which good ethnographic descriptions could be crafted.

Example 1: Where Does Mathematical Authority Reside?

As indicated in Part I, mathematical truth or correctness is a delicately grasped object. One might say that the ultimate authority is the mathematics itself: False proofs are still false, even if people believe them, for example (and ultimately, one expects, the flaws in them will be uncovered). Nonetheless, mathematical authority is, in practice, exercised by human hands and minds. There are, of course, collective standards for mathematical correctness, for example, the review process, in which experts certify (to the degree they can; cf. the proof of the four-color theorem) that an argument is correct. Through such processes the mathematical community implements mathematical authority with consistency and (in general) with accuracy. This public process both is based in, and provides substance for, individuals' mathematical knowledge and authority. Mathematicians, having internalized the standards of correctness in their mathematical communities, apply those standards to what they know as individuals. In turn, the application of that abstract mathematical authority results in a very powerful personal ownership of the mathematics they can certify. To put it another way, arriving at mathematical certainty is the very personal process of applying an internalized impersonal standard.[1] In that sense, ultimate mathematical authority resides deeply in individuals, and collectively in the mathematical community.

Now, contrast this view of where authority resides with the typical student's view. Most college students possess little of the sense of personal knowledge or internal authority just described. They have little idea, much less confidence, that they can serve as arbiters of mathematical correctness, either individually or collectively. Indeed, for most students, arguments (or purported solutions) are merely proposed by themselves. Those arguments are then judged by experts, who determine their correctness. Authority and the means of implementing it are external to the students. Students *propose;* experts judge and *certify*.

One explicit goal of my problem-solving courses is to deflect inappropriate teacher authority. I hope to make it plain to the students that the mathematics speaks through all who have learned to employ it properly, and not just through the authority figure in front of the classroom. More explicitly, a goal of instruction is that the class becomes a community of mathematical judgment which, to the best of its ability, employs appropriate mathematical standards to judge the claims made before it.

In the course discussed here, the explicit deflection of teacher authority began the second day of class when a student volunteered to present a problem solution at the board. As often happens, the student focused his attention on me rather than on the class when he wrote his argument on the board; when he finished he waited for my approval or critique. Rather than provide it, however, I responded as follows:

"Don't look to me for approval, because I'm not going to provide it. I'm sure the class knows more than enough to say whether what's on the board is right. So (turning to class) what do you folks think?"

In this particular case the student had made a claim that another student believed to be false. Rather than adjudicate, I pushed the discussion further: How could we know which student was correct? The discussion continued for some time, until we found a point of agreement for the whole class. The discussion proceeded from there. When the class was done (and satisfied) I summed up.

This problem discussion illustrated a number of important points for the

---

[1] One is likely to get to this point via interactions with others, of course.

students, points consistently emphasized in the weeks to come. First, I rarely *certified* results, but turned points of controversy back to the class for resolution. Second, the class was to accept little on faith. That is, "we proved it in Math 127" was not considered adequate reason to accept a statement's validity. Instead, the statement must be grounded in mathematics solidly understood by this class. Third, my role in class discussion would often be that of a Doubting Thomas. That is, I often asked, "Is that true? How do we know? Can you give me an example? A counterexample? A proof?," both when the students' suggestions were correct and when they were incorrect. (A fourth role was to ensure that the discussions were respectful—that it is the mathematics at stake in the conversations, not the students!)

This pattern was repeated consistently and deliberately, with effect. Late in the second week of class, a student who had just written a problem solution on the board started to turn to me for approval, and then stopped midstream. She looked at me with mock resignation and said, "I know, I know." She then turned to the class and said, "O.K., do you guys buy it or not?" [After some discussion, they did.]

The pattern continued through the semester. It was supplemented by overt reflections on our discussions that focused on what it means to have a compelling mathematical argument. The general tenor of these discussions followed the line of argumentation outlined in Mason, Burton, and Stacey's (1982) *Thinking Mathematically:* First, convince yourself; then, convince a friend; finally, convince an enemy. (That is, first make a plausible case and then buttress it against all possible counterarguments.) In short, we focused on what it means to truly understand, justify, and communicate mathematical ideas.

The results of these interactions revealed themselves most clearly in the following incident. Toward the end of the semester I assigned the following problem.[2]

The Concrete Wheel Problem

You are sitting in a room at ground level, facing a floor-to-ceiling window which is 20-feet square. A huge solid concrete wheel, 100 miles in diameter, is rolling down the street and is about to pass right in front of the window, from left to right. The center of the wheel is moving to the right at 100 miles per hour. What does the view look like, from inside the room, as the wheel passes by? (See Fig. 3.1.)

This problem tends to provoke immediate and widely divergent intuitive reactions, among them:

1. The room will go (almost) instantaneously dark as the wheel first passes the window. It will stay dark for a short while and go (almost) instantaneously light as the wheel leaves.

---

[2]The problem is borrowed from diSessa (and borrowed in turn, I believe, from Papert).

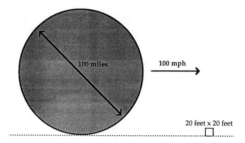

FIG. 3.1. The situation described in the concrete wheel problem.

2. Same as (1), but the room stays dark for a relatively long time.
3. The room darkens slowly, as though a large window shade was being pulled more or less:

   a. horizontally from left to right, as follows:

   b. diagonally from the upper left corner as follows:

   c. vertically downward as follows:

   The room then stays dark for a short/long period of time, after which it lightens in a way complementary to the way it darkened.

When the problem was posed, students made many of the conjectures listed above. As usual, the class broke into groups to work on the problem. One group became the staunch defenders of one conjecture, while a second group lobbied for another. The two groups argued somewhat heatedly, with the rest of the class following the discussion. Finally, one group prevailed, on what struck me as solid mathematical grounds.[3] As is my habit, I did not reveal this but made my usual comment: "O.K., you seem to have done as much with this as you can. Shall I try to pull things together?" One of the students replied, "Don't bother. We got it." The class agreed.

While one might dismiss this event as being trivial (the students simply indicated that they had understood the material, and the class progressed; what's

---

[3]I shall refrain from giving the answer in order not to spoil readers' possible pleasure in determining it themselves.

the big deal?), or even see their rejection of my offer to pull things together as being somewhat abrasive (I had signaled my intentions, and they told me not to bother), either view misses the significance of the event. First, it is important to note that the classroom was functioning as a mathematical community. Various points of view were advanced and defended mathematically. The arguments in favor of different positions were made on solid mathematical grounds, and ultimately the correct view prevailed, for good reason. One could ask for no better at a meeting of professional mathematicians. Second, and more importantly, the locus of mathematical authority had shifted radically. From the student's point of view, I was no longer needed as an authority figure to provide external *certification* of results. As in the mathematical community at large, the mathematics spoke through the students. It did so collectively, in the dialogue that took place in the community; it did so individually, in that the students demanded the appropriate mathematical standard of argumentation, and then believed the results. This was *their* mathematics. They had ownership of it, not only in the motivational sense, but in the deep epistemological sense that characterizes the true mathematical knowing and understanding possessed by mathematicians.

Example 2: Who Can *Do* Mathematics?

Speaking broadly, research mathematics is one thing, and classroom mathematics is something else altogether. When they are *doing* mathematics as researchers, mathematicians are pushing the boundaries of knowledge—not only their own, but that of the mathematical community. Publishable research consists, in essence, of results that (a) are new to the community of mathematicians, and (b) deemed of sufficient merit or interest to warrant distribution. In contrast, classroom mathematics generally consists of the distillation and presentation of known results to be "mastered" by students. The implicit but widespread presumption in the mathematical community is that an extensive background is required before one can *do* mathematics. Undergraduates who publish mathematics are exceedingly rare, and even graduate students with publications prior to their thesis work are relatively uncommon. Until students get to the point of doing research (typically in the third year of graduate school), learning mathematics means *ingesting* mathematics.

There are, of course, exceptions to this rule. There is, for example, the Moore method. The *Journal of Undergraduate Research* has, for half a century, published student work in mathematics. Occasionally classroom work produces results of professional quality. For example, an article by Banchoff and student associates (1989) appeared in a recent *UME Trends,* and a discussion in one of my problem-solving classes not long ago led to a publication in the *College Mathematics Journal* (Schoenfeld, 1989a). But the threshold of research *qua* research is unreasonably high for most undergraduate courses.

Here I wish to pursue an alternative perspective, one based on the notion of *intellectual community.* In the introduction to this chapter I indicated that I work to make my problem-solving courses "microcosms of selected aspects of mathematical practice and culture," in that the classroom practices reflect (some of) the values of the mathematical community at large.

Part I of this chapter presented one mainstream view of mathematics, as the *science of patterns*. I described this elsewhere by saying that the business (and pleasure) of mathematics consists of perceiving and delineating structural relationships. Suppose we add to that the notion, as suggested above, that research—what most mathematicians would call *doing* mathematics—consists of making contributions to the mathematical community's knowledge store. And finally, one adds part of the mathematician's aesthetic, that making such contributions is part of the mathematician's intellectual life, and something of intrinsic value.

My goals for my problem-solving courses are to create local intellectual communities with those same values and perspectives. The notion of localization works as follows: A contribution is significant if it helps the particular intellectual community advance its understanding in important ways.

Elsewhere I (Schoenfeld, 1990) described one of my class's discussions of the Pythagorean Theorem. Here I review that discussion from the perspective of social and epistemological engineering. The initial problem posed to the class was very broad, in essence: "What can we do with the Pythagorean Theorem?"

In its discussion of the result (well known to all of the students), the class began by proving the theorem a variety of different ways. It explored three- and *n*-dimensional analogues of the theorem; it pursued geometric extensions and analogues. Then it began to focus on the diophantine equation

$$a^2 + b^2 = c^2.$$

Could we find all positive integer solutions to this equation?

Now, any mathematician can tell you there is a general solution to this problem. A triple of integers (a, b, c) with the property that $a^2 + b^2 = c^2$ is called a Pythagorean triple. Every Pythagorean triple (a, b, c) can be shown to be of the form

$$a = k(M^2 - N^2), b = 2k(MN), c = k(M^2 + N^2),$$

where k is the largest common factor of a, b, and c, and M and N are relatively prime integers. In a content-oriented course (e.g., elementary number theory), one would typically present the proof of this result in about 10 minutes, and then move on to another result. But part of the engineering effort in teaching this course consists of seeding classroom dialogue with problems at the appropriate level for community discourse, and then holding back as the community grapples with those problems to the best of its ability.

In this case, the students began working on the problem by generating some of the whole-number Pythagorean triples they knew; (3, 4, 5), (5, 12, 13),

(6, 8, 10), (7, 24, 25), (8, 15, 17), (9, 40, 41), and (10, 24, 26). On the basis of these empirical data they made the following observations:

1. Integer multiples of Pythagorean triples are Pythagorean and hence of little intrinsic interest. (If you can generate all the relatively prime Pythagorean triples, then you can generate all the rest.)
2. In every relatively prime triple, the hypotenuse was odd.
3. In every relatively prime triple where the smaller leg was odd, the hypotenuse exceeded the larger leg by 1.
4. In the relatively prime triple where the smaller leg was even, the hypotenuse exceeded the larger leg by 2.

As a result of observation (1), the class restricted its attention to relatively prime Pythagorean triples. They conjectured that observation (2) was always true, and proved it. On the basis of observation (3), they conjectured that there are infinitely many Pythagorean triples of the form $(2x + 1, 2y, 2y + 1)$, and proved it. On the basis of observation (4), they conjectured that there are infinitely many Pythagorean triples of the form $(2x, 2y - 1, 2y + 1)$, and proved it. On the basis of those two results, and the fact that they knew of no other triples, the class conjectured that all relatively prime triples are of the types described in (3) and (4). They began their work on this conjecture by proving it for the first relevant case: They proved there are no relatively prime Pythagorean triples of the form $(x, y, y + 3)$. At that point a student asked: If they were successful in proving their conjecture, did they have a publishable theorem?

The answer, of course, was no. As noted above, the complete solution to the problem they were working on is a standard result presented in elementary number theory courses. Nonetheless, neither the student's question nor the class's achievements should be discounted. The individual student's comment indicated that he, at least, thought that the class might be at the frontiers of knowledge—a far cry from what happens in most classrooms.

And, in two significant ways, the students were. First, three of the results they proved:

There are infinitely many triples of the form $(2x + 1, 2y, 2y + 1)$;
There are infinitely many triples of the form $(2x, 2y - 1, 2y + 1)$; and
There are *no* relatively prime triples of the form $(x, y, y + 3)$,

were new to me and (although easily proven) are a surprise to many mathematicians. Hence, the product of their labors was not inconsequential. But more importantly, these students, in their own intellectual community, were *doing* mathematics. They were, at a level commensurate with their knowledge and abilities, truly engaged in the science of patterns.

FINAL COMMENTARY

In Part I of this chapter I tried to portray mathematics as a living, breathing discipline in which truth (as much as we can know it) lives in part through the individual and collective judgments of members of the mathematical community. I suggested that

1. Mathematicians develop much of that deep mathematical understanding by virtue of apprenticeship into that community—typically in graduate school and as young professionals.
2. In standard instruction students are typically deprived of such apprenticeships, and hence of access to doing and knowing mathematics.

In Part II of this chapter I tried to convey some of the character of my problem-solving courses. In essence, I create artificial communities in them—communities in which certain mathematical values, consistent with some of those of the mathematical community at large, predominate. The following is a more precise delineation of some of the main themes of those courses:

1. Mathematics is the science of patterns, and relevant mathematical activities—looking to perceive structure, seeing connections, capturing patterns symbolically, conjecturing and proving, and abstracting and generalizing—all are valued.
2. Mathematical authority resides in the mathematics, which—once we learn how to heed it—can speak through each of us and give us personal access to mathematical truth. In that way mathematics is a fundamentally human (and for some, aesthetic and pleasurable) activity.

I hope to have illustrated, in examples 1 and 2, how students, by living in such artificial microcosms of mathematical practice, come to develop as mathematical doers and thinkers. I conclude this chapter with the comment that in a very serious sense, these artificial environments provide students with a genuine experience of *real* mathematics. By that standard, conventional mathematics instruction is wholly artificial.

ACKNOWLEDGMENTS

This research was supported by the U.S. National Science Foundation through NSF grants MDR-8751520, MDR-8550332, and BNS-8711342. The Foundation's support does not necessarily imply its endorsement of the ideas or opinions expressed in this chapter.

The author owes thanks to many colleagues for their comments and criticisms, and wishes in particular to thank Cathy Kessel, Bob Scher, and Elliot Turiel.

REFERENCES

Balacheff, N. (1987). *Devolution d'un probleme et construction d'une conjecture: Le cas de "la somme des angles d'un triangle"* [The evolution of a problem and the construction of a conjecture: The case of the "sum of the angles of a triangle"]. (Cahier de didactique des mathematiques No. 39). Paris: IREM Universite Paris VII.

Banchoff, T., & Student Associates. (1989, August). Student generated iterative software for calculus of surfaces in a workstation laboratory. *UME Trends, 1*(3), 7–8.

Benacerraf, P., & Putnam, H. (1964). *Philosophy of mathematics: Selected readings*. Englewood Cliffs, NJ: Prentice-Hall.

California Department of Education. (1989). *A question of thinking*. Sacramento, CA: California Department of Education.

California Department of Education. (1992). *Mathematics framework for California public schools: Kindergarten through grade twelve*. Sacramento, CA: California Department of Education.

Carpenter, T. P., Lindquist, M. M., Matthews, W., & Silver, E. A. (1983). Results of the third NAEP mathematics assessment: Secondary school. *Mathematics Teacher, 76*(9), 652–659.

Collins, A., Brown, J. S., & Newman, S. (1989). Cognitive apprenticeship: Teaching the craft of reading, writing, and mathematics. In L. B. Resnick (Ed.), *Knowing, learning, & instruction: Essays in honor of Robert Glaser* (pp. 453–494). Hillsdale, NJ: Lawrence Erlbaum Associates.

Fawcett, H. P. (1938). *The nature of proof (1938 Yearbook of the National Council of Teachers of Mathematics)*. New York: Columbia University Teachers College Bureau of Publications.

Hoffman, K. (1989, March). *The science of patterns: A practical philosophy of mathematics education*. Paper presented to the Special Interest Group for Research in Mathematics Education at the 1989 Annual Meeting of the American Educational Research Association, San Francisco.

Kitcher, P. (1984). *The nature of mathematical knowledge*. New York: Oxford University Press.

Kuhn, T. S. (1962). *The structure of scientific revolutions*. Chicago: University of Chicago Press.

Lakatos, I. (1977). *Proofs and refutations* (revised ed.). Cambridge: Cambridge University Press.

Lakatos, I. (1978). *Mathematics, science, and epistemology*. Cambridge: Cambridge University Press.

Lampert, M. (1990). When the problem is not the question and the solution is not the answer: Mathematical knowing and teaching. *American Educational Research Journal, 27*, 29–63.

Lave, J., Smith, S., & Butler, M. (1988). Problem solving as an everyday practice. In R. Charles & E. Silver (Eds.), *The teaching and assessing of mathematical problem solving* (pp. 61–81). Hillsdale, NJ: Lawrence Erlbaum Associates.

Lave, J., & Wenger, E. (1989). *Situated learning: Legitimate peripheral participation* (IRL Rep. 89-0013). Palo Alto, CA: Institute for Research on Learning.

Mason, J., Burton, L., & Stacey, K. (1982). *Thinking mathematically*. New York: Addison-Wesley.

National Council of Teachers of Mathematics. (1989). *Curriculum and evaluation standards for school mathematics*. Reston, VA: NCTM.

National Council of Teachers of Mathematics. (1991). *Professional standards for teaching mathematics*. Reston, VA: NCTM.

National Research Council. (1989). *Everybody counts: A report to the nation on the future of mathematics education*. Washington, DC: National Academy Press.

National Research Council. (1990). *Reshaping school mathematics*. Washington, DC: National Academy Press.

Pólya, G. (1954). *Mathematics and plausible reasoning* (two vols.). Princeton: Princeton University Press.

Popper, K. R. (1959). *The logic of scientific discovery*. London: Hutchinson.

Schoenfeld, A. (1987). What's all the fuss about metacognition? In A. Schoenfeld (Ed.), *Cognitive science and mathematics education* (pp. 189–215). Hillsdale, NJ: Lawrence Erlbaum Associates.

Schoenfeld, A. (1989a). The curious fate of an applied problem. *College Mathematics Journal*, *20*(2), 115–123.

Schoenfeld, A. (1989b). Ideas in the air: Speculations on small group learning, environmental and cultural influences on cognition, and epistemology. *International Journal of Educational Research*, *13*(1), 71–88.

Schoenfeld, A. (1990). On mathematics as sense-making: An informal attack on the unfortunate divorce of formal and informal mathematics. In J. Voss, D. Perkins, & J. Segal (Eds.), *Informal reasoning and education* (pp. 311–343). Hillsdale, NJ: Lawrence Erlbaum Associates.

Schoenfeld, A. (1992). Learning to think mathematically: Problem solving, metacognition, and sense-making in mathematics. In D. Grouws (Ed.), *Handbook for research on mathematics teaching and learning* (pp. 334–370). New York: Macmillan.

Steen, L. (1988). The science of patterns. *Science, 240*, 611–616.

Webster's new universal unabridged dictionary (2nd ed.). (1979). New York: Simon & Schuster.

A Discussion of Alan Schoenfeld's Chapter

Leon Henkin
Judah L. Schwartz

> **Leon Henkin,** *University of California, Berkeley*
> **Alan H. Schoenfeld,** *University of California, Berkeley*
> **Judah L. Schwartz,** *Harvard Graduate School of Education*

Leon Henkin. The position that some mathematical truths are socially negotiated is a dangerous one, but I couldn't possibly deal with it in the amount of space I have here. I think, Alan, that what you were telling us in your description of your course is about how you were withholding the authority to tell your students what was true and what was false. However, you *were* setting standards. I think that if the students interpreted your standards in a way that led them to something that you thought was wrong, and they were about to go away with the wrong belief, you wouldn't be doing your proper job as a teacher if you let them get away with it.

But let me get to my main comments. This question of what mathematics is has tremendous social force behind it. I've worked with some teachers in elementary schools, got them keyed up to want to bring new ideas into the classroom, and the thing that stymies them is the parents. When the kids come home with these new exciting things, the parents come and protest to the teacher and to the principal, "That isn't mathematics. I want my children to learn mathematics!" meaning what they think mathematics is, because of what they learned when they were students.

And so, I believe that when you are talking about changing the concept of what mathematics is, you are scrapping all of the old curriculum. Of course, it *has* to be scrapped, but there's no way to snap our fingers, even with such an august company as in this room, and get it scrapped. Making changes in something with the enormous social inertia of a school system is incredibly difficult. It has to be looked at as a very long-term and political event, not just as an educational event.

I think that we ought to bring in the *processes* of mathematics right from the beginning, that is, in kindergarten exposure and certainly from first grade on. We know that the elementary school curriculum has focused on computation and so has the high school curriculum and lower division college curriculum. We compute with algebraic expressions maybe, instead of with numbers, but we're still just getting the answer rather than reasoning about the underlying material. In fact, reasoning is certainly possible at the lowest grade levels. I know there are Piagetian theories asserting that only at a later stage of development can people reason, because it's such an abstract activity. But I don't think that's correct. I have been involved in efforts with my student Nitsa Hadar to show the contrary, that I think are successful. The fact is that every kid who learns language between the ages of 2 and 6 is learning an enormous number of abstractions. Every noun is an abstraction. Cat is an abstraction, blue is still more of an abstraction. If kids can use language without being far down the line developmentally, they can do the abstraction that is needed in mathematics.

Finally, you were talking about truth—truth in science, truth in mathematics. There is indeed a social dimension to truth, but there's also a logical dimension to truth, and as a logician I am entitled to say something about it. The big difference between science and math is that in science you're talking about one structure: The structure that you're looking at with your senses. Whereas the great difference in mathematics is that we use the same language to talk simultaneously about quite different structures, that is, the nature of abstraction. Abstraction is the heart of mathematics, and the corresponding built-in ambiguities in the use of language are the heart of mathematical language.

Alan Schoenfeld. Let me respond briefly to Leon's comment. In my presentation I was deliberately provocative about the nature of mathematical "truth" as being socially defined. In my chapter I am more circumspect, and in the long run, I suspect I shall leave such issues to philosophers. My main purpose in raising epistemological issues was to incite people to think about the ways in which we as members of the mathematical community decide what we will (individually and collectively) accept as true. Thinking about these issues sets the stage for considering alternative pedagogical practices such as those described in my chapter.

Let me now turn to the big practical issue Leon raises—that if we take the message of my chapter seriously, we need to scrap the current curriculum. I believe we do need to scrap it—but having been one of the authors of the new California Mathematics *Framework* (1992), I am all too aware of the political, practical, and intellectual shoals on which such an enterprise can founder. Some of the staunchest opposition to the *Framework* came from professional mathematicians, who were dismayed that they could not identify the places where specific content items were to be found, in a curriculum organized around large, thematic, problem-based units. In addition, parents often feel uncomfortable when their children are in experimental programs with unproven track records, especially those that feel unfamiliar and leave the parents unable to help their children with their work. And the state superintendent of education felt uncomfortable about the lack of explicit teacher authority in the new pedagogy suggested by the framework: In a student-centered, collaborative classroom, he worried, just what *is* the teacher's authority?

Then there are the bottom-line issues. Would a curriculum along the lines envisioned in the California Mathematics *Framework* be teachable by the current teaching force? Gaea Leinhardt's contribution to this volume strongly suggests the answer is "no." Indeed, in recognition of the large amount of in-service work required to help teachers teach in the ways suggested, members of the *Framework*'s writing committee wanted to specify minimum in-service requirements. "Don't," they were instructed. "You can't tell the state, or the school districts, how to spend their money." But without such expenditure, reform may fail. In sum, there is good reason to be concerned about the issues Leon raises. Orchestrating change is a long-term intellectual and highly political process. And there are a myriad of ways the process can go awry.

Judah Schwartz. Let me first say that I subscribe almost totally to the spirit of Alan's talk. But this last exchange is one that I would like to underline, because it says to me that there really is some powerful political mileage to be gotten out of the Trojan Mouse strategy (see Schwartz's chapter in this volume). If you use mathematical objects that are recognizable to parents, to school boards, and to state superintendents of education, you might get past level zero and maybe even level one of criticism and get far enough along that you won't be interfered with at the outset. *Then* you have an opportunity to make change.

Now let me turn to the observation about mathematics being the science of patterns. The evidence Lynn Steen cited was really about new content. Alan then went on to talk about new content not being the issue so much as a focus on the processes of mathematics—and getting kids to engage in the processes of mathematics. I would like to endorse that perspective very strongly. I think that it is consonant with the notion of school being the place to develop good habits of mind. In fact, I would go further and say that in *every* domain that a child is asked to study in school, the child ought to be asked to create in that domain, at some level. History isn't all done, either; neither is literature. No subject matter is all done, and one of the glories of the human spirit is that ability to create in all kinds of domains. And school ought to make that very, very obvious.

Alan, you talk about mathematics as being an issue of making sense of the surroundings, not simply accumulating facts. Yet, when I listen to your description of your problem-solving courses, which I think are spectacular, it seems to me that they are at least in part mislabeled. That is, because they are, in some substantial measure, problem-posing courses. They are courses that help people learn how to *pose* problems. This in some sense is harder, and it's certainly a different task. When Bruce Reznick spoke about the fact that one knows that Putnam exam problems already have solutions, and more generally, that students are almost always asked to solve problems they know have solutions, he made an important point. I once asked the scientific attaché in the French Embassy in Washington, at a meeting about nuclear weapons, "What help did the French get in developing hydrogen fusion nuclear weapons from the U.S.?" He said something very interesting: "The fact it could be done." That was his characterization of the help that they got. So I would like to argue that at least in part your courses are about problem posing, and that that is a very important aspect of the processes of mathematics that you are referring to. And maybe suggest that when you speak about these things that you include that in a more salient way.

You spoke about truth as socially negotiated in mathematics, that one's not going to get by me any more than it got by Leon. Paulos, in his book *Innumeracy* (1988), talks about some people who are listening to the weather man on television who predicts that there's a 50% chance of rain on Saturday and a 50% chance of rain on Sunday, and therefore it is certain to rain on the weekend. Now, if truth is socially negotiated in mathematics, the question is: socially negotiated by *whom?* If it's up to anybody to socially negotiate truth, then any group of people could publish their own journal, and *caveat lector.* Then you buy the journal, you read it, and you decide whether you believe it—that is, you get into the act of negotiating truth. This is dangerous territory, and you want to be careful here—but you did warn us that you were going to be provocative, and you were.

Let me turn to the NAEP busing problem as a way of introducing the next section of my comments. The physicist in me argues that people should not be taught the mathematics of number, but they should be taught the mathematics of quantity. There is a question of to what extent the barbarism "31 *r* 12" in answer to the busing problem is a consequence, not of classroom discourse practices, but just simply a consequence of teaching the wrong mathematics. I'm not trying to argue that this would explain all the variance, but I am trying to suggest that there might be another source for barbarisms of this sort. This is not to say that we set up reasonable intellectual climates in our mathematics classrooms. We don't, and in fact I think your list of student beliefs is accurately descriptive of the sorts of things that go on. But students might do better if what we taught was more sensible, too.

This leads me to the issue of beliefs in general. One of the things I learned from recent presidential campaigns is the power of the sound bite. These weren't campaigns of issues; people threw 10-second barbs at one another. Now, if that's the way it is, then that's the way it is, and the question is, can we arm ourselves with sound bites to counter the Neanderthals of the world? So (this is addressed particularly to Tom Romberg, because I hope he can get it into the NCTM machinery) here is a sound bite that I want to propose to NCTM: "Mathematics should never pose a problem that has only one right answer." This is crafted deliberately as a sound bite and should be understood as such. But I think it stands on its own. People don't quite know how to ask questions of that type, and I think that we can help them by giving examples of how to do so in all kinds of domains of mathematics. And, I think the entailments of dealing with different kinds of questions really address your list of beliefs in an essential way.

Finally, on the last transparency of your talk, you made a very strong statement. The transparency read as follows:

Theorems

1. Mathematics is a living, breathing, and exciting discipline of sense-making.
2. Students will come to see it that way if and only if they experience it that way in their classrooms.

Sharing your predilection to make exaggerated statements, I sympathize. But, I have to tell one more story. Last week I had the tremendous privilege of being visited by I. M. Gelfand. Gelfand had talked to a friend of mind in Moscow who told him about the work I was doing. Gelfand walks in and he says, "I hear that you are doing wonderful things in geometry, but I have a very strong argument with you, I will tell you later." The conversation goes on. And in the conversation, he was asking me about the kinds of things I do. At one point I said, "Look, I really think that making conjectures and exploring them is a very important piece of doing mathematics and should be part of everybody's mathematical education." He says: "Now I withdraw my objection. And I will tell you what this is about." He says: "Your friend told me that you thought that conjecturing in mathematics was necessary and sufficient, but I now hear you saying only that it is necessary. I withdraw my objection." Perhaps you should withdraw the "only if" from Theorem 2. You see, this is a room full of people who probably didn't learn their mathematics with the only if.

REFERENCES

California Department of Education. (1992). *Mathematics framework for California public schools*. Sacramento, CA: California Department of Education.

Paulos, J. A. (1988). Innumeracy: Mathematical illiteracy and its consequences. New York: Hill and Wang.

4 Democratizing Access to Calculus: New Routes to Old Roots

James J. Kaput
University of Massachusetts, Dartmouth, MA

> *What we call the beginning is often the end*
> *And to make an end is to make a beginning.*
> *The end is where we start from.*

T. S. Eliot's words, redirected from *Four Quartets,* Little Gidding V, where they were referring to problems of poetics, describe the standard school and university approach to calculus (Eliot, 1962, p. 144). Our introduction to the massive intellectual edifice called *calculus* is often the students' end experience in mathematics, but somehow it is to be the beginning of their careers as users of mathematics. But at another level, pedagogically, the historical end is where we start from.

The immediate occasion for writing this chapter was a conference that brought people together whose interests include mathematics, problem solving, and mathematics learning and cognition. Given that the university mathematics community is struggling to come to grips with its collective responsibility to teach mathematics and contribute to the nation's mathematical and scientific infrastructure, and given that this struggle has focused on the need to teach calculus much more effectively than in the past, this chapter reexamines our views of calculus. In particular, I reassess its proper place in the curricular path that so many students seem unable or unwilling to complete.

In order to do this I am forced to examine the nature of the mathematical content of calculus: What is it? What are its objectives, methods, and, especially, its representations? And in order to do this, I am further required to look back historically at how we came to be in the situation we now find ourselves. How did calculus come to be defined in the way it has? Our consensual, largely tacit, view of the subject has defined the curriculum and helped hold it in place. I especially examine the twin assumptions, first, that calculus is a "capstone" *course,* and second, that calculus needs to be introduced in the language of algebra. Regarding the first assumption, I look closely at the origins of the major underlying ideas of calculus for clues regarding how calculus might be regarded as a web of ideas that should be approached gradually, from elementary school onward in a longitudinally coherent school mathematics curriculum. Regarding the second assumption, I look closely at dynamic graphical means for representing important calculus ideas in ways that reflect their origins in the study of change and the critical roles of motion imagery in guiding this study.

SETTING THE PERSPECTIVE

> Without a well-developed notation the differential and integral calculus could not perform its great function in modern mathematics. The history of the growth of the calculus notations is not only interesting, but it may serve as a guide in the invention of fresh notations in the future. The study of the probable causes of the successes or failure of past notations may enable us to predict with greater certainty the fate of new symbols which may seem to be required, as the subject gains further development. (Cajori, 1929, p. 196)

Why the extraordinary stability of the standard approach to calculus across the past two centuries, despite the profound changes in the contexts in which calculus would be used, in the theoretical foundations of the subject, in the extraordinary growth of the mathematical ideas related to or built on the basic ideas of the subject, in the technological media in which the ideas of calculus are represented, and in the populations who are expected to learn those ideas? To begin to answer this question, we must look beyond the relatively superficial matters of texts, publishers, and syllabi that often dominate discussions of calculus reform and ask what has *not* changed over the years.

Certain interlocking cultural background factors did not change. The following factors comprise an essentially unquestioned, tacit background against which the "figures" of our understandings of calculus as a discipline and our relations to it are formed and sustained:

1. The language, primarily algebraic, in which the content of the subject was written has remained relatively constant. This helps hold a strong curricular prerequisite structure in place.
2. The static inert media in which the languages of calculus have been instantiated remained fixed across the years.
3. Calculus has come to be regarded as a *course* (or sequence of such) rather than a collection of ideas that can be approached gradually, involving many different perspectives, levels, and representational tools, as is the case with geometry, for example.
4. Calculus has remained the unquestioned province of the intellectual

elite—despite large-scale changes in the demographics of school and university populations and even changes in the role of schools and universities in society.

The stability of the first two factors has reinforced the second two. Algebra and knowledge of important classes of functions as algebraically defined objects have acted as an insurmountable barrier to calculus access for all but the intellectual elite. The second factor has limited the forms in which the ideas of calculus can be expressed, whether they be strictly algebraic or coordinate graphical, again, in a very quiet way, helping hold the other factors in place. These two factors also fit with the ways we typically translate mathematical content into mathematical curriculum: A presumed logical order drives the curricular order, the prerequisite structure (Kline, 1968); and major ideas are identified with mathematics "courses," which are inevitably experienced by students in a layered structure (Steen, 1990).

Today, all these background factors either are changing or are newly susceptible to change. As a result, the potential exists for a fundamental and systemic change in our individual and collective relationship with calculus. Thus, I am not concerned with rethinking certain topics in a canonical calculus syllabus, although the central ideas of calculus are examined in some detail, but rather with rethinking the entire enterprise of calculus as a school or university subject.

PREVIEW OF THE CHAPTER

The chapter is in three parts: Part I, Background on our notational framework; Part II, A curriculum- and pedagogy-sensitive historical overview of calculus; and Part III, Implications and future directions. The notational framework is intended to alert the reader to aspects of the mathematical experience that are often taken for granted but which play a major role in defining that experience. Further, having become aware of these notational aspects and how they have functioned in the past, we are in a better position to dream alternatives. The historical material that comprises a major portion of the chapter in Part II is intended, on the one hand, to do what history is always supposed to do—to give us a better understanding of the present, especially relative to issues of curriculum and pedagogy. On the other hand, it is also intended to help expose opportunities for major alternatives to the current tacit strategy of treating calculus as a target body of ideas that must wait on a long apprenticeship with arithmetic and algebra. Thus, I give more attention than usual to what are normally thought of as precursors to calculus and to larger themes and less attention to the historical development of technique and theory. Also, I give detailed attention to the representational struggles of Oresme, including efforts on his part that were decidedly confused and even unsuccessful, in order to provide a sense of the interaction between a developing understanding and the developing means for the expression of that understanding, as further outlined in Part I. He was in the same position as today's students, needing to develop both an understanding of the mathematics of change and notational systems for expressing that understanding.

The third part of the chapter examines what the new technologies can offer, particularly given a broader place for calculus in the school curriculum, and what new questions are raised by the alternative approaches suggested. I look at possibilities for an approach to the mathematics of change for all students, which begins in elementary school and builds gradually toward the formal system that is now identified as *calculus*.

PART I: A FRAMEWORK FOR THINKING ABOUT NOTATIONS

> In signs one observes an advantage in discovery which is greatest when they express the exact nature of a thing briefly, and, as it were, picture it; then indeed the labor of thought is wonderfully diminished. (Leibniz quoted in Cajori, 1929, p. 184)

Structuring Experience: A General Perspective on the Expression of Mathematical Knowledge

I take the point of view that we organize the flow of experience jointly using two sources of structure, one mental—the structures of mind—and the other material—the material artifacts, including spoken and written language, produced and used in accordance with our cultural inheritance in one or more physical media. Confining our attention to mathematical experience, the mental structures comprise our mathematical knowledge, formal and informal, and the material artifacts are the mathematical notations that we are able to relate to that knowledge. We relate our conceptions to physical notations in two ways as depicted in Fig. 4.1, each cyclic and each with both a deliberate and an automatic or "preconscious" aspect.

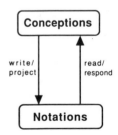

FIG. 4.1. Two interacting sources of structure.

4. DEMOCRATIZING ACCESS TO CALCULUS   81

Consider first the cyclical acts associated with reading, which begins with our attention directed toward an existing physical notation. Notations operate on our mental structures by perceptually evoking them (it is useful to think of resonance phenomena), and we can then deliberately respond, or read them. As we do so, we further organize our attention being applied to the notations in acts of parsing and thus reflect our own mental structures back upon them. This organizes still further the reading process at a higher level of organization in a cyclic pattern as indicated by the two arrows in Fig. 4.1. In sum, this amounts to the act of "building meaning" from the notations.

The second type of interaction involves producing, modifying, or elaborating notations. Here the emphasis is downward rather than upward, but it is again cyclic—think of it as beginning with the left arrow. We begin with the projection of mental contents through an act of writing in some notation system, which might mean a traditional form of writing, but can vary according to the system and the media involved. For example, it could involve putting cardboard puzzle pieces in place, putting sticks or objects together to denote numerical computations, and so on. With each "chunk" of such outward action comes an interpretive, reading of the result, which further informs the next writing act (both consciously and preconsciously).

Apart from the differences in starting point and the presence or absence of notations at the start of the process, the difference between reading and writing acts is mainly one of emphasis associated with the goal of the activity. While we have referred to the reading processes as *cyclic,* in fact, they are not, if we take into account the changes in mental state resulting from the sequence of perceptual and interpretive acts. And for writing acts, both the notation and the mental state change. In terms of Fig. 4.1, the up and down arrows do not return to the same places within the two boxes.

But many significant representation acts involve more than one notation system, for example, algebraic symbols and coordinate graphs. Figure 4.2 helps describe a horizontal dimension of referential relationship, which, as a species of representation, is quite different from the vertical one between conceptions and material notations. Here, we express relationships between physical notation A and physical notation B, in which each (and perhaps even the correspondence) is

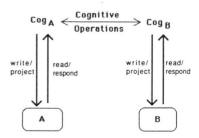

FIG. 4.2. "Horizontal" reference—Between notations.

expressible in material form. By Cog_X I mean a mental structure built using the system X as the primary organizing structure. Note that the actual referential relationship between A and B exists only as mental operations of a person for whom the notations are interpretable. Such reference exists only by way of composite actions that "pass through" the mental world of conceptions for example, as combined acts of interpretation, mental translation-operation, and projection; it does not exist apart from the actions of interpretants, although members of a community may share the conceptions and even a referential relationship in the sense of being able to generate the interpretive acts as needed. It is in this sense in which I speak of "mathematical objects" later, as sharable conceptual experiences among a given community, for example, a community of mathematicians (see Kaput, 1992a, for further discussion).

The directionality of the reference in general depends on the cognitive operations involved, which in turn depend on the context and hence is not fixed. At one time a coordinate graph B may act as a referent for an algebraic equation A, where A is conceived as representing B, and at another the situation may be reversed, with the graph conceived as representing the equation. As depicted in Fig. 4.2, the correspondence between notations is not provided physically. However, such a physical connection can be provided, for example, by a computer program, which links equations and graphs in one or both directions. While such physical connections may influence acts of reference, they do not constitute them.

The same scheme shown in Fig. 4.2 is intended to include the situation, in which, for example, B is a situation that is being mathematically *modeled* using the notation system A (Kaput, 1989, in press). In this case, Cog_B is based on one's understanding of situation B. An example might involve a body rolling down an inclined plane and an attempt to model its displacement graphically based on a sequence of measurements. Here there is an immediate contrast between, say, a physics class and a mathematics class. In the former, B exists, and actual actions involving B take place, leading to the construction of A. In a mathematics class, B does not exist, while Cog_B is assumed to exist. Instead, B might be represented by a surrogate representation, say a table of data. (This will be elaborated later.) When I speak of notations being used to represent phenomena or even other mathematical entities, I often refer to them as *representations*, rather than merely notations.

The examinations of two important distinctions will be discussed later. One involves the kinds of notation systems that are used—the distinction between display and action notations. The latter did not develop until the rise of algebra in the 17th century but had profound impact on the nature of mathematics. The second concerns the distinction between static/inert media and dynamic/interactive media. By *medium* we mean the physical material in which the information-carrying dimensions of a notation system can be instantiated (Kaput,

1987). I discuss this issue in the context of potential computer contributions in the third part of the chapter.

PART II: HISTORICAL BACKGROUND ON THE EVOLUTION OF CALCULUS

Identifying Taproots

Three Root Aspects of Calculus.

There seem to be three very broadly reaching roots that feed into what we now call *calculus,* each reaching back to antiquity, but with differing degrees of relative importance and forms of expression through the ages. One has concerned itself with geometric issues related to computations of areas, volumes, and tangents. It was primarily practical, as reflected initially in the work of Archimedes, for example. A second was a mix of practical and theoretical interest involving the characterization and theoretical exploitation of continuous variation of physical quantities. While the second did not begin in earnest until the Scholastics' attempts to mathematize change and did not thrive until the decades before and including Newton, it had a history going back to the Greeks' notion of generating geometric objects through continuous motion. It took more abstract form in the hands of the mathematician/physicists in the 18th and 19th centuries who quantified ever more physical phenomena. A third root was inherently theoretical, beginning with Zeno's ancient motion paradoxes and continuing through to the development of a formal theory of limits in the 19th century that now logically frames the other two strands. It includes more recent efforts that resulted in a complete theory of infinitesimals (Keisler, 1976; Robinson, 1966). Obviously, these roots interweave complexly, and the story of their interweaving is still being written.

Beginning with Root Problems Instead of Methods:
Applied Phylogeny.

The cognitive application of "ontogeny recapitulates phylogeny"—the development of a person's understanding of an idea parallels, or should parallel, the historical development of that idea—is always tempting, particularly as a curriculum design principle. But there are many reasons to be careful in using it. Beyond a certain inefficiency in historical process (resulting from the "blindness of the watchmaker") and the differences between a collective historical enterprise and an individual's learning, we need to consider the irregularity of historical developments in a subject such as calculus over the years, for example, the fact that early Scholastic attempts to exhibit changing quantities graphically preceded coordinate graphs, and the fact that the latter came into existence before the idea of function was in place.

The perspective of Piaget and Garcia. Perhaps the deepest and most complete analysis of the parallels between historical and individual development has been made by Piaget and is described in Piaget's last book,[1] coauthored with Garcia (Piaget & Garcia, 1989). Their thesis offers three main stages for the development of any major idea in increasing levels of structure: (a) the intraoperational, (b) the interoperational, and (c) the transoperational, together with recursive mechanisms effecting the transitions from one to the next. Oversimplifying drastically, one might say that in the first stage one performs actions within the objects, with attention to the properties of the objects themselves. Next, one shifts attention to relationships and transformations between objects and invariances across objects. Lastly, one builds a higher level structure that embodies these relationships as its elements, and one attends to the properties of this structure.

The defining feature of Piaget's approach, however, is that the stages and mechanisms that he postulates are not psychological, or historical (so he is not "reporting" an accidental parallel between the two), but rather, epistemological—this is how knowledge is inherently constructed. Piaget thinks of history and the individual as comprising parts of the same "epistemological laboratory." Further, each stage transforms the character of the objects and actions that preceeded it, which come to exist in a new way as participants in a higher level structure. And, the stages iterate. The structures in the third, transoperational stage can be taken as objects in which actions can be taken in a new intraobject stage, and so on.

Piaget and Garcia trace this process for the fields of mechanics, geometry, and algebra. The easiest to summarize is geometry, when jumping in at one intraobject level of this process, the initial objects are particular geometric figures, and one examines their internal properties. But at the interlevel, relationships among these are of interest, as in projective geometry, in which transformations play a key role in formulations and proofs of results. As is widely appreciated, this step was very much facilitated by the appearance of analytic geometry, which allowed statements to be made directly about general classes of figures, rather than particular figures. However, those transformations were not themselves objects of study, but rather served as the means by which geometry is done. Later, however, in the 19th century, as the attention shifts upward in abstraction, these transformations themselves become the elements of structures, namely groups, and these structures in turn allow us to see geometries in an entirely new, wholistic way, as culminated in Klein's Erlanger Program. Similar things happen in algebra, in which substitution transformations in equation solving came to

---

[1]The author is indebted to Ed Dubinsky for bringing this work to his attention.

have a group structure in their own right that in turn fundamentally transformed the way we understand equation solving and led to a new theory, that of groups and fields. This is a glimpse of what Piaget and Garcia (1989) see as a very general process:

> Abstract mathematical notions have, in many cases, first been used in an instrumental way, without giving rise to any reflections concerning their general significance or even any conscious awareness of the fact that they were being used. Such consciousness comes about only after a process that may be more or less long, at the end of which the particular notion used becomes an object of reflection, which then constitutes itself as a fundamental concept. (p. 105)

An analysis of the idea of derivative that shares these features of Piaget's has been offered by Grabiner (1983). She defines four periods associated with the derivative: (a) when it was first used (to determine tangents and quadratures before Newton and Leibniz), (b) when it was discovered (by Newton and Leibniz), (c) when it was explored and developed (in the 18th century), and (d) when it was finally defined (in the 19th century). In her terms:

> That is, examples of what we now recognize as derivatives first were used on an ad hoc basis in solving particular problems; then the general concept lying behind all these uses was identified (as part of the invention of the calculus); then many properties of the derivative were explained and developed in applications both to mathematics and to physics; and finally, a rigorous definition was given and the concept of derivative was embedded in a rigorous theory. (p. 195)

We shall refer to Piaget's perspective from time to time to help isolate major transitions in the historical development of calculus that, because they also represent significant reorganizations of knowledge, are relevant to the design of curriculum and instruction. However, while it helps identify major transitions, it does not fully explicate the role of such factors as notation development (Gardner, 1979).

Notations, Purposes, and Methods.

A close look at the events alluded to earlier will reveal that some of the greatest innovations took place in association with the application of or the development of notations, especially when the amount of forward "propellant" of the innovations is considered. In the case of geometry, it was the generality-elevating role of algebraic description, and in the case of derivatives, it was the potent new notations introduced by Leibniz.

To help clarify matters, I distinguish between what each of the three root aspects of calculus identified earlier is about—its purposes, animus, and problems—and its substance—its notations, theory, and its methods.

From a modern instrumental (or technique-oriented) perspective, purpose, notation, and method of calculus are almost the same: Efficiently encoded methods for the efficient handling of problems have organically grown over the problems themselves, so that now one sees mainly methods. The original problems have long since composted and have fertilized the growth of luxuriant methods. Problems now only appear in synthetic form, as chemical fertilizer, to be sprinkled on a highly organized mass of methods. Indeed, one could identify the contemporary calculus as experienced by students as being entirely a set of methods, with a bit of theory perhaps thrown in or pointed to so that the students realize they are not being told the whole story.

But looking historically, there are problems attacked before efficient methods were developed for their solutions. Moreover, history reveals several layers of growth that separate today's methods from those that were developed when the root problems were at the center of attention.

Hence, our first application of calculus's phylogeny is mainly to guide a choice of root problems with which to begin the study of the subject, in particular, the quantification of variable quantity, rather than to determine the details of an historically organized curriculum, for example, as in Priestly (1979), in which all the background assumptions about calculus remain intact. Nevertheless, certain rather large historical struggles and transformations in the nature of legitimate mathematical activity will appear, struggles and transformations that connect across the different layers of methods and representations that developed across the centuries.

In terms of Grabiner's four stages (use, discovery, development, and formal definition) I concentrate mainly on the first two periods, but with some attention also to the third and fourth.

Each of the three broad root patterns described above, the geometrical, the characterization of continuous variation, and the theoretical, has weaved around a few central problems, despite the radical changes in technique associated with changes in the languages used to encode those problems and their solutions. Of these root problems, the geometric and the continuous variation roots, are especially of interest. More particularly, I am more interested in those root problems associated with describing change and accumulation of continuously variable quantity, and especially the relation between change and accumulation as represented geometrically and kinematically. I make this choice, because (a) they seem to have the greatest relevance to curriculum design for younger students, (b) they seem to have been lost in the ubiquitous curricular press for computationally efficient algebraic methods, and (c) continuous variation can now be richly represented in dynamic electronic media.

This is, in part, a return to what mathematicians of the 13th–14th centuries, and then the decisive 17th century, were attempting to achieve—the mathematization of variation—prior to the 18th-century crowning work by Euler and Lagrange that exploited those earlier constructions algebraically. This in turn preceded 19th-century attempts, especially those of Bolzano, Cauchy, and Weierstrauss, to put the whole enterprise on a sound logical footing.

Our approach, beginning with the characterization of variation, thus puts in an ancillary position certain issues that most mathematicians today regard as central to calculus, limits and continuity, which turn out to be rather recent concerns when compared to geneology of the root problems.

It will be useful to review a bit of the history of the notion of continuous variation. The basic sources are Boyer (1959), Boyer and Mertzbach (1989), especially Clagett (1968), Edwards (1979), and Struik (1986).

Early Understandings of Continuous Variation.

Boyer (1959) noted that while the Greeks "had made some qualitative speculations on the subject of motion, . . . the idea of continuous variation by means of geometrical magnitude or of studying it in terms of the discreteness of number does not seem to have arisen with them" (p. 71). It was not until the 14th century:

> that a theoretical advance was made which was destined to be remarkably fruitful in both science and mathematics, and to lead in the end to the concept of derivative. This consisted in the idea—often expressed, to be sure, in terms of dialectical rather than mathematical method—of studying change quantitatively, and thus admitting into mathematics the concept of variation. (p. 71)

Edwards (1979) emphasized the limitations of Greek mathematical thought imposed by their insistence on absolute logical rigor:

> Although the Greek bequest of deductive rigor is the distinguishing feature of modern mathematics, it is arguable that, had all succeeding generations also refused to use real numbers and limits until they had fully understood them, the calculus might never have developed, and mathematics might now be a dead and forgotten science. (p. 79)

A parallel between the Greeks and the students of today. The discussion by both Boyer and Edwards of the reasons why the Greeks did not begin a serious study of variation, especially involving motion, is suggestive of a gap that exists today in the usual approach to the mathematics of motion. First, the Greeks kept their discrete and continuous worlds separate, with a high demand for rigor in mathematical statements that prevented arguments from mixing the two worlds. Second, they had no concept of acceleration—all astronomical motions of interest were either uniform or circular, whereas the motion of everyday objects was highly nonuniform. "Motion was, it appeared, a quality rather than a quantity; and there was among the ancients no systematic quantitative study of such qualities. . . . In general, Greek mathematics was the study of form rather than of variability" (Boyer, 1959, p. 72). Boyer goes on to note that "the quantities entering into Diophantine algebraic equations are constants rather than variables, and this is true also of Hindu and Arabic algebra" (p. 72). Finally, according to Van der Waerden (1963), the actual learning of Greek mathematics took place in a strongly oral tradition, so that when social and political changes cut off the oral transmission based in the Greek academies, and their mathematics was represented only in books, progress ceased. Succeeding generations could not master the highly formal technique based on the study of text alone, especially the key, but extremely difficult, work by Archimedes.

I am struck by a parallel between certain aspects of Greek mathematics and aspects of students' mathematical condition as they traditionally enter the study of calculus. Students are primarily (discrete) arithmetic creatures (Kaput & West, 1993), whose understanding of algebraic literals is as unknowns rather than as variables. (To be sure, today's students have a much easier arithmetic of number with which to arithmetize their world as compared to the Greeks' cumbersome geometric representations.) They are usually without a concept of acceleration, and the only kinds of motion that they see depicted within mathematics are those describable via algebraically simple formulas—entirely unlike the irregular motion they experience in their daily lives—and hence they do not cognitively connect these school applications to their everyday experience.

Further, the view of rigorous mathematical proof that students inherit from their study of Euclidean geometry (Schoenfeld, 1986), coupled with the views of formal proof inherent in the typical calculus text, leave them with a profound schism between what might need to be mathematically explained in their daily experience of variable quantities and the world of deductive mathematical argument, including its formal definitions. With this starting point, neither Aristotle nor Archimedes could even begin to think of the calculus of variation. Should we now expect more of the average student, especially based on only 100–200 hours of often poorly designed and executed instruction? Instead, the student attempts to memorize as much of the barrage of algebraically encoded technique as possible. In fact, given their prior instruction in manipulative algebra, in which the only functions are given by closed-rule formulas, and in which no significant attention is given to the fact that functions have domains and ranges, they approach the calculus with something of the same practical epistemology as was prominent in the 18th century, when all the theoretical problems that were dealt with in the next century did not exist (Fraser, 1988).

Mathematizing Variation: The Discursive Beginnings

Gradually, in the 13th–15th centuries, the classic Peripatetic doctrine, that all motion is the result of some external force, was replaced by the impetus doctrine, that a body once set in motion holds some tendency to remain in motion. This made mathematically possible the idea of velocity at a point, instantaneous velocity. A precise mathematical language suitable for the expression of this idea was centuries in coming. Beginning in the 13th century, probably in the work of

Duns Scotus and others, it took the form of "a discussion of the latitude of forms, that is, of the variability of qualities" (Boyer, 1959, p. 72). One basic distinction the Scholastics made, based on the Aristotelian distinction between quantity and quality, was between *quantities,* which can vary in *extent,* such as weight, distance, length, or volume, and *qualities,* which can vary in intensity, such as brightness of illumination, density, velocity, or even certain attributes of the soul (which we might characterize in psychological terms, e.g., intelligence, happiness, etc.).

A *form* refers to "any quality which admits of variation and which involves the intuitive idea of intensity—that is, to such notions as velocity, acceleration, density. . . . In general, the latitude of a form was the degree to which the latter possessed a certain quality, and the discussion centered about the *intensio* and the *remissio* of the form, or the alterations by which this quality is acquired or lost" (Boyer, 1959, p. 73). This is in contrast to an accumulation style of increase and decrease of quantity. This difference, of course, is the same difference we continue to use today to distinguish between intensive and extensive quantities. And the relationship between intensity and extent is at the conceptual foundation of the calculus. Given something that possesses a variable quality, how does one define the total extent of that quality? And vice versa?

The writers of the 14th century, when the discussions intensified, distinguished between uniform and nonuniform change (*latitudo uniformis* and *latitudo difformis*) of forms. They also distinguished between two types of *latitudo difformis,* namely, *latitudo uniformiter difformis* and *latitudo difformiter difformis:* linear versus nonlinear change. Occasionally, the latter was further distinguished into two types: *latitudo uniformiter difformiter difformis* and *latitudo difformiter difformiter difformis.* I mention the terminology to give a sense of the difficulty experienced by those who are attempting to develop a coherent mathematical theory of variation before a systematic language for the expression of a theory was available.

Richard Suiseth, an early Oxford scholar of the first half of the 14th century, who was known as the Calculator—in the same sense that Aristotle was known as *the* Philosopher, and Paul, *the* Apostle—helped, together with his slightly older contemporary, Bradwardine, to lay the basis for an increasingly sophisticated analysis of variation. He managed to formulate and argue in the context of thermal phenomena the basis for a mean value theorem for a linearly changing quantity, that is, the idea of average of *latitudo difformis.* The arguments were long and discursive, sprinkled with numerical examples, but, more importantly, they were not geometric—not a single diagram ever appeared. Even when dealing with infinite intensities, he appealed to intuitions regarding uniform change. It is helpful to quote Boyer at length:

> To Calculator we owe perhaps the first serious effort to make quantitatively understandable these concepts of mathematical physics. His bold study of the change of such quantities anticipated not only the scientific elaboration of these, but also

adumbrated the introduction into mathematics of the notions of variable quantity and derivative. In fact, the very words *fluxus* and *fluens,* which Calculator used in this connection, were to be employed some three hundred years later by Newton, when in his calculus he spoke of such a variable mathematical quantity as a fluent and called its rate of change a fluxion. Newton apparently felt as little need as Suiseth for a definition of this notion of fluxion, and was satisfied to make a tacit appeal to our intuitions of motion. Our definitions of uniform and nonuniform rate of change, are, as Suiseth anticipated, numerically expressed; but their rigorous definition could only be given after the development, to which Newton contributed, of the limit concept. This latter arose out of the notions of the calculus, which, in their turn, had evolved from the intuitions of geometry. The prolix dialectic of Calculator made no appeal to the geometrical intuition, which was to act as an intermediary between his early attempts to study the problem of variation and the final formulation given by the calculus. This link between the interminable discursiveness of Suiseth and the concise symbolism of algebra was supplied by others of the fourteenth century who studied the latitude of forms. (1959, p. 79)

Mathematizing Variation: Oresme's Struggle for Representations

Most notable among those who studied the latitude of forms was Nicole Oresme, who recognized the need for geometric representation of variation:

> The work of Oresme therefore makes most effective use of geometrical diagrams and intuition, and of a coordinate system, to give his demonstrations a convincing simplicity. This graphical representation given by Oresme to the latitude of forms marked a step toward the development of calculus, . . . [I]t was the study of geometrical problems and the attempt to express these in terms of number which suggested the derivative and the integral and made the elaboration of these concepts possible. (cited in Boyer, 1959, p. 80)

Precursor Representations.

Others before Oresme, including Duns Scotus, had helped prepare the representational path from Aristotle. In particular, Roger Bacon in the mid-13th century introduced a one-dimensional approach to thinking about variable intensities of qualities: "Every inherent form receives intension and remission, on account of which it becomes understandable when set forth as a line that is called the line of intension and remission. And since every inherent form has a contrary and a mean, that same line will be imagined as containing contrary forms. Suppose a hotness is placed in any place whatever on the aforesaid line. . . . " (cited in Clagett, 1968, p. 57).

He went on to discuss the relation between the "longitude" of the mean (on that line) and the contraries. An illustration of how Bacon and others of his time dealt with the relation between quality, which can have varying intensity, and the quantity of that quality is illustrated in the following passage, in which the author (whom Clagett argues is probably Bacon) is dealing with the mixture of two units of water hot in the "sixth degree" and one unit of "twelfth degree":

> For example, let there be given water of two weights hot in the sixth degree in respect to some point contained in the same line; let there be given another water of one weight hot in the twelfth degree with respect to the same point; a mixture of the two waters having been made, the hotness of the mixture will be raised in a line of intension through eight degrees, with respect to the aforesaid point, since the distance that is between six and eight is one-half of the distance that is between eight and twelve, just as the water of one weight is half the water of two weights. (cited in Clagett, 1968, p. 58)

Here, as elsewhere among Scholastics of the 13th and early 14th centuries, there is a weighted average of intensity being computed arithmetically and then being placed on a single line. Such lines are essentially abstract conceptual aids, because they do not appear in the manuscripts before Oresme but serve as ordered placeholders for quantities taken to be linearly ordered. They are likely based in Aristotle's comparisons between "contrary qualities" such as hotness and coldness, which represented extremes.

In the *Liber calculationum* of the Calculator, dated from the 1340s, there is some presaging of the use of two dimensions to express variable intensities as he and other Schoolmen attempted to come to grips with the "extent" or "quantity" of a quality. In particular, he gropes with how to increase a quality's extent without increasing its intensity via an analogy that increases a rectangular shape in one dimension but not the other:

> If a length of a foot be taken, and if there were added to the side of it a length as great or smaller, the whole length would not be increased, because [the second length] is added in such a way that it cannot increase that dimension (i.e., length) as such. . . . Thus in the question at hand, that acquired quality is acquired in such a way that it does nothing intensively to the whole quality, but only [affects it] quantitatively. (Clagett, 1961, pp. 336–337)

Oresme's Early Diagrams of Qualities Possessed by Linear Objects.

Because Oresme's approach is of special interest, I look more closely at his work itself, as included in the aforecited lengthy tract by Clagett (1968), who offers a translation and extensive commentary on Oresme's *De configurationibus*, "published"[2] in the mid-1350s. Oresme's approach was to represent the "subject," which possessed the quality, as a horizontal line—a *longitudo*. For

---

[2]This was prior to the printing press, so it was a handwritten manuscript circulated among interested peers and students.

each point of the *longitudo* was determined a *latitudo,* a perpendicular straight line whose length represented the intensity of the subject's quality. Actually, he used this system only for "linear subjects" or what we might term one-dimensional subjects. An example of such a subject might be a rod whose temperature or illumination varies from one end to another. He also discussed the representation of the qualities of planar regions in analogous fashion, with perpendiculars at each point, thereby generating a solid. He even attempted to discuss qualities of three-dimensional objects, while attempting to avoid a new dimension.

The global representation of the latitudes of forms was repeatedly used in the special cases of *latitudo uniformis* and *latitudo difformis*—indeed, the title of Oresme's chief work uses the phrase *De configurationibus,* emphasizing his attention to the global forms associated with different types of variation. In particular, uniform velocity over a time interval, that is, constant velocity, is represented by a rectangle whose width represents that interval and whose height represents the velocity; and uniformly difform velocity (linearly changing velocity) is represented by a right triangle. In a sense, *the geometric object (rectangle or triangle) was being used to represent a total event as a global representation,* in which the subject was the time interval over which the motion occurred. This is not quite the same as thinking of the curve, that is, the "summit," of the rectangle or triangle, as representing the event. The latter conceptualization was not possible until after Fermat and Descartes. Nonetheless, Oresme used the shape of the entire figure (its *configurationibus*) as the basis for talk about varying intensities. He also described variation of quality in terms of the "summit line," which expresses the essential information, because the perpendicularity of the intensity lines determined the remaining sides of the figure.

He actually characterized *latitudo uniformiter difformis,* linearly changing intensity, in terms of what we now call slope, in Part I, Chapter xi of *De configurationibus*[3]:

> If any three points [of the subject line] are taken, the ratio of the distance between the first and the second to the distance between the second and third is as the ratio of the excess in intensity of the first point over that of the second point to the excess of that of the second point over that of the third point, calling the first of those three points the one of greatest intensity. (in Clagett, 1968, p. 193)

Oresme went on to repeat his statement using the following very familiar Fig. 4.3. Incidentally, his language revealed that he treated the figure in a right to left orientation, and throughout his work, there is no indication of a fixed directionality that would indicate a directionally ordered number line conception behind any of the horizontal lines. He read other figures from left to right.

In addition to thinking of an intensity's configuration as both a region with

---

[3] I refer to Oresme's work from Clagett's translation (1968) in terms of Roman numeral part and chapter, so the following is from I.xi.

4. DEMOCRATIZING ACCESS TO CALCULUS   93

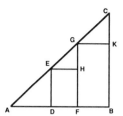

FIG. 4.3. Oresme's "graph" of linearly increasing intensity.

a certain shape and as a summit line, Oresme also used motion imagery to describe varying intensity. What he did not say is as revealing as what he did say, given the temptation succumbed to by some historians to ascribe to Oresme much more modern mathematics than he actually created. He imagined a point moving regularly along a subject line, as follows:

> Again, we can be led to a knowledge of the differences which have been premised by the imagery of motion. For let point d be imagined as moving regularly on line AB and in such a way that any point of line AB over which d comes will be equal and similar in intensity to that same point d. [See Fig. 4.4 below.] If, therefore, in the beginning of the motion the point d has a certain degree or some intensity and it continually remains in that same degree without alteration throughout the motion, then it will describe in line AB a uniform quality. But if in the beginning of the motion, point d has none of the quality and during the motion point d is continually altered and regularly increased in intensity, then it will describe a quality uniformly difform terminated at no degree. If, moreover, d is regularly increased in intensity, but in the beginning of the motion has some quality or intensity, then it will describe a quality uniformly difform terminated in each extreme at [some] degree. Similarly, if in the beginning of the motion d has some quality and it is regularly decreased in intensity to the end of the motion, then d will describe a quality uniformly difform terminated in both extremes at [some] degree. If the quality of d is decreased in intensity to no degree, then the quality described will be uniformly difform terminated at no degree. But if d is irregularly moved and regularly increased or decreased in intensity, or even conversely, it will describe a quality difformly difform. However, it could happen that point d would be irregularly moved and irregularly altered in such a compensatory or equivalent fashion, that it would then describe a quality uniformly difform. But whenever there would be no such compensation, then it would describe a quality difformly difform. (I.xii, in Clagett, 1968, pp. 195–197)

At this point, he did not think of the point d as generating a summit line above the subject line or as the top of a perpendicular line sweeping out a region. His representation called for point d somehow to "carry" the intensity of the quali-

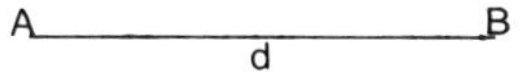

FIG. 4.4. A point moving along segment AB.

ty—with no geometric feature indicated to do so! I thought of the point as having a varying brightness. Sixth-grade students attempting to build graphs of motion described by diSessa (diSessa et al., 1991) struggled with some of the same representational problems, although they exhibited surprising competence and creativity. Much later, he suggested that, as a point traverses the linear subject, we could think of it as turning the part traversed white. He then went on to take an even more generative approach—in the sense of Pappus:

> The intensive acquisition of punctual quality is to be imagined by the motion of a point continually ascending over a subject point and by its motion describing a perpendicular line imagined [as erected] on that same subject point. But the intensive acquisition of a linear quality is to be imagined by the motion of a line perpendicularly ascending over the subject line and in its flux or ascent leaving behind a surface by which the acquired quality is designated. For example, let AB be the subject line. I say, therefore, that the intension of point A is imagined by the motion, or by the perpendicular ascent, of point C, and the intension of line AB, or the acquisition of the intensity, is imagined by the ascent of line CD. [CD is the top of a rectangle with base AB.] Further, the intensive acquisition of a surface quality is in a similar way to be imagined by the ascent of a surface, which (by its motion) leaves behind a body by means of which that quality is designated. (III.i, in Clagett, 1968, p. 395)

Some insight into the overall situation is provided by a closer look at the referential relationship between Oresme's "configurations" and the qualities thereby represented—in terms of Fig. 4.2, the A of his representation and the B it was used to model. It was an issue to which he paid explicit attention in his struggle to express himself. The following key statement reveals that he intended a particular unit length reference to be chosen for the vertical lines to represent the quality under consideration, reflecting a basic principle of measurement:

> Although some linear quality can be correctly imagined by any plane figure other than those mentioned before, still not any quality can be imagined by any figure. Indeed no linear quality is imagined or designated by any figure except the ones in which the ratio of the intensities at any points of that quality is as the ratio of the lines erected perpendicularly in those same points and terminating in the summit of the imagined figure. (I.vi, in Clagett, 1968, p. 179)

Thus here, as elsewhere, he maintained there is a separation between the representation and the thing being represented—the A and the B in Fig. 4.2. However, there is a difficulty with the role of the horizontal line as sometimes representing the subject directly, while the vertical lines are viewed as entirely imaginary, as aids to the imagination—the notations for X and Cog_X are not always clearly separate. Two distinct forms of reference are embodied in the same system, which yields special difficulties in the case of velocity as shall be seen presently. Nonetheless, he knew well that the representation should be faithful to variation in the quality being represented, although the second sentence does not quite express this as well as he did in a later discussion, in which he used specific numbers and a picture, appropriately generalized by a follow-up statement. He thus distinguished between suitable and nonsuitable configurations (I.vii, in Clagett, 1968). He did this, however, without specification of a unit of measurement for the quality being represented. It is quantified, but not in the modern sense (Luce & Narens, 1987). Nonetheless, by focusing on preserving ratios of pairs of intensities at arbitrary points of the subject, he forced faithfulness.

If one reviews Oresme's language, he frequently and explicitly addressed "the imagination" and the functionality of his notations: "The aforesaid differences of intensities cannot be known any better, more clearly, or more easily than by such mental images and relations to figures, although certain other descriptions or points of knowledge could be given which also become known by imagining figures of this sort" (I.xi, in Clagett, 1968, p. 193).

Oresme's Characterization of Variable Motion.

Oresme knew that uniform acceleration yielded uniformly difform velocity, a fact argued discursively by Suiseth and others in the previous generation at Oxford. In fact, his teacher Jean Buridan at Paris was a leading developer of the impetus theory of motion (Dugas, 1988, chap. 4). Oresme also proved the mean value result for uniformly difform velocity, sometimes known as the Mertonian Rule, geometrically using some version of Fig. 4.5 below, variants of which appeared frequently in his work (see also Struik, 1986, p. 137). Indeed, scribes in Paris occasionally added Oresme's figures to earlier documents by Suiseth and others as they copied them (Clagett, 1968, p. 61)—creative copy machines!

In particular, he said that over a given interval of time, a body starting from rest and moving with (positive) uniform acceleration would travel as far as one with uniform velocity equal to half the final velocity of the accelerating body. While Oresme did not state explicitly that the areas of the rectangle ABGF and triangle ABC represent distance traveled, his proof concluded that the distances were equal based on the fact that triangles AEF and CEG are congruent (III.vii, in Clagett, 1968). Boyer noted (1959), "This is perhaps the first time that the area under a curve was regarded as representing a physical quantity, but such inter-

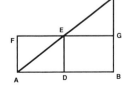

FIG. 4.5. Oresme's Mean Value Theorem—Geometrically presented.

pretations were to become before long commonplaces in the application of the calculus to scientific problems" (p. 84).

A sense of how far Oresme had to come is illustrated by his starting point early in the part of his book devoted to characterizing velocity, wherein he separated out three features of motion that need to be accounted for—distance, time, and velocity:

> Every successive motion of a divisible object has parts and is divisible in one way according to the division and extension or continuity of the mobile, in another way according to the divisibility and duration or continuity of time, and in a third way— at least in imagination—according to the degree and intensity of velocity. From its first continuity motion is said to be 'great' or 'small'; from its second, 'short' or 'long,' and from its third, 'swift' or 'slow.' And so motion has two extensions, one that pertains to the subject and the other that pertains to time, and one intensity. Now the two extensions can be imagined in a certain way as mutually intersecting at right angles in the manner of a cross, so that the extension of duration ought to be said to be 'longitude' and the extension in subject ought to be 'latitude,' while the intensity could be called the 'altitude' of this motion or velocity. (II.i, in Clagett, 1968, p. 271)

An attempt is made to create axes for time and for the (linear) object itself. He then struggled with two ways of displaying motion, one with time as the horizontal axis and the intensity of velocity as vertical (yielding a two-dimensional figure), while the other had the horizontal axis standing for the object itself, which then moved, acting as a surrogate object, to generate another two-dimensional figure as a trace of the motion—he referred to this in terms of velocity having a "double longitude."

This creates what he referred to as a "small difficulty concerning uniformity" that relates to the issue of describing motion either by time or by "subject":

> Punctual velocity according to subject (as well as punctual quality) is to be imagined by a straight line, and linear velocity according to subject is to be imagined by a surface or by a surface figure while the velocity of a surface is to be imagined by a body in completely the same way as was described in connection with the figuration of qualities in I.Iv. Instantaneous punctual velocity, however, is to be imagined by a straight line while punctual velocity enduring in time is to be imagined by a surface; moreover the velocity of a line with respect to time is to be imagined by a body and similarly the velocities of a surface and a body are imagined by bodies. . . . And thus it is evident that subject difformity of a line in motion is designated by a surface figure, and similarly the temporal difformity of the velocity of a point in motion is [also] designated by a surface figure. And since we can learn about corporeal figures from a knowledge of surface figures, therefore a comparison of other difformities could be discerned from the comparison of the subject difformity of the velocity of a line in motion to the temporal difformity of velocity of a point in motion. (II.ix, in Clagett, 1968, p. 293)

He thus recognized the difference between instantaneous velocity and velocity across time (although in his universe, all "continuous" entities had some minimal grain size, albeit indefinite). He attempted to deal with a shortage of representing dimensions and especially with respect to representing time, in which he must use a static medium. He was further handicapped by a lack of a fluent use of such idealizations as point masses. Nonetheless, he (and other Schoolmen) thus struggled to describe motion of bodies "with extent" and even bodies which are elastic, so that different parts begin moving at different times. They also used this parallel approach to discuss angular velocity, in which linear velocity varies proportionally with distance from the center of rotation.

Moreover, he and others struggled with different ways of describing increase and decrease of velocity, as echoed later in the writings of Galileo: Should velocity of a falling body be described in terms of ratios of distances traveled and ratios of times elapsed, or should it be described in terms of ratios of distances from the destination and ratios of times elapsed (which was sometimes taken to be the center of the earth)? (Many of these examples are found in II.iv, in Clagett, 1968.) Similarly, he noted that in comparing a large tree which grows 2 inches with the growth of a small tree that grows 1 inch, one might regard the larger tree as being augmented more slowly than the smaller (II.v, in Clagett, 1968), if one takes into account the ratios of starting and ending magnitudes.

Once again, I am moved to wonder how much of this struggle to represent and understand variable motion is important to experience before one moves on to the idealizations that are the starting points for contemporary student mathematical experience with motion. The sixth-grade students mentioned earlier worked very hard before they were able to come up with a representation of velocity against time rather than position (diSessa et al., 1991).

Oresme's Influence on the Development of Calculus.

The magnitude and subsequent import of Oresme's advances are not universally appreciated. For example, Kline (1972) says, in reference to Oresme's technique, that while it was "a major technique in the Scholastic attempt to study physical change, was taught in the universities, and was applied in efforts to revise Aristotle's theory of motion, its influence on subsequent thought was minor" (p. 211). On the other hand, while a paper trail is hard to track from Oresme to Galileo, the "Third Day" of Galileo's *Discorsi* in 1638 bears a striking resemblance to Oresme's treatment of variable motion, including the diagrams (Clagett, 1968, p. 71). In fact, Clagett argues (pp. 103–110) that Galileo's early writings contain quotes from Calculator and probably from Oresme, thereby revealing direct influence.

Boyer (1959) suggests that Oresme, in keeping with the mathematical atomism of his times, thought of lines as having some width, so the areas could be thought of as made up of large numbers of vertical lines—which dovetailed with his notion of instantaneous velocity, a velocity for a very short interval. These notions came to be shared by others in Paris, in Italy, and at Oxford in the latter 14th and in the 15th and 16th centuries, but achieved their boldest formulation by Galileo, whose debt to Oresme may be significant. Oresme also, when analyzing *latitudo difformiter difformis,* which he represented as a flat-bottomed semicircular region, "remarked that the rate of change of an intensity is least at the point corresponding to the maximum intensity" (in Boyer, 1959, p. 85).

While Boyer cautions against reading into this observation the kind of thinking captured in the calculus of three or more centuries later (as many historians apparently have), it is of interest to examine what kind of thinking this is. After all, it does not involve even the description of a curve, let alone an algebraically defined one—it is not analytic geometry in the sense of Fermat and Descartes. But Boyer noted, "His remarks show clearly, however, how fruitful the idea of the latitude of forms was to be when it entered the geometry and algebra of later centuries, to become eventually the basis for calculus. The Scholastic philosophers were striving to express their ideas in words and geometrical diagrams, and were not so successful as we who realize, and can make use of, the economy of thought which mathematical notation affords" (p. 86).

By describing variable intensities in terms of the varying heights of the rectangles and regarding the area of the rectangle as representing the *quantity* of the quality, he also presaged later thinking. In a sense, the later use of rectangles under a curve amount to an attempt to recover this representation in a true coordinate system, in which the objects are curves and points on those curves.

Oresme, as had Suiseth, also worked with simple infinite series. Oresme did so in the context of the latitude of forms, especially velocities and distances. He examined a variety of cases in which velocity was constant for the first half of a time interval and then increased (via uniform acceleration) over subsequent shrinking subintervals. For example, he considered the case in which it doubled in the next quarter interval, tripled in the next eighth, quadrupled in the next sixteenth, and so on. In this way, by thinking in terms of concrete distances covered, which were also represented by areas of rectangles, he concluded that the total distance covered would be four times that covered in the first half of the interval.

Boyer remarked that had these Scholastics been more closely attuned to the work of Archimedes rather than Aristotle, they may have been able to build on their results more fruitfully, as happened much later, in the work of Stevin and others in the 17th century. While Galileo, Descartes, and Leibniz had the highest respect for these men and their work, Boyer (1959) noted that:

> the type of work they represented was not destined to be the basis of the decisive influence in the development of the *methods* of the calculus. The guiding principles were to be supplied by the geometry of Archimedes, although these were to be

modified by kinematic notions derived from the quasi-peripatetic disputations of the Scholastic philosophers on the subject of variation. (pp. 88–89; emphasis mine)

I emphasized the word "methods" in this quote in order to make the point that the objectives of calculus, its reasons for being, were already established by these men. Further, their willingness to deal with the infinite and infinitesimal prepared the way for work to come. And a gradual philosophical shift during the 15th and 16th centuries away from the tight influence of Aristotle and toward Platonism fostered a growing assumption that the propositions of mathematics are established by the intellect. This in turn helped free mathematical thinking from strict adherence to the messages of the senses. Additionally, the relative loss of rigor (compared to Euclidean geometry) helped even more to allow the infinite and infinitesimal into mathematics proper. A level of rigor comparable to Euclid was not again achieved until the 19th century.

By the middle of the 15th century, Nicholas of Cusa had begun to include methods of Archimedes in geometric problems, and Boyer (1959) noted:

> as Plato's thought may have been instrumental in the use of infinitesimals made by Archimedes in his investigations preliminary to the application of the rigorous method of exhaustion, so also the speculations of Nicholas of Cusa may well have induced mathematicians of a later age to employ the notion of the infinite in conjunction with the Archimedean demonstrations. (p. 92)

And so the gradual merging of the static geometry of the Greeks with the Scholastics' interest in mathematizing variation set the stage for the spectacular developments of the 17th century.

Representing Variable Magnitudes by Line Segments Rather Than Coordinatized Points.

Coordinatization of space proved not to be easy across history in the sense that, despite the use of coordinates in ancient Babylonian astronomy, the Greeks never used them in mathematics. Indeed, the Greeks never even coordinatized the one-dimensional line. While Aristotle had noted that time could be thought of as analogous to a line, the Greeks never developed a calendar that linearly ordered events. They specified dates via laborious reference to events known to the listener or reader (Bochner, 1966).

Further, it is important to realize that the main contribution of coordinate geometry, outside the generative role associated with the area and tangent problems, occurred mainly in later years. Bochner remarked that coordinate geometry came into serious functional use only in the second half of the 18th century and played a remarkably small role even in Newton's *Principia* and *Opticks,* in the sense of providing the setting in which one thinks about phenomena.

We recall that the *intensio* of an intensive quantity such as temperature (for a linear subject) was represented in the 14th and 15th centuries by a linear object, the vertical line. In a sense, Oresme can be seen as adding a second dimension to the thing posessing the quality itself to express the intensity of that quality. It acted as a kind of narrow-bar bar graph over the subject.

It is also important to note that not only did Oresme fail to provide a modern Cartesian graph of the intensity of a quality, neither did Descartes. As is well known by historians, both Descartes and Fermat, who are usually credited with the simultaneous independent invention of analytic geometry (Descartes gets chief billing, because his work saw publication first; Fermat's was published posthumously), developed what might be called an "ordinate geometry" based on lengths of line segments. For each, an algebraic equation in two variables was associated not with a set of coordinate points, but rather with a set of line-segment pairs, produced as follows: mark off a segment of some length on a horizontal axis to the right of a reference point (this represented the value of one of the variables), and then place a second perpendicular line segment at the end of the first. The endpoints of the second line segments represented the equation. While Descartes started with geometric figures and translated them to equations, Fermat tended to go in the other direction. "Thus the works of Descartes and Fermat, taken together, encompass the two complementary aspects of analytic geometry—studying equations by means of curves, and studying curves defined by equations" (Edwards, 1979, pp. 96–97). But neither actually used coordinate axes. Although we normally think of Descartes in particular as providing an analytic approach to geometry, I examine key differences in his conception of quantity and operations on quantities that both distinguish his thinking from his predecessors as well as from us.

Parallel behavior by students. Vergnaud and Errecalde (1980) studied student translation of quantities given arithmetically to graphs using line segments (students 10–13 years of age). They discovered that the process of relating numbers to points on an ordered line was highly nontrivial, and that most students tended to express numbers as intervals rather than as points. In effect, the early attempts to graph intensities of qualities were similar, except that the intervals used by Oresme were set perpendicular to the subject they were describing. The elements of the object that possessed the varying quality were designated by points, but not the intensities, which were designated by segments.

The standard approach to teaching children the beginnings of graphical representation of quantities and relationships among them has not systematically used bar graphs, despite the fact that such graphs are frequently used to represent categorical or discrete data. Indeed, they are barely mentioned in an extensive review of children's learning of functions and graphing that catalogues a wide range of student difficulty in constructing and interpreting coordinate graphs

(Leinhardt, Zaslavsky, & Stein, 1990). (However, bar graphs have occasionally been criticized as not being an appropriate lead-in to *coordinate* graphing.) It seems as if, both historically and psychologically, quantity seems more easily represented by line segments than by points, or, put slightly differently, by lengths rather than locations.

The Development of Action Notation Systems in the 17th Century

Returning to the historical development of calculus, I have already noted that the Scholastics' work proved not to be decisive in getting calculus underway, especially with respect to calculus as method. Several ingredients needed to be added beyond the infusion of the problems and techniques of Archimedes. Key among these was the development of an algebra that on one hand would support symbol manipulation, and on the other, could be used to represent variables and relations among these, including the emerging idea of function. The idea of variation that Oresme and others were able to cognize could only be expressed notationally in natural language and in his diagrams—no notation for variable yet existed.

Action vs. Display Notation Systems.

Notation systems differ drastically in the form and explicitness of the information that they afford their users—by differing in the conceptions and mental operations that they either support or supplant (Salomon, 1979, chaps. 3, 5).

A difference that is especially important for our purposes lies in the extent to which different notation systems serve primarily either to display information for the user to read or respond to—a *display system*—or to provide systematic means for the user to act on it physically—an *action system*. For example, a coordinate graph of a function in $\mathbb{R}^2$ has historically been used as a display notation—it provides an organized display of a quantitative relationship that simultaneously displays an infinite ordered set of numerical relationships, provided, of course, the user has the appropriate prior conceptions available to read the graph. A similar statement applies to tables of data. In stark contrast, an algebraic description of a quantitative relationship using alphanumeric characters is designed to be manipulated according to certain rules of syntax. This is to be distinguished from what we have termed a "procedure-representing object" such as an arithmetic or algebraic expression that can be read as specifying a concatenation of products, sums, powers, and so on (Kaput, 1989).

Certain major historical developments bearing on the conceptual development of calculus can be interpreted in these terms. In particular, the great movement to a system of algebra in the 16th and 17th centuries changed algebra from a rhetorical shorthand display system for representing arithmetical procedures to an action system, in which derivations and arguments could occur.

In describing the mathematics of the Renaissance as mainly arithmetic and algebra, Bochner (1966) commented on its impact:

> It brought about momentous innovations which gradually wrought a changeover from the inertness of traditional syllogistic schemata of Greek mathematics to the mobility of symbols and functions and mathematical relations. These symbols and functions and relations have penetrated into many areas of systematic thinking, much more than sometimes realized, and they have become the syllables of the language of science. (pp. 185–186)

He also noted:

> In mathematics, the 16th century was the century of the rise of so-called algebra, whatever that be. Not only was this algebra a characteristic of the century, but a certain feature of it, namely the "symbolization" inherent to it, became a profoundly distinguishing mark of all mathematics to follow. . . . [T]his feature of algebra has become an attribute of the essence of mathematics, of its foundations, and of the nature of its abstractness on the uppermost level of the "ideation" à la Plato. (pp. 38–39)

In Mahoney's terms (1980), this development made possible an entirely new mode of thought "characterized by the use of an operant symbolism, that is, a symbolism that not only abbreviates words but represents the workings of the combinatory operations, or, in other words, a symbolism with which one operates" (p. 142).

Bochner (1966), again, commented on the difference between using a notation to recognize similarity and difference versus using it to perform actions:

> that various types of "equalities," "equivalences," "congruences," "'homeomorphisms," etc. between objects of mathematics must be discerned, and strictly adhered to. However this is not enough. In mathematics there is the second requirement that one must know how to 'operate' with mathematical objects, that is, to produce new objects out of given ones. Plato knew philosophically about the first requirement for mathematics, but not at all about the second. And Greek mathematics itself never developed the technical side of the second requirement, and this was its undoing. (p. 313)

We recall Piaget's three developmental stages and the role of actions, in particular, transformations on objects, in signaling the move from intra- to interobject, relational thinking (Piaget & Garcia, 1989). From this perspective, the operant nature of the algebraic symbol system provided the means for the transition from the former stage to the latter for quantitative relationships, particularly those that were now representable by curves. It then became possible to operate on the entities that the curves represented. But there is more involved— arithmetic needed to become more flexible as well.

Syntax, Semantics, and the Loosening of Referential Constraints.

Whenever one speaks of operations in two systems, A and B, one must be concerned with how they correspond—how objects and relations correspond under the allowable actions in either system. (This includes the case when one notation system is acting as a model of some aspect of a nonmathematical situation.) Elsewhere (Kaput, 1989, 1992), I discussed these complex issues of syntax and semantics in a bit more detail, in which I distinguished between *syntactically guided actions* within a notation system apart from constraints outside the system and *semantically guided actions*, where actions in A, say, are guided by presumed linkages to their referents in B (in the sense of Fig. 4.2), where, as I have emphasized, the linkages are in the cognitive realm.

A chief reason that quantitative Greek mathematics did not develop was the combination of two constraints: (a) the underlying semantic linkage to a static geometric model for quantities and operations on them, and (b) a stress on logical rigor. The first limited the notion of quantity that was possible (ratios of "like" measures) and operations on quantities (those that fit the allowable three physical dimensions). The second constraint forced strict adherence to semantically guided actions on quantities. No action was allowable that did not have a well-defined geometric referent. This amounted to a strong ontological commitment to the geometric reality of all intermediate steps in a proof, for example. It also was behind the reluctance to admit the infinite into mathematical arguments.

This is in strong contrast to modern (post-17th-century) algebraic manipulations, where one can, at least temporarily, fly free of ontological commitments during a derivation and operate on the basis of (one's conception of) the syntax of the algebraic system. The big steps toward operational freedom were taken by Vieta and Descartes (Mahoney, 1980, pp. 143–146). In particular, Descartes' trick for performing all "arithmetic" operations in terms of a simple line-segment proportion that no longer required one to keep track of the dimensions of the operands and results, together with his symbolism for exponents (which was based on his line system rather than on dimensionality), led to a new conception of number and variable that opened a whole new world of possibilities. Boyer and Mertzbach (1989) put it as follows:

> In one essential respect he broke from Greek tradition, for instead of considering x^2 and x^3, for example, as an area and a volume, he interpreted them also as lines. This permitted him to abandon the principle of homogeneity, at least explicitly, and yet retain geometric meaning. Descartes could write an expression such as $a^2b^2 - b$, for, as he expressed it, one "must consider the quantity a^2b^2 divided once by unity (that is, the unit line segment), and the quantity b multiplied twice by unity." It is clear that Descartes substituted homogeneity in thought for homogeneity in form, a step that made his geometric algebra more flexible—so flexible indeed that

today we read "xx" as "x squared" without ever seeing a square in our mind's eye. (pp. 377–378)

"Homogeneity" here refers to homogeneity in geometric dimension, assuming each factor in an expression accounts for one dimension. Up to this time, based on the assumption that all quantities were based on geometric entities, terms could be additively combined or compared only if they had the same dimension. This circumvention of the homogeneity requirement was a major step in moving toward the modern concept of number, a step that was required before numbers could easily be identified with points on a line or before pairs of numbers could be identified with points in the plane, rather than with segments or pairs of segments, respectively. It represented both a change in the referent for the symbols and a weakening of the correspondence, in the sense of Fig. 4.2. Thus, while Descartes maintained a geometric referent for his quantities and operations, it was no longer the confining one of the Greeks. Furthermore, once having established the existence of referents, he was not bound to use them as semantic constraints for all his actions, but rather he leaned on the increasingly powerful syntax of his notation system, as a syntactic guide. (This will be elaborated on further later in the chapter.)

One can regard the earlier work of Oresme and the Scholastics as a sustained attempt to understand situations B involving variation—building Cog_B—and representing aspects of such situations in one or another graphical forms A, from which one could then reason in graphical terms—in Cog_A. However, the system A was primarily a display system, to display variation in the intensity of qualities via the configurations and, via areas, the extent of qualities—it was not designed to be manipulated. Although I have not focused on it, the Scholastics also used their system to deal with elementary arithmetic and geometric infinite series (e.g., see Oresme, III.viii–xiii, in Clagett, 1968, pp. 413–435) by creatively subdividing the horizontal "longitude" line into finer and finer intervals and placing narrow rectangles on the subdivisions to represent the terms of the series. Again, however, their system allowed only the display of series and not the manipulation or combination of such. Later, in the hands of a Gregory or Newton, all manner of transformations on series became common, and the series themselves jumped up a level in abstraction, where they could then be compared with one another in an interoperational way.

While I have no simple or direct curriculum or instruction lesson to draw from this aspect of the history of calculus, it should sensitize us to the laborious and complex constructions that were necessary and the various constraints to overcome before the calculus could even become *possible* in the sense that we now know it.

Continuity, Motion, and Indivisibles.

While the Scholastics loosened constraints on the consideration of the infinitely small and large, they also played an important role in planting the seed of variable quantity, the "generation of magnitudes" that carried through to the thinking of Newton and, in a slightly different way, of Leibniz. Descartes also helped modernize the concept of quantity and operations with quantities in order to make it more flexible and useful. Moreover, the increasing attention to Hindu-Arabic algebra, which had a more flexible set of conceptions of number and arithmetic that allowed irrationals and even negatives, helped prepare the way for what was to come.

During the 16th and 17th centuries, a vague but persistent notion of continuity, of continuously generated magnitudes, of continuously generated geometric figures, seemed to lie behind increasingly intense efforts to bridge the rectilinear and the curvilinear efforts expressed using imprecise concepts of infinitesimals. Thus thinkers like Kepler regarded a circular region as made up of an infinite collection of infinitesimal triangles, a sphere made up of cones, and generated other curvilinear figures by rotating figures about axes in the style of Pappus. As Boyer (1959) said:

> This striving for an expression for the idea of continuity constantly reappears throughout the period of some fifty years preceding the formulation of the methods of the calculus. Leibniz himself, like Kepler, frequently fell back upon his so-called law of continuity when called upon to justify the differential calculus; and Newton concealed his use of the notion of continuity under a concept that was empirically more satisfying, though equally undefined—that of instantaneous velocity, or fluxion. (p. 110)

While there is not enough space to recount the details of the lengthy struggle with continuity, infinitesimals, and indivisibles (Boyer, 1959, chap. iv, is an excellent source), I must note that it lay behind the development of representational tools in this period. And there was a complex interplay between conceptions based on kinesthetics and those based on geometric increments.

For example, by observing a difformly difform variable quality displayed by a semicircle, Oresme noted, using the language of "change," that near the maximum intensity the change for a given change in the (horizontal) subject was least. And Kepler (in 1615), using data tables to express the relation between the dimensions of solid figures (in his case, Austrian wine casks), remarked that the increment in volume for a given size increment in dimension was smallest near the largest volume. But none of these men or their contemporaries was able to come up with techniques to find maxima other than to observe them in static displays of variable quantities. Later, Fermat actually came up with a technique based on an algebraic description of infinitesimals based on increments, although still without either the idea of limit or function behind it. Further, his technique was written in terms sufficiently general as to provide the language for finding the tangent to a curve as well.

During the first half of the 17th century, Galileo using kinesthetic metaphors, his student Cavalieri using infinitesimals combined with dynamic imagery associated with geometric figures of rotation, and especially Torricelli who used this idea as the basis for much of his work found application of the yet incomplete idea of "flowing indivisibles" that could generate lines, surfaces, and solids. Indeed, Torricelli was the first to integrate motion imagery, especially instantaneous velocity, into what he termed "geometric demonstrations" and was convinced that Archimedes thought this way (Boyer, 1968, p. 132). He obtained tangents to "generalized parabolas" (what we now would call the graphs of x^n, using composite motions consisting of a uniform horizontal velocity and a variable vertical velocity to generate the curves of interest). Or, in the case of various spirals, he used points sliding with variable velocity along a uniformly rotating line. Others apparently employed this approach, including Roberval and even Descartes, but Torricelli's work enjoyed great popularity and hence influence. He also managed to use indivisibles to solve various area problems. Interestingly, not only did he fail to see generality in his methods (a feature common to all the work before Newton and Leibniz), his motion-generated curves could only be created if one knew the motions in advance. They did not apply to curves associated with arbitrary equations. Nonetheless, they borrowed from the common experience of continuous motion a way of thinking about geometric objects and relationships that proved highly productive. In a sense, he never ascended beyond the intraoperational stage of Piaget and Garcia (1989).

This way of thinking that developed during the first two-thirds of the 17th century amounted to the introduction into mathematics of a reliable way of representing continuous variables, a way of representing the notion that the Scholastics had introduced centuries earlier. It yielded display representations for continuous variables. As stated by Edwards (1979):

> The notion of a variable, as first emphasized by Descartes and Fermat, was indispensible to the development of the calculus—the subject can hardly be discussed except in terms of continuous variables. Moreover, analytic geometry opened up a vast virgin territory of new curves to be studied, and called for the invention of algorithmic techniques for their systematic investigation. Whereas Greek geometers had suffered from a paucity of known curves, a new curve could now be introduced by the simple act of writing down a new equation. In this way, analytic geometry provided both a much broadened field of play for the infinitesimal techniques of the seventeenth century, and the technical machinery needed for their elucidation. (p. 97)

On the other hand, despite the new representations, none of these mathematicians was able to come to terms geometrically with the fundamental logical difficulties of infinitesimals: After all, points have no length, but somehow they make up a line; lines have no width, but they were thought to comprise a region with area; planes have no thickness, but somehow they were thought to comprise a solid with volume. And if they were to have length, width, or thickness, respectively, then only a finite number of them would be needed to comprise the object in question. But if an infinite number of them was present, then they must be infinitely small, and so on. Cavalieri tried clever analogies, including threads of cloth comprising a region and pages of a book comprising a volume (pun intended?). While they all continued to use indivisibles on the grounds that they got the right results (Torricelli even wrote about their paradoxes), it would be more than two centuries until things would get straightened out logically. Philosophers were more troubled by these difficulties than mathematicians.

Boyer (1959) suggested that "the lack of a suitable basis for indivisibles was perhaps more serious than the omission of instantaneous velocity, for whereas intuition may serve to clarify the use of velocity as an undefined element, there is no such safe guide in the use of indivisibles" (p. 134). Somehow, thanks to our stronger intuitions, motion is a safer haven from the dangers of absurdity than the geometry of indivisibles.

The French, including Pascal and Roberval, and Wallis, made stronger use of numerical argument, although in the end inevitably resorted to geometric considerations. Yet, they sometimes argued in ways that helped justify arithmetic steps using differences in dimensionality to justify discarding a term, presaging Barrow, Leibniz, and others who would discard infinitesimals of higher order. Indeed, the loose analogies between arithmetical statements and geometric ones proved to be a potent heuristic, because one could lean on the reasonably well-defined actions on arithmetic statements to manipulate these statements into various forms that had transparent interpretation. These forms, particularly involving series, were then interpreted in quasi-empirical fashion to produce more general insights expressed algebraically, for example, that the quadrature of the general parabola x^n was $x^{n+1}/(n + 1)$. This general trend toward the arithmetization of infinitesimal methods was exemplified by Wallis, who lost no opportunity to jump to conclusions inductively on the basis of arithmetic results generalized to algebraic form, including the extension of the above result to rational and even irrational exponents (Edwards, 1979, pp. 113–115).

However, none of these mathematicians, thinking geometrically, arithmetically, or kinematically, resolved the problem of infinitesimals. The intractability of the continuity–infinitesimal problem is of critical importance for anyone expecting to teach students the basic ideas of calculus. The fact that analogous arguments were repeated for centuries is testimony to styles of thinking that should be expected from students, provided that they have experiences analogous to those who produced those arguments. To help emphasize how little progress had been made, the following quotation from Barrow's famous geometry lectures explains how the area under a time-versus-velocity graph represents distance:

> To every instant of time, or indefinitely small particle of time (I say instant or indefinite particle, for it makes no difference whether we suppose a line to be composed of points or indefinitely small linelets; and so in the same manner, whether we suppose time to be made up of instants or indefinitely minute timelets);

to every instant of time, I say, there corresponds some degree of velocity, which the moving body is considered to possess at the instant. . . . If through all points of a line representing time are drawn . . . parallel lines, the plane surface that results is the aggregate of the parallel straight lines, when each represents the degree of velocity corresponding to the point through which it is drawn, exactly corresponds to the aggregate of the degrees of velocity, and thus most conveniently can be adapted to represent the space traversed also. (from Boyer, 1959, p. 180)

He also noted that the lines could be replaced by narrow rectangles, but insisted that it did not make any difference. The similarity with Oresme and Galileo is clear. Of course, Barrow is the last step before Newton, his student, and the Newton–Leibniz invention of calculus as a general method.

The Evolution of Representations of Abstract Relations.

The major ingredient lacking in the early 17th century was a good algebraic action representation system in which the concepts that *display* systems *exhibited* could be fruitfully manipulated to go beyond the achievements of the Greeks, especially the great Archimedes. As already noted, development of such a system began with the literals of Vieta and the use of continuous variables by Fermat, Descartes, and others such as Wallis who moved to arithmetize the geometric methods. In significant measure, the early actions on symbols were initially arithmetic, with generalizations expressed algebraically—algebra served as a language for expressing the initial curve via an equation and a language for expressing the general result. Much argument came to take the form of a sequence of arithmetically computed results followed by an inductive leap to an algebraically formulated generalization. This was an approach used by Newton and Leibniz as well, although in Newton's case the computations were increasingly algebraic in character, for example, his extensive work based on the binomial expansion.

In Boyer's (1959) words, "This literal symbolism was absolutely essential to the rapid progress of analytic geometry and calculus in the following centuries, for it permitted the concepts of variability and functionality to enter algebraic thought" (p. 98). But the profound tendency to regard geometry as the referential bedrock for all argument made arithmetization and algebraization a very slow process during the 17th century. Mahoney (1980) pointed out the tension between thinking in the geometric and algebraic modes, respectively:

For example, one can often see this tension in Pierre Fermat: on the one hand, he consciously solves problems that the ancient mathematicians were powerless to confront or that they could not even have posed; on the other hand, he maintains that his solutions carry on the traditions of ancient mathematics, even though these solutions employ mathematical tools and concepts with which an Archimedes or an

Appolonius would hardly have agreed. One senses the same tension in Fermat's contemporary and rival, Rene Descartes, who on the one hand holds his algebraic universal mathematics to be a reconstruction of those general methods that underlay Greek mathematics and that the Greeks meanly withheld from later generations, and who on the other hand praises himself for having created a mathematical method that the Greeks had never possessed. (p. 141)

Mahoney and others have pointed out that accompanying the move to algebra and arithmetic was a continuing relaxation of the demands of rigor. One way to deal with the problem of infinitesimals was to change the forms of discourse and argument, which is what the introduction of algebra accomplished. This was an explicit goal of such men as Descartes in his *Discourse on Method,* and, according to Mahoney (1980), was rooted in the birth of pedagogy in the 16th century, in the writings of Ramus and the new textbooks of Outred, Herigone, Recorde, Harriot, and others: "One characteristic of the intellectual world of the sixteenth and seventeenth centuries is precisely the extension of the school and university system to include broader segments of society" (p. 150). This process was abetted by the introduction of the new technology of the printing press and the introduction of ever more easily useable notations—not too unlike the circumstances today. Two of the background factors mentioned earlier were, in fact, changing in the 16th century (the school population and the medium in which knowledge was externalized).

Descartes was very conscious of the role of symbolic algebra in helping support his method, as he stated in Rule XVI of his *Regulae,* quoted in Mahoney (1980):

> Those things that do not require the present attention of the mind, but which are necessary to the conclusion, it is better to designate by the briefest symbols [nota] than by whole figures: in this way the memory cannot fail, nor will thought in the meantime be distracted by those things which are to be retained while it is concerned with other things to be deduced. . . . By this effort, not only will we make a saving of many words, but, what is most important, we will exhibit the pure and bare terms of the problem, such that while nothing useful is omitted, nothing will be found in them which is superfluous and which vainly occupies the capacity of the mind, while the mind will be able to comprehend many things together. (p. 150)

As put by Kline (1972), Descartes "sees in algebra a powerful method wherewith to carry on reasoning, particularly about abstract and unknown quantities. In his view algebra mechanizes mathematics so that thinking and mathematical transformations become simple and do not require a great effort of the mind. Mathematical creation might become almost automatic" (p. 281). This goal was even more important to Leibniz, who sought a general formal language that would serve all reasoning in a similar fashion, an idea that grew in currency, capturing the imagination of mathematicians, scientists, and philosophers in the

17th and 18th centuries. For example, Lavoisier, while introducing his *Nomenclature of Chemistry,* pointed out the difference between his nomenclature, as a systematic abbreviation system, and the real mathematical symbolism of algebra: "Algebra is the analytical method par excellence; it was invented to facilitate the labors of the mind, to compress into a few lines what would take pages to discuss, and to lead, finally, in a more convenient, prompt, and certain manner to the solution of very complicated questions" (quoted in Gillespie, 1960, p. 245).

Furthermore, the development of computations with infinite series provided an alternative representation system that supported an even wider class of actions, again, without logical foundation. The standard approach, beginning in the 17th century mainly with Newton and continuing for more than a hundred years, was to treat infinite series as polynomials, first for purposes of algebraic operations and later for the operations of calculus. In fact, as noted by Kline (1972), "Both Newton and Leibniz regarded calculus as an extension of algebra; it was the algebra of the infinite, or the algebra that dealt with an infinite number of terms, as in the case of infinite series" (p. 324). (For an excellent review of Newton's remarkable development and application of the binomial series, see Edwards, 1979, pp. 178–187.) In Newton's own words (quoted in Edwards, 1979, p. 187): "[W]hatever common analysis [i.e., ordinary algebra] performs by equations made up of a finite number of terms [i.e., polynomials] (whenever it may be possible), this method may always perform by infinite equations [i.e., infinite series]."

Just as analytic geometry widened the playing field (after all, any equation led to a curve), so did infinite series, because now a whole new class of "functions" could be dealt with.

The Invention of Calculus: Underlying Conceptions

The reason why the invention of calculus should be attributed to Newton and Leibniz, despite the fact that quadrature and tangent problems had been solved in particular cases by so many others, and despite the fact that Barrow had even exhibited the inverse relation between them, was put forth by Edwards (1979):

> What is involved here is the difference between the mere description of an important fact, and the recognition that it is important—that is, that it provides the basis for further progress. In mathematics, the recognition of the significance of a concept ordinarily involves its embodiment in new terminology or notation that facilitates its application in further investigations. As Hadamard remarks, 'the creation of a word or a notation for a class of ideas may be, and often is, a scientific fact of very great importance.' (p. 189)

To this we would add the fact that the general method was interoperational, whereas all the operations on specific curves and *equations* (as their algebraic expression was known) were intraoperational in the sense of Piaget and Garcia

(1989). The intraoperational processes associated with finding tangents and quadratures could now be seen from a new level of abstraction, as inverses.

Newton and the Role of Motion Imagery.

The language of algebra and its use in coordinate geometry, by the third quarter of the 17th century, were quickly becoming standard tools for the development of new mathematics. This increased both the types of problems that were addressed and the methods for their solution. Increasingly, the justification for arguments was cast in loose algebraic and arithmetic terms, with frequent use of empirical induction from particular calculations. The underlying notion of number had gradually become more abstract and general during the previous hundred years. And in Newton's mind, numbers included what we would now call negatives and irrationals. (The story of this important matter is too complex to be dealt with here, but an interesting analysis of the interaction between the development of algebra and that of number is given by Klein, 1968.) Finally, the concept of continuous variable entered the mainstream of thought. In fact, the brilliant young Scott James Gregory introduced an explicit definition of function in 1667: "A quantity obtained from other quantities by a succession of algebraic operations or by any other operation imaginable" (in Kline, 1972, p. 339). The young Newton mastered all this work, especially the new algebraic techniques, and quickly exploited and extended them beyond any level seen before.

More importantly for our purposes, the underlying kinetic imagery associated with continuous motion was integrated with all these developments in Newton's mind. In his words of 1676, in the *De quadratura curvarum* included in the *Sourcebook* (Struik, 1986):

> I consider mathematical quantities in this place not as consisting of very small parts, but as described by a continued motion. Lines [curves] are described, and thereby generated, not by the apposition of parts but by the continued motion of points. . . . These geneses really take place in the nature of things, and are daily seen in the motion of bodies. And after this manner the ancients, by drawing moveable right lines along immoveable right lines, taught the genesis of rectangles. (p. 303)[4]

This dynamic way of viewing curves enabled him, when coupled with his invention and use of binomial series, to approach the area problem in a way fundamentally different from his predecessors, not as a limit of a sum of infinitesimal areas, but rather in terms of the rate of increase in area, in the "moment" of area. As put by Boyer (1959):

> In other words, what we should now call the derivative is taken as the basic idea and the integral is defined in terms of this. Mathematicians from the time of

---

[4]The larger selection from which this important quotation is taken can be found on pp. 303–309.

Torricelli to Barrow had in a sense known of such a relationship, but Newton was the first man to give a generally applicable procedure for determining an instantaneous rate of change and to invert this in the case of problems involving summations. Before this time the tendency had been rather in the opposite direction—to reduce problems, whenever possible, to the determination of quadratures. With this step made by Newton, we may consider that the calculus has been introduced. (pp. 191–192)

In three works over a period of about 10 years from the mid-1660s to the mid-1670s (all published considerably later), Newton attempted to come to terms with the underlying ideas of the infinitely small. In his earliest writings (*De analysi per aequationes numero terminorum infinitas*), there were traces of infinitesimal geometric quantities. But as noted by Boyer (1959) in discussing Newton's demonstration that the area under $y = ax^{m/n}$ is $z = [n/(n + m)]ax^{(n+m)/n}$:

The ordinate y seems to represent the velocity of the increasing area, and the abscissa represents the time. Now the product of the ordinate by a small interval of the base will give a small portion of the area, and the total area is only the sum of these moments of area. This is exactly the infinitesimal conception of Oresme, Galileo, Descartes, and others, in their demonstrations of the law of falling bodies, except that these men had found the area as a whole through the addition of such elements, whereas Newton found the area from its rate of change at a single point. It is difficult to tell in exactly what manner Newton thought of this instantaneous rate of change, but he very likely accepted it as similar to the conception of velocity which Galileo had made so familiar but had not defined rigorously. (p. 193)

In his second approach (*Methodus fluxionum et serierum infinitarum*), Newton continued to think of the geometric entities as generated by the continuous motion of points, lines, and planes rather than as aggregations of infinitesimal elements. This had the effect of using time as a mediating or auxiliary variable, which was allowed to exist in infinitesimally small increments. And so the quantities that were treated as being generated in the continuous flow of time could take on corresponding increments. It is here where he introduced his "dot" notation for fluxions, the instantaneous rates of change of quantities. Thus, he could write a dot over the literal standing for a variable quantity to represent the rate of change of that quantity—as an implicit function of time, what we might write in Leibniz's notation as dx/dt. Indeed, when dealing with the area problem, he even normalized his representation by taking x-dot to be one. That is, he regarded x as representing time itself. He seemed to regard the idea of continuous motion and continuous change as so clear and compelling that no further definition was required. As quoted in Edwards (1979) from the *Methodus fluxionum et serierum infinitarum:*

I consider time as flowing or increasing by continual flux & other quantities as increasing continually in time & from ye fluxion of time I give the name fluxions to the velocity wth wch all other quantities increase. . . . I expose time by any quantity flowing uniformly & represent its fluxion by an unit, & the fluxions of other quantities I represent by any other fit symbols. . . . This method is derived immediately from Nature her self. (p. 210)

In the third stage of his thinking (*De quadratura curvarum*, written in the early 1690s and quoted earlier), he tried to purge all mention of the infinitely small, leaning even more heavily on continuous-motion imagery. Here, after using the binomial expansion as before in determining the fluxion of x^n, he formed the ratio of the change in x to the change in x^n (the reciprocal of the usual ratio). He then cancelled the change in x as we would before taking the limit as the change approaches 0. He referred to the result as the "ultimate ratio" of the changes—he thought of it as a ratio of changes rather than as a change in ratios, and hence he thought of it as a single number. This is likely linked to his underlying conception of number as a ratio, an underlying conception that may have helped fuel the subsequent extended confusion about the nature of fluxions and differentials as quotients.

Newton developed a product rule for fluxions, from which he derived the power rule for positive integers and then extended it to negative and then rational exponents (which were called *indices* at that time). He used infinite series to represent the trigonometric functions and their inverses for the first time, to determine the calculus of these functions, and to apply all this work to solve an immense array of a particular area, tangent, arc length, and optimization problems. In a blizzard of creativity accomplished mainly while he was only in his 20s and early 30s, he established most of what we now take as the basic calculus of a single variable.

Leibniz: The Power of Notations.

The story of Leibniz is the story of the development of notation in concert with the development of concepts.

> His infinitesimal calculus is the supreme example in all of science and mathematics, of a system of notation and terminology so perfectly mated with its subject as to faithfully mirror the basic logical operations and processes of that subject. It is hardly an exaggeration to say that the calculus of Leibniz brings within the range of an ordinary student problems that once required the ingenuity of an Archimedes or a Newton. Perhaps the best measure of its triumph is the fact that today we can scarcely discuss the results of Leibniz' predecessors without restating them in his differential notation and terminology. (Edwards, 1979, p. 232)

Anyone doubting the power of notation as a cognitive technology (Pea, 1987) need only consider the role of the Hindu–Arabic placeholder system in changing the mental operations required to perform complex computations. What once was possible only by accomplished scholars in ancient Rome can be done by ordinary students today.

Leibniz began thinking about areas and tangents in the early 1670s and, with the advice of other mathematicians and scientists of the day, notably Huygens (20 years his senior), became acquainted with much of the work done up to that time, much of the same work that Newton had learned of.

A significant long-term difference between him and Newton, however, seems to be in Leibniz's early and continuing interest in formal and combinatorial arithmetic and especially in his quest for a formal language in a "certain sensible and palpable medium, which would guide the mind as do the lines drawn in geometry and the formulas for operations which are laid down for the learner in arithmetic" (quoted in Baron, 1969, p. 9).

Leibniz's Starting Points.

We are aided in our review of Leibniz's work by his own history of his discoveries written some 30 years later (in his *Historia et origo*) and a persistent self-consciousness of the importance of his discoveries that led him to describe frequently to others the rationale for the work leading to and sustaining those discoveries. He acknowledges that he began with the simple background metaphor of sums and differences of sequences of numbers. For example, given a finite sequence $a_0, a_1, a_2, \ldots, a_n$, one can form the sequence $d_1, d_2, \ldots, d_n$ of consecutive differences $a_i - a_{i-1}$. If one were to sum the differences, all intermediate terms cancel, leaving $a_n - a_0$. He applied this idea to the sequence of perfect squares, from 0 to n^2, to obtain, first, the sequence of odd integers up to $2n - 1$ as its sequence of consecutive differences. From this he is able to conclude that the sum of the odd integers up to $2n - 1$ must be n^2.

He extended this result to infinite sequences where, as we would now put it, the limit of the nth term as n grows without bound is 0. For such sequences the limit of the sum of differences would amount to the first term of the sequence, because the partial sums all have the form "first-term-minus-last-term," and the last term goes to 0. From this type of investigation he went on to work with modifications of Pascal's arithmetic triangle and figurate numbers, partially instigated by problems put to him by Huygens. Instead of building a triangle beginning with the consecutive integers and taking sums to get the usual Pascal triangle (actually known well before Pascal), he began with their reciprocals and took differences, referring to the result as his *harmonic triangle*. By applying his simple result on sums-of-differences of infinite sequences to the infinite rows of his harmonic triangle, he was able to derive a variety of numerical infinite series results, including solutions to Huygens's problems.

According to Edwards (1979), the inverse relationship between the standard sum-based Pascal triangle and the difference-based harmonic triangle "implanted in Leibniz's mind a vivid conception that was to play a dominant role in his development of the calculus—the notion of an inverse relationship between the operation of taking differences and that of forming sums of the elements of a sequence" (p. 238). Indeed, it was only many years later that he was persuaded by John Bernoulli to change his names for the two branches of calculus from "differential and summative" to "differential and integral."

We would add that this early episode reveals another important aspect of Leibniz's style of mathematical thinking. He had a strong tendency to be guided by the formal results of syntactical manipulations of symbols, but at the same time, an eagerness to tie these results to previously established systems of reasoning—in this case, the system of reasoning is that of the arithmetic of numbers and sequences of numbers. A last element of his work revealed in this early episode, shared by many in this era, is a ready tendency to extrapolate from the finite to the infinite on the basis of the reasonableness of the results from the finite case.

Ironically, it was another of Pascal's triangles that set Leibniz on his way toward his general solution of the quadrature and tangent problems. Pascal had, about 20 years earlier, constructed a right triangle on a circle in such a way that its hypotenuse was tangent to the circle and the corresponding radius through the point of tangency passed through its right angle on the interior of the circle as in Fig. 4.6.

In the analogous differential figure on the right-hand side of Fig. 4.6, the circle's radius becomes the normal to a more general algebraically given curve, and the difference between the arc *ds* and the tangent *EF* is ignored. Triangle *EFK* is similar to triangle *ADG,* from which it follows that the ratio of sides *AD* to *EF* equals the ratio of sides *DG* to *FK*, so *DG*EF = AD*FK*. In the analogous differential-labeled figure, this equation becomes *yds = ndx*, from which Leibniz was able to derive an amazing number of results that solved and then went beyond most of the quadrature and tangent problems of his day.

While Barrow and many others previously had used this "characteristic triangle" for more limited purposes, Leibniz used it much more generally as the basis for his notions of differential, derivative, and integral. Indeed, in a remarkable series of manuscripts written over a period of a few months late in 1675, he translated his geometrically motivated reasoning into the algebraic notation that has served us so well since. (For slightly differing accounts of the way he pulled the algebra off the geometric figures and then refined his notation, see Cajori, 1929, and Burton, 1985.)

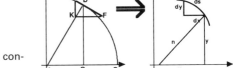

FIG. 4.6. Leibniz's basic geometrical construction.

The Interplay of Geometric and Formal Algebraic Reasoning.

While the full details are beyond the scope of this chapter, Leibniz used a variety of ingenious elaborations of the basic geometry that he increasingly tied to algebraic notations. The foundation of his bridge from geometry to algebra was to think of the curve as made up of infinitesimal segments and the y-values as an infinite sequence indexed (or ordered) by the associated x-values. This was an outgrowth of his earlier work with arithmetic sequences. That earlier work had, it appears, oriented his thinking toward the inverse relationship between sums and differences.

In remarkable bootstrapping of meaning, he described his geometric constructions algebraically, successively refined his algebraic descriptions in order to simplify and streamline them, and in turn applied them to an ever wider array of geometrical problems. It was in this most productive period, October and November 1675, that he developed the differential notation—to describe those successive differences in y-values—and the integral notation—to describe sums of these.

In an October 29th manuscript in which he first introduced the elongated S for sum of infinitesimals (cited in Edwards, 1979), he attacked the inverse tangent problem (where l stands for the successive y-differences):

> Given l, and its relation to x, to find $\int l$. This is to be obtained from the contrary calculus, that is to say, suppose that $\int l = ya$. Let $l = ya/d$. Then just as $\int$ will increase, so d will diminish the dimensions. But $\int$ means a sum, and d a difference. From the given y, we can always find y/d or l, that is, the difference of the y's. Hence one equation may be transformed into the other. (p. 253)

Three days later, the y/d notation becomes dy, when he recognized that the integral did not actually change dimensions, because it was merely a sum of areas. But with his acute sensitivity to patterns in formal symbols, he created ever more systematic formulas generalized from particular results. By November 11, he was asking whether the differentials of a product (respectively, quotient) is the product of the differentials (respectively, quotients). He was able to answer in the negative by applying his method to x^2, and went on to determine the "antiderivative" of x^3, checking his results against the results of Sluse's geometrically based tangent rule. By the following July, he had the power rule for differentials and "antiderivatives" for nonintegral exponents and even composite functions (a form of the Chain Rule), for example, he could determine the differential of the square root of $ax^2 + bx + c$. By November 1676, he had returned to Sluse's tangent rule, previously accepted on Sluse's geometric basis, and showed how it follows from his calculus of differentials by straightforward computations and the assumption that products of differentials can be ignored in the final result. In July 1677, he had the product rule worked out in an argument that would be familiar to a calculus student of today:

4. DEMOCRATIZING ACCESS TO CALCULUS

$$d(xy) = (x + dx)(y + dy) - xy = xdy + ydx + dxdy$$

He then noted that "the omission of the quantity $dxdy$, which is infinitely small in comparison with the rest, for it is supposed that dx and dy are infinitely small, will leave $xdy + ydx$" (quoted in Edwards, 1979, pp. 255–256).

He lost no opportunity to tie such formally based results to geometric arguments. For example, he pointed out that by applying the integral sign to the terms of $xdy + ydx$, to get $\int xdy + \int ydx$, one is modeling the addition of areas as in Fig. 4.7. (He does not mention here, but definitely understood, the meaning of the product $dxdy$ as the negligible area of the upper right "corner" rectangle of the shaded region—not shown in the figure.)

Leibniz went on, generalizing to moments about the x and y axes, to assert that his calculus shows these sorts of results without reference to *any* figure (quoted in Edwards, 1979), "so that now there is need for no greater number of the fine theorems of celebrated men for Achimedean geometry, than at most those given by Euclid in his Book II or elsewhere, for ordinary geometry" (p. 256).

By mid-1677, he wrote the now standard $\int ydx$ for the area under the curve regarded as a polygon made up of infinitesimal segments, where the area was based on an approximation using infinitesimally wide rectangles bounded on top by the curve segments. Again, he discarded the area of the triangles formed at the top "since they are infinitely small compared with said rectangles" (Edwards, 1979, p. 257). He went on, in the same 1677 manuscript, "mounting to greater heights, obtain the area of a figure by finding the figure of its summatrix or quadratrix" (p. 257), essentially by antidifferentiation. Indeed, he has developed the definite integral form of the Fundamental Theorem of Calculus.

Over the next few years he applied his method, his "calculus," to solve a wide range of particular problems, mainly geometric, while at the same time expanding the general formal theory, including higher differentials.

On the question of whether differentials of any order actually exist as *quantities*, Leibniz seemed to be an agnostic. He did not feel a need, at least as a mathematician, to plumb the philosophical questions surrounding the infinitely small or infinitely large. Rather, he was pragmatic, arguing that "it will be sufficient simply to make use of them as a tool that has advantages for the purpose of the calculation, just as the algebraists retain imaginary roots with great profit" (quoted in Edwards, 1979, p. 265). He also felt confident that one

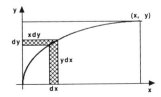

FIG. 4.7. Leibniz's integral of a product.

could make rigorous along the lines of Archimedes's exhaustion arguments, using only finite quantities, the proofs involving differentials, including the elimination of differentials of higher dimension.

Acceptance of infinitesimals was highly irregular over the following century as the nature of the geometric continuum evolved from a particulate geometric object to a more modern numerical continuum and as the underlying views of what constitutes mathematics and proof evolved (see especially Boyer, 1959, chap. 6). One can see analogies between the role of notations in the acceptance of differentials as mathematical objects and the role of notations in the acceptance of other mathematical entities such as negative numbers, irrationals, and complex numbers. Syntactically driven operations with symbols led to combinations that initially had no well-established referent or even had an interpretation that was inconsistent with prior interpretations of the symbols. Furthermore, in each case many years intervened between the appearance of the objects as the results of operating with accepted objects and their acceptance as objects that could be used in operations in their own right. For example, negatives occurred as the result of computations with positives for more than a century (in Europe) before Harriot in the early 1600s put a negative number as an entity alone on one side of a conditional equation (Kline, 1972).

Perhaps the most famous case of symbolic operations leading eventually to entities involves the appearance of roots of negative numbers, *imaginary numbers* as they were later termed by Descartes. This occurred in calculations by Cardan in 1545, who was attempting to divide 10 into two parts whose product is 40, that is, to solve the equation $x(10 - x) = 40$. He obtained the roots $5 + \sqrt{(-15)}$ and $5 - \sqrt{(-15)}$. He said "Putting aside the mental tortures involved," multiply them, whereby one gets $25 - (-15)$, which is 40 (in Kline, 1972, p. 253). It was not until Argand had developed the complex plane that they eventually were fully accepted as mathematical objects. Later, we shall see a similar occurence in the evolution of the concept of function in the 18th century.

Implications for Curriculum Design.

At a most general level, this pattern of syntactically driven extensions of symbol systems leading to conceptual extensions, in particular, to the construction of conceptual entities, raises important curriculum design questions that are ripe for research. How much and what kinds of actions on formal symbol systems are needed to support the mental construction of objects that can in turn serve as referents for new symbols and systems of reasoning (Greeno, 1983; Harel & Kaput, 1991; Sfard, 1991)? In the critically important case of functions, the work of Dubinsky and colleagues (Ayers, Davis, Dubinsky, & Lewin, 1988; Breidenbach, Dubinsky, Hawks, & Nichols, 1991) has shown both that the passage from process conception to conceptual entity is difficult, and that it is traversable with appropriate forms of deliberately designed experience, in their case, experience with defining and manipulating wide varieties of functions in a computer environment (Dubinsky & Schwingendorf, 1991).

In the particular case of calculus, especially in the contexts of differential, derivative, (definite) integral, antiderivative, infinite series, and limits, specific instantiations of these research questions abound. In each case, we need to know more about students' root conceptions of the continuum, of limits and the infinitely large and small, and about students' operant assumptions regarding the nature of symbols and actions on them. Work in progress by Dubinsky (1991) and colleagues with college students offers promise. (See Mamona-Downs, 1990, for a recent review of other research. For a European perspective, see Cornu, 1983; Tall, 1986, 1987, among others.) Furthermore, many of these questions need to be asked in a developmental framework, not merely of college age students.

The highly successful approach taken by Leibniz may have major implications regarding appropriate early "precalculus" experiences for students. In particular, he began with a calculus of finite differences: The fact that they were termed *infinitesimals* seems to play a relatively small role in the actual form of the early results. He initially developed the Fundamental Theorem as a relation between sums and differences. And he (and most others of his time, with the exception of Newton) was guided by intuitions rooted in a particulate conception of geometric curves, coupled with an understanding of functional relationships between independent and dependent variables, where the independent variable acted as a discrete index for the dependent variable's values (this will be elaborated on in Part III. Incidentally, Leibniz exploited the available symmetry between independent and dependent variables to develop very effective substitution routines.)

Leibniz's Calculus as an Action Notation System.

I have described earlier the importance of the development of algebra as an *action* notation system, as a system on which one acts directly and which has a well-defined syntax—rules defining allowable transformations and rules for defining allowable objects and equivalences. In accordance with the usage of the day, Leibniz referred to this action system as a *calculus,* for example, in a letter to l'Hospital he refers to learning the "calculus of Descartes" (in Edwards, 1979, p. 244). Of course, Leibniz had as his central aim the development of a calculus for his new analysis. And he was so successful that the word *calculus* has become narrowed to refer to *his* action system and its associated conceptual and representational structures.

His strong tendency to search for and then immediately exploit patterns in formal symbols led him quickly to treat his integral and differential as *operators,* indeed, as inverse operators. They were not merely descriptors of geometric relationships, or even descriptors of relations between symbols, but were *operators that acted on other symbols to produce new symbols.* This is a critical difference. Repeatedly, he used actions on the symbols to lead the way to new results (usually checking for consistency with what was already known) or to ask new questions. In this sense, his manipulation of notations drove his discovery, or, to paraphrase the quote from him earlier in this section, it "guided his mind as do the lines drawn in geometry and the formulas for operations which are laid down for the learner in arithmetic."

It should be noted, however, that the view of calculus as an extension of algebra was not unique to Leibniz and came to be widely shared—finding an antiderivative, signified by $\int$, was just another operation, similar to, say, finding a square root, signified by $\sqrt{}$. How often do students today who have no sense of the difference among changing the form of $\sqrt{x}$ to $x^{1/2}$, determine the derivative of the function expressed by the latter symbolism and change the form of the result from $(1/2)x^{-1/2}$ to $1/(2x^{1/2})$? Of course, Leibniz *would* know the difference, and his distinction would be based on his idea of the functional relationship between variables. (He was the first actually to use the word *function* in the modern sense of quantitative dependency relationship).

I now turn to the last part of the historical review—the "algebraic century" of calculus following its invention.

Analysis As Algebra—The Age of Euler and Beyond

The Algebraic Universe.

In the 18th century, calculus had two faces: its algebraic face, reflected in the monumental work of Euler and his contemporaries, and its metaphysical face, reflected in an ongoing uncertainty regarding its foundations, especially the ideas of infinitesimals, function, and limit. I deal with each only briefly. (A more complete story of the algebraic developments appears in Edwards 1979, chap. 10 entitled "The Age of Euler," and the story of the metaphysical struggles appears in Boyer, 1959, chap. 6, entitled "The Period of Indecision.")

Kline (1972) provided some advice:

> It will be helpful, in appreciating the work and arguments of the eighteenth century thinkers, to keep in mind that they did not distinguish between algebra and analysis. Because they did not appreciate the need for the limit concept and because they failed to recognize the problem introduced by infinite series, they naively regarded the calculus as an extension of algebra. (p. 401)

In order to give the flavor of Euler's style and that of others of his time, I borrow Edwards's account (1979) of Euler's handling of the relationship between the natural logarithm and natural exponential (pp. 272–273).

Euler begins by defining what we would call the logarithm of x base a as that exponent y such that $a^y = x$. He notes that $a^0 = 1$, and defining ϵ to be an infinitely small number, he writes $a^\epsilon = 1 + k\epsilon$ (where k is a constant that depends on a).

4. DEMOCRATIZING ACCESS TO CALCULUS   121

Now, given a finite number x, he introduces the infinitely large number N defined by $N = x/\epsilon$. Then he performs the following series of computations:

$$a^x = a^{N\epsilon} = (a^\epsilon)^N = (1 + k\epsilon)^N \text{ (since } a^\epsilon = 1 + k\epsilon)$$
$$= (1 + kx/N)^N$$
$$= 1 + N(kx/N) + [N(N-1)/2!](kx/N)^2$$
$$+ [N(N-1)(N-2)/3!](kx/N)^3 + \ldots \text{ (binomial series)}.$$

Thus, simplifying, he writes

$$a^x = 1 + kx + (1/2!)[N(N-1)/N^2]k^2x^2$$
$$+ (1/3!)[N(N-1)(N-2)/N^3]k^3x^3 + \ldots$$

But now, because N is infinitely large, he assumes that:

$$1 = (N-1)/N = (N-2)/N = \ldots$$

so he can simplify the previous representation of a^x to:

$$a^x = 1 + (kx)/1! + (1/2!)k^2x^2 + (1/3!)k^3x^3 + \ldots$$

For $x = 1$, he concludes that:

$$a = 1 + (1/1!)k + (1/2!)k^2 + (1/3!)k^3 + \ldots$$

He now introduces the now familiar number e as that value of a for which $k = 1$:

$$e = 1 + (1/1!) + (1/2!) + (1/3!) + \ldots$$

He then identifies e as the base for natural or hyperbolic logarithms and writes out its value to 23 places. Now he can write the prior equation $a^x = (1 + kx/N)^N$ as:

$$e^x = (1 + x/N)^N, \text{ or}$$
$$e^x = 1 + (1/1!)x + (1/2!)x^2 + (1/3!)x^3 + \ldots$$

What if a freshman calculus student performed these maneuvers? The student would likely get clobbered for making enormous and unjustified algebraic leaps, but would get some encouragement for being clever. Nonetheless, in many instances it is a matter of direct translation to render certain statements from Euler to the modern notation. For example,

$$e^x = (1 + x/N)^N$$

can be rewritten as the limit as n grows without bound of $(1 + x/n)^n$:

$$\lim_{n \to \infty}(1 + x/n)^n$$

But, of course, the notation translations replacing the "infinite N" by limit statements reflect deeper conceptual transformations. Euler and his contemporaries did not have the conceptual apparatus denoted by the limit notation. The fact that they managed to achieve such success without falling into utter chaos reveals the coherence of the mathematical system that they were operating within, a coherence ultimately born out by the nonstandard analysis model developed by Robinson (1966). It also reveals the utter faith that they had in their system.

Another illustration from Edwards (1979, p. 278) is Euler's alternative development of the differential of the natural logarithm of x, beginning with the previously derived formula (where N is as before, an infinitely large number defined in terms of the infinitely small number ϵ by $N = x/\epsilon$):

$$log(1 + x) = N[(1 + x)^{1/N} - 1].$$

By definition of N and substituting x for $1 + x$,

$$log\ x = (x^\epsilon - 1)/\epsilon.$$

But now, based on the power rule,

$$d(log\ x) = (\epsilon x^{\epsilon-1} dx)/\epsilon$$
$$= (x^\epsilon dx)/x.$$

Then he concludes that this last expression is equal to dx/x, because ϵ is infinitely small!

As put by Fraser (1988), "The 18th century faith in formalism, which seems today rather puzzling, was reinforced in practice by the success of its methods. At base it rested on what was essentially a philosophical conviction" (p. 331).

But there is more to his view of the calculus as embodied in the algebraically based study of algebraic formulas. Leibniz had suggested that the differentiation rule for natural logarithms was valid only for positive (real) values of x. Euler objected, observing in 1751:

> For, as this [differential] calculus concerns variable quantities, that is, quantities considered in general, if it were not generally true that $d/x = dx/x$, whatever value we give to x, either positive, negative, or even imaginary, we would never be able to make use of this rule, the truth of the differential calculus being founded on the generality of the rules it contains. (quoted in Fraser, 1988, p. 331)

As quoted from Volume 1 of his *Introductio* in Youschkevitch (1976, p. 61), Euler insisted that variables could take on *any* numbers as values, which, among other things, resulted in his failure to distinguish significant difference between real and complex analysis, a distinction that had to wait for Cauchy:

> The calculus of Euler and Lagrange differs from later analysis in its assumptions about mathematical existence. The relation of this calculus to geometry or arithmetic is one of correspondence rather than representation. Its objects are formulas constructed from variables and constants using elementary and transcendental operations and the composition of functions. When Euler and Lagrange use the term

"continuous" function they are referring to a function given by a single analytical expression; "continuity" means continuity of algebraic form. A theorem is often regarded as demonstrated if verified for several examples, the assumption being that the reasoning in question could be adapted to any other example one chose to consider. (Fraser, 1988, p. 328)

This is a perspective not unlike that of most undergraduates today, for whom any talk of limits and discontinuity falls on deaf ears (Dreyfus & Vinner, 1983; Vinner, 1987). Such talk is from another universe, another century, that of Cauchy and Weierstrauss:

> In the modern calculus attention is focussed locally, on a curve near a point or on a neighborhood about a number. By contrast, the algebraic viewpoint of Euler and Lagrange is global. The existence of an equation among variables implies the global validity of the relation in question. An analytical algorithm or technique implies a uniform and general mode of operation. In Euler's and Lagrange's presentation of a theorem in the calculus, no attention is paid to considerations of domain. The idea behind the proof is always algebraic. It is invariably understood that the theorem in question is generally correct, true everywhere except possibly at isolated values. The failure of the theorem at such values is not considered significant. The primary fact, the meaning of the theorem, derives always from the underlying algebra. (Fraser, 1988, p. 329)

Indeed, any contemporary teacher of calculus will recognize this algebraic view of the mathematical world at work when students are asked to prove the existence of a limit, where, for the students, the essence of the process is in the algebraic maneuvers relating the "given epsilon" to the "needed delta" (Williams, 1990, 1991).

One should be careful, however, not to understate Euler's brilliant and far-reaching accomplishments, accomplishments with long-term theoretical implications. As put by Boyer (1959):

> Although his views on the fundamental principles of the calculus lacked all semblance of the precision and rigor which entered calculus in the following century, the formalistic tendency which his work inaugurated was to free the new analysis from all geometric fetters. It also made more acceptable the arithmetic interpretation which was later to clarify the calculus through the limit concept which Euler himself neglected. (p. 246)

Another example of Euler's larger and indirect influence in opening up the concept of function is shown in the following section. But, before closing this section, I should mention the work that took the algebraic perspective to the extreme, Lagrange's attempt to formulate the integral and differential calculus without any use of infinitesimals, differentials, or limits. He tried to use infinite series, in particular, power series, to define the derivative and from there go on to develop much of the theory and methods of the calculus. While his attempt failed, mainly because he did not sufficiently attend to issues of convergence, he spun off some valuable results along the way, including new forms for the remainder term in Taylor's formula. He also was the first to use the word *derivative* and the *prime* notation, $f'(x)$, $f''(x)$, and so on (Cajori, 1929). In fact, he was among the first to treat the derivative as a function in its own right rather than as a relation between variables based on a ratio. Furthermore, his development yielded $f'(x)$ as a particular coefficient in an infinite series, entirely independent of differentials, geometry, or of motion metaphors—it was strictly an abstract algebraic object, as were all higher derivatives.

It may be speculative, but one interpretation of Lagrange's work is as an attempt to move beyond the interoperational development of calculus to a transoperational one (Piaget & Garcia, 1989). Here his aim was to create a structure that captures all the known procedures in a complete system. In his unsuccessful attempt, he was nonetheless led to think about derivatives and integrals in a more abstract way than others, particularly to think of the derivative as a "derived function," as "just another function."

The Slow, Irregular Evolution of the Concept of Function in the 18th and 19th Centuries.

The complex process of moving from this stable and algebraically self-contained world to the modern one is described by Grattan-Guinness (1970a, 1970b) and Grabiner (1981). It involves subtle interactions among understandings of the physics of vibrating strings, heat flow, and other phenomena, the differential equations used to model them, and the convergence of the infinite series that appeared as solutions to these equations. Interestingly, none of these appears in the experience of beginning calculus students today.

It also revolved around a gradual widening of the concept of function, which initially involved the question of what kinds of constructions were allowable in building a function. Euler began with a concept of function similar to that of Leibniz, but broadened it in his work on the vibrating string problem to include functions defined piecewise by different rules (but continuous, indeed piecewise smooth in the modern sense). He called these "mixed functions." A famous controversy with Jean d'Alembert ensued over whether the initial position of a vibrating string (fixed at its endpoints) need be "continuous" in the Euler sense, or whether it could be a "mixed function," as is the case if the string is plucked from its midpoint. Euler took the latter position, while d'Alembert, a highly respected colleague, disagreed. (d'Alembert, among his other achievements, was the first to realize, in 1754, that Newton's fluxions could actually be described as the limit of what are now the familiar ratios $\Delta y/\Delta x$ of finite differences, where y and x are related by some equation.)

Over the next several decades the discussion encompassed most of the mathematical community and was at the heart of the expansion and deepening of analysis. Does a function need to be explicitly representable, or could it be given implicitly by an equation that might in turn not be algebraically solvable? Or might it be describable over some interval only as the limit of a convergent series? And what functions are so describable anyway? And what of uniqueness of formulaic or series representations? In the latter part of the century it was realized that certain functions seemed to be determined everywhere by their series expansion in a small interval, almost as if there were a formula hiding behind them.

Then there were those new representations involving sines and cosines that Fourier exploited so effectively in the early part of the 19th century to model heat transfer that seemed to generate the most peculiar "functions." (Trigonometric series were used by Euler and Lagrange earlier, but not systematically so.) Indeed, even curves "not determined by any definite equation, of the kind wont to be traced by a free stroke of the hand" (Euler quoted in Youschkevitch, 1976, p. 68), sometimes called "mechanical curves," seemed to make sense as potential descriptions of quantitative relationships, because such could be imagined as modeling physical phenomena. These curves were frequently called "discontinuous" by Euler and others. Indeed, by 1807, Fourier showed that even "mixed functions" could be described by single trigonometric series over their entire domain of definition.

Interestingly, Euler would regard the hyperbolic function $y = 1/x$ as *continuous* and a smooth, connected, but quantitatively arbitrary hand-drawn curve as *discontinuous*. Surprisingly, the now common absolute value function was not used until Cauchy described it in the 1820s in the context of exhibiting a function that was both *mixed* and *continuous*—he noted that it could be expressed piecewise as minus x for negative x and x for nonnegative x, but could also be described as the square root of the quantity x squared for all x. (Similar examples were actually offered in the 1780s as well, shortly after Euler's death.) These sorts of examples, but especially those by Dirichlet (who in 1829 gave us the everywhere discontinuous function defined to be one constant at rationals and a different constant at irrationals) and by Fourier, put decisive stress on both the prevailing notion of continuity as "algebraic" continuity and the algebraically based notion of function.

Euler himself broadened his notion of function as he worked, and as early as 1755 at the beginning of his major exposition of differential calculus he offered the following definition:

> If some quantities so depend on other quantities that if the latter are changed the former undergo change, then the former quantities are called functions of the latter. This denomination is of broadest nature and comprises every method by means of which one quantity could be determined by others. If, therefore, x denotes a variable quantity, then all quantities which depend on x in any way or are determined by it are called functions of it. (Euler, quoted in Youschkevitch, 1976, p. 70)

But of special interest is the fact that Euler did not include such more general functions in his studies, only continuous or mixed functions; in fact, only analytic functions in the modern sense of "analytic" are considered in the book where this definition appeared. In fact, some version of "analytic" seemed to be inherent in most functions studied up until Fourier, even in Cauchy's foundation-setting work in 1821. But then, rather quickly, Euler's general definition came to have influence, probably because there was little alternative, given what the trigonometric series investigations were turning up. Thus, Fourier wrote, in 1821, in his famous book on the theory of heat, "In general, a function $f(x)$ represents a set of values or ordinates, each of which is arbitrary" (quoted in Youschkevitch, 1976, p. 77, in French, translated by the author). Fourier went on to emphasize the arbitrariness of these successive values, suggesting that each might be given individually.

It is commonly thought that Dirichlet is the author of the general concept of function, but in fact he (in 1837) and Lobatchevsky (in 1834) defined what amounts to arbitrary, but *continuous,* functions, as evident in Lobatchevsky's definition:

> General conception demands that a function of x be called a number which is given for each x and which changes gradually together with x. The value of the function could be given either by an analytical expression, or by a condition which offers a means for testing all numbers and selecting one of them; or, lastly, the dependence may exist but remain unknown. (quoted in Youschkevitch, 1976, p. 77)

This retrogression to conditions narrower than that offered by Euler some 80 years previously may be somewhat surprising, particularly because both Dirichlet and Lobatchevsky used discontinuous functions in their important work with trigonometric series (including the former's now-famous "Dirichlet conditions" for representability by Fourier series). Moreover, other mathematicians, including Stokes much later in 1880, continued to identify continuity as defined by Cauchy with algebraic continuity as defined by Euler (Youschkevitch, 1976). But from this data, it seems that we can conclude that until these functions were constituted as conceptual objects, their formal expression carried no weight in the mathematicians' actions—no matter how brilliant they were. To expect formal definitions to be meaningful to ordinary students seems all the more farfetched. Nonetheless, Hankel in 1870 offered an essentially general definition of function in which he also explicitly included discontinuous functions (in the sense of Cauchy) but inappropriately attributes the definition to Dirichlet.

Actually, the process of generalization of the concept of quantitative function (as opposed to functions on more general sets of objects) continued into the 20th century as new constructs came to be developed by such people as Baire, Borel, Cantor, and others. Inherent in all these progressions, reaching back even before Euler, were feedback relations between representability and constructability on the one hand and admissibility on the other. If in some way it could come to be represented in a way that connected with other well-accepted objects, it would be allowed to exist, but not before. Thus, Euler's general definition was not applied even by him, and then for a century hence, because there was no way to represent—and hence meaningfully study, compute, or apply—such general objects. On the other hand, new representation systems, such as those provided by trigonometric series, produced objects that forced a widening of the concept of function. Youschkevitch (1976) makes the point that Euler's premature general definition of function seemed to open the world to the wider conceptions that came later. See also Thompson's discussion of the concept of function (in press).

We should not be surprised that students might not appreciate the generality of our definitions, particularly given the (early) Euler-like algebraic universe that we provide for them in their prior courses.

The Evolution of Rigor in the Basic Concepts and Methods of Calculus—Cauchy's Role.

The concepts of continuity, limit, derivative, and integral evolved slowly from the middle of the 18th through the 19th centuries. As already indicated, much of this conceptual structure was entirely irrelevant to mathematicians before the second half of the 18th century, despite the fuss stirred by Bishop Berkeley's famous polemic attacking Newton and his followers for muddle-headedness relative to the underlying ideas of fluxions and fluents (Boyer, 1959; Edwards, 1979). It was not until the success of the methods of the calculus was in jeopardy that basic notions became an issue. And it was more than half a century until the matter came under significant conceptual control, thanks to Cauchy's work early in the 19th century, which provided a substantially rigorous foundation for basic concepts. This was given in three great texts published in 1821, 1822, and 1829, respectively.

Cauchy's work was preceded by earlier efforts to characterize continuity. One of these was by Louis Arbogast in 1791, a little over 200 years ago, who came up with a characterization of function continuity (over an interval) that amounted to the Intermediate Value Theorem condition; however, he also held that functions defined by different algebraic rules over different subintervals were discontinuous—the hold of the algebraic content of functions still held sway as a determining factor in his characterization. He used the term *discontiguous* for those cases "when the different parts of a curve do not join together" (Edwards, 1979, p. 304). Then in 1817, the deep-thinking Bohemian priest Bernhard Bolzano provided another effort that was quite successful and which was paralleled by Cauchy. Basically, he said that a function $f(x)$ is continuous at a set of numbers x if $f(x + \omega) - f(x)$ can be made as small as one wishes by choosing ω as small as required for all such x. (Some controversy exists as to whether Cauchy knew of Bolzano's work, which was circulated in a privately published pamphlet; Grattan-Guinness, 1970a.)

Another predecessor was d'Alembert, whose characterization of the derivative was mentioned earlier. He also had the following to say in the middle of the 18th century about differentials, which, as we have seen, Euler had regarded as zeros whenever it was convenient to do so: "A quantity is something or nothing: if it is something, it has not yet vanished; if it is nothing, it has literally vanished. The supposition that there is an intermediate state between these two is a chimera" (quoted in Boyer, 1959, p. 248).

Others tried, with varying degrees of success to come to terms with differentials and infinitesimals, especially, L'Huilier, Lacroix, and Carnot, whose approaches, despite wide popularity well into the 19th century, essentially reverted to Leibnizian ideas. Most shared with Lagrange's approach an inability to deal with convergence of the series that they developed and manipulated, or they exhibited other fundamental difficulties. Zeno's challenge, from two millennia before, remained unanswered.

Out of all this confusion, Cauchy's achievement stands even more eminent by its depth and comprehensiveness. His ambitious agenda was to put the whole of calculus as then understood on a firm logical basis, an agenda that he accomplished within the bounds of contemporary standards of proof and available number systems. He began with a definition of limit: "When the successive values attributed to a variable approach indefinitely a fixed value so as to end by differing from it by as little as one wishes, this last is called the limit of all the others" (quoted in Boyer, 1959, p. 272).

Cauchy then went on to define that most difficult of ideas, that of infinitesimal. Under the influence of the widespread success of the idea of variable, he broke with the patterns of the past that tended to treat infinitesimals as fixed quantities and instead made them variables, on a par with any other variables: "One says that a variable quantity becomes indefinitely small when its numerical value decreases indefinitely in such a way as to converge toward the limit zero" (quoted in Boyer, 1959, p. 273). He carefully defined infinitesimals of higher order and even infinities of higher order, again as variables, but in this case variables that grow without bound. From here he went on to define derivatives as the limit of the now standard ratio and treated them as did Lagrange, not as quotients, but as derived functions in their own right, even using the $f'(x)$ notation. He proved the Chain Rule and other basic properties of derivatives (although his proof of the Chain Rule contains the familiar error involving zero denominators).

Finally, he defined differentials in terms of derivatives, as is the case in modern treatments, which are, of course, modeled on his. In particular, if dx is a finite constant quantity, say h, then the differential of $y = f(x)$ is defined as $f'(x)dx$. Now the ratio of the differentials dy/dx tautologically coincides with $f'(x)$, which is also the limit of $\Delta y/\Delta x$ as $\Delta x \to 0$. From this base he was able to define differentials of higher order. He also defined continuity of functions at a point in terms of limits, that is, the limit as $x \to a$ of $f(x)$ must be $f(a)$.

With the differential calculus under conceptual control, he departed from the pattern that had prevailed since the breakthrough by Newton and Leibniz that enabled one to treat the integral as an antiderivative and returned to its definition over an interval as a limit of a sum—in this case of finite products $f(x_i)\Delta x_i$ in the usual sense, where it is assumed the limit is taken as the $\Delta x_i \to 0$. He called the limit a *definite integral* and cautioned against regarding it as a sum instead of a limit. He was then able to state and prove the version of the Fundamental Theorem of Calculus that asserts that the derivative of the definite integral of a continuous function $f(x)$ from a fixed point to x is $f(x)$. During the remainder of the 19th century, others (of course, Riemann, Darbaux, Peano, and Lebesgue) refined this basic idea of the integral as a limit of products, but its basic form was put in place by Cauchy.

Cauchy also defined convergence of infinite series in terms of limits of sums. He ran up against the inadequacies of conceptions of the number system when he tried to develop what are now known as Cauchy conditions for convergence—the conditions that state that convergence of a sequence is equivalent to the difference of its terms eventually getting infinitely small. The problem was that to guarantee convergence to a number, one needs the existence of the number! But if the number is defined in terms of convergence to itself, one has logical circularity. The geometric intuitions of continuity of the line, which had carried thinking in one form or another since the Greeks, indeed, since the Pythagoreans, no longer sufficed as a foundation for calculus. Cauchy had gone as far as he could, as far as the geometric continuum could take him.

His formulations gained wide, although not quite universal, acceptance in the following decades. He made a few well-known errors along the way, based on his tacit acceptance of the long-standing *principle of continuity* that Leibniz and many others had leaned on, the principle that properties of elements used in a limit will hold for the limit itself. This caused him to believe that the limit of continuous functions is continuous, and more generally, to ignore the concept of uniform convergence.

The Final Steps to Logical Rigor—The Number System and Weierstrass.

In addition to the logical shortcomings of the continuum that began to be evident by the middle of the 19th century, there were also problems with the relation between continuity and differentiability. Again, a geometric intuition—that a smoothly drawn (continuous) curve ought to have tangents—led to difficulties. Bolzano in the 1820s and then Weierstrass in 1872 exhibited a function that was continuous on a bounded interval, but which failed to have a derivative at infinitely many points. But Weierstrass did much more, for he put the finishing touches of logical rigor on what Cauchy had developed. His major achievement was to remove the last vestiges of continuous motion from the concepts of variable (now merely a letter representing any of a set of quantities), limit, continuity, and hence derivative. His $\epsilon - \delta$ definition, the one now standard in textbooks, was a static concept, replacing the motion metaphor used in Cauchy's definitions, as indicated previously—no longer did values of a variable "approach indefinitely a fixed value." Weierstrass rendered limits as atemporal concepts. Furthermore, he had no need for infinitesimals. His logical structure rendered them superfluous.

Lastly, not only Weierstrass, but also Dedekind, Heine, Meray, and Cantor (and actually, even Russell), constructed a genuine arithmetic continuum in a structure consistent with that of the concepts of limit and of function continuity. Nor did they employ geometry. In the words of Boyer (1959), "Geometry, having pointed the way to a suitable definition of continuity, it was in the end excluded from the formal definition of this concept" (p. 292).

Of interest is the view of the historians themselves on the achievements of the 19th century. Boyer (1959) exults in the abstractness of the formulations. "The introduction of uniform motion into Newton's method of fluxions was an irrelevant evasion of the question of continuity, disguised as an appeal to intuition. There is nothing dynamic in the idea of continuity" (p. 294). As argued in Kaput (1979), whether or not a logical structure can be imposed on a concept, its actual learnability and comprehensibility depends on its interpretation in the stable conceptual structures at hand, be they, in this case, embedded in motion metaphors or closeness metaphors. Furthermore, these conceptual foundations make themselves known in the languages and notations used to express the concept, despite claims of logical purity. Thus, we still speak of limits, variables, and constants, and we use limit notations with arrows, and so on. These are the means by which the concepts are learnable, knowable, and applicable.

More Recent Events.

While the calculus of functions of a real variable had thus been put in logical order, not all questions of logical consistency had—or have—been resolved, of course, because new questions arise that are associated with the theory of infinite sets that was needed to create the reals. Indeed, the foundation issues have not been removed but only moved to questions of decidability, consistency, and completeness.

Furthermore, the formalization achieved by Weierstrass turns out to be only one of perhaps many possibilities. The nonstandard analysis of Robinson (1966) amounts to a much more direct formalization of the work of Leibniz, Euler, and Lagrange. Here the complete, ordered field of reals is extended to the (incomplete) ordered field of *hyperreals*, which includes infinitesimals and infinite numbers. The idea of infinitesimal is used to model the notion of "infinitely close," which in turn serves as the basis of definitions of limit, continuity, differential, derivative, integral, series, and all the rest. The beauty of Robinson's approach is that it logically "legalizes" the facile computational techniques of the 18th century, including the kinds of maneuvers illustrated previously that Euler used so productively. Some evidence is accumulating that infinitesimals may provide accessible objects and language by which students can deal with the key ideas of calculus (Frid, in press). It also helps answer a question that puzzled so many mathematicians in the 19th century: Given the lack of foundation for so much of what was done previously, why was it so correct so much of the time? The answer is that there *are* consistent models of numbers and functions in which those maneuvers *are* logically legitimate. The models simply had not been formalized in the 18th century.[5]

It is important to point out that the process of growth in mathematics is continuing at an ever rapid pace, even in nontechnical areas that affect foundations once thought to be settled in the sense of not requiring further elaboration. A good example is provided by the new notion of hyperset (Barwise & Moss, 1991), which subsumes the old idea of set in the same way that the hyperreals subsume the reals—hypersets can be members of themselves in ways that (well-founded) sets cannot. But they allow the modeling of a large and growing class of circular phenomena in computer science, logic, and cognitive science that were previously out of bounds. And, of course, new iterative operations made feasible by computers open yet another large arena to analysis that is directly accessible to students (Bridger, 1989; Devaney, 1990; Peitgen & Richter, 1986; Morrison, 1991). Of course, these have yet to be systematically studied regarding their cognitive content and structure.

Reflections on the Significant Historical Factors

Given that the syntax of (single-variable) calculus is coherently and consistently constructed in its operative notation, we can be confident that the outcome of syntactically guided actions on symbol strings will be semantically correct (Kaput, 1987, 1989, 1993). This is the genius of Leibniz's contribution. One can mechanically "ride" the syntax of the notation without needing to think through the semantics.

This is perhaps among the greatest achievements of the human mind, although, to be sure, the real achievement was that of the masters who built the concepts initially and wrote them in language that embodied the organizing syntax that we consumers now can use with confidence. While it is commonly said that the storage and accumulation of human knowledge across generations has been made possible by the invention of writing, I would suggest that the real *generative* power of knowledge stored in writing follows from the storage of the

---

[5]There is a well-known, but not widely used, textbook based on this approach (Keisler, 1976), and other more recent materials have likewise taken the nonstandard approach (Bishop & Bridges 1985; Tall, 1981).

syntactically organized systems created by the giants. Using these, others of us less powerfully endowed can recreate their ideas and, more importantly, act on or with those ideas without the need to understand each step of the action. It suffices to make the semantic connection at strategic junctures in the process, especially at the beginning and the end.

However, it is all too easy to confuse acting by the rules with real mathematical activity. This is the major failing of our curriculum—in Skemp's terms (1987), we teach "instrumental understanding" instead of "relational understanding." We are able to get some of the students to write some of the notes without them ever hearing the larger melody, let alone the interplay of melodies. So their achievements are in small pieces, based on remembering rules for the manipulation of character strings. In the longer run, the capability of memory to hold the rules and their various combinations and restrictions is overwhelmed. The entire experience is, ultimately, alienating and debilitating for the great majority of students—including the majority who never get close to that experience called "the calculus sequence."

Part III is a response to this intolerable circumstance. It is based on the following four major historical factors identified in the historical review, each of which will be woven into recommendations:

1. The Scholastics' early attempts at mathematizing genuinely experienced variation before algebra,
2. Newton's (and others') heavy reliance on motion imagery to conceptualize continuous phenomena,
3. Leibniz's heavy reliance on the forms of symbols and the conceptual role of finite difference calculus in his independent development of calculus, and
4. The subtle shifts in both the semantics of mathematics and the nature of justification as the roles of geometry and algebra shifted in the period between the 16th and 19th centuries.

PART III: IMPLICATIONS AND FUTURE DIRECTIONS

The most important messages of this last part of the chapter are its twin recommendations drawn from the historical review and a survey of current conditions and possibilities:

1. Calculus needs to be studied across many years of school, from early grades onward, much as a subject like geometry should be studied. Hence, its many purposes should be examined, not merely its refined methods.

But most especially, its root problems should take precedence as the organizing force for curriculum design.
2. The power of new dynamic interactive technologies should be exploited in ways that reach beyond facilitating the use of traditional symbol systems (algebraic, numeric, and graphical), and, especially, in ways that allow controllable linkages between measurable events that are experienced as real by students and more formal mathematical representations of those events.

I discuss the rationale for these recommendations, offer suggestions on their implementation, including concrete examples, and discuss implications for needed research and development activity.

The Beginnings of a Longitudinally Coherent Calculus Curriculum

> I think the great flaw . . . is the assumption that you can build a house from the roof down. . . . Can you really hold conferences and talk seriously about calculus at the university level and not spend an equal amount of time on the secondary mathematics curriculum? It seems inconceivable to me. (Starr, 1988, p. 36)

Under current views, calculus should not be taught unless it is taught well, which means that it should be taught in ways that match the ideals of good introductory university courses. It is widely acknowledged that the standard casual cookbook introduction to calculus, called "crap calculus" by Kenelly (quoted in Cipra, 1988, p. 99), does more harm than good. Indeed, the Mathematical Association of America (MAA) recommends that, unless a rigorous AP calculus course is offered, the student would be better off taking solid preparatory courses or other topic alternatives. But, of course, the underlying view of "calculus" in that recommendation, and in virtually all public discussion of the subject, is not the one I am dealing with here—the view is essentially some local variation on the canonical syllabus that accepts, without question, the current place of calculus in school mathematics. The "calculus" that I refer to in my first recommendation is much broader in scope. Yet, the best way to show what I mean is to offer some particular illustrations of an alternative view. I take President Starr's suggestion quoted earlier from his talk at the "Calculus for a New Century Conference" a step further, going back to the elementary level.

Early Experiences.

After a student in elementary school can form, conceptualize and compute sums and differences of whole number quantities, the student is ready to begin analyzing increase and decrease patterns in numerical data, both pure numerical data and data arising from counts and measures, especially sequential data, that is, data occurring in temporal sequences—tracking growth of plants, classmates, points scored in games, money won and lost, numbers of people in the school building on an hourly basis, and so on. This same approach is being taken in a curriculum development project at the Technical Educational Resource Centers (TERC) (Nemirovsky & Tierney, 1992; Russell, 1991) and in an associated research project (Nemirovsky, 1993; Rubin & Nemirovsky, 1991).

Issues of representation of quantity and change in quantity are explicitly addressed, using both children's idiosyncratic notations as well as conventional notation systems such as tables and graphs, including graphical representations of all kinds.

Only gradually can quantity changes be considered apart from the quantities that gave rise to them. Thompson (1993) has examined the difficulties that fifth graders have in understanding differences as quantities in their own right, as quantities that can participate in additional operations and comparisons, as in "Whose growth rate is bigger?" Or, "What would the result be if we had a growth rate double this one?" Disentangling change in x from x is very difficult, but evidence exists that it can be done if students have sufficient experience, and if it is of an appropriate kind (Tierney & Nemirovsky, 1991). If this were done, these students would be in a much better position to distinguish $f'(x)$ from $f(x)$. In effect, I am suggesting that a developmental stage jump needs to be accomplished for young students, from treating differences at the intraoperational level, where their cognitive resources are consumed in conceiving and computing differences, to being able to treat those differences, those derived quantities, as entities in their own right, capable of participating in higher level actions (Breidenbach et al., 1991; Piaget & Garcia, 1989).

The kinds of activities that would accomplish this are not at all well represented in today's curriculum, and their systematic organization has yet to be developed. Yet their objective is the ability to deal with discrete change situations that amount, in their more sophisticated form, to solutions of difference equations, precursors to differential equations. For example, a standard starting-point problem of this genre might involve being given the record of the numbers of students entering and leaving a given room over time, and then being asked the actual number in the room after a certain period, in which one is given the number initially present in the room. Other problems involve tracking the number in the room at each countpoint, given various initial conditions. This style of thinking, not now part of the 4th- through 6th-grade curriculum, would likely help lay a base for the kinds of thinking needed in understanding both integration as well as the solution of differential equations. After all, the student is solving simple difference "equations" with boundary conditions (Nemirovsky et al., 1993).

The Middle Grades.

These kinds of activity for younger students involve quite simple computations. They need not involve rule-based functions, but can involve the irregular data collected from actual situations. But, nonetheless, they can lead to increasing formalization through constant, linear, and, eventually, exponential change situations at higher grade levels. They should also involve work with mean values—what constant growth yields the same total result as some variable growth rate? Important ties with primitive ideas of fairness, multiplication, division, and average come into play. This work can also involve interesting investigations of number patterns, sums of integers, figurate numbers, and so on, even, say, the sums of odd integers. Of course, they all have graphic representations. Studying the mathematics of change offers much opportunity to integrate ideas that do not currently cohere in our curriculum.

Further, one can become quite systematic in this approach, as Strang (1990a, 1990b) has illustrated. He has shown how one can use discrete versions of all the basic function types to develop the inverse relationships embodied in the Fundamental Theorem of Calculus. There is a large body of mathematics that can be approached in this style (Richardson, 1954; Smith, 1984; Zia, 1991). One might even consider gradually pulling students toward the standard differential and integral notation, although there is a clear danger of premature algorithmization. Ultimately, the goal should be to render the Fundamental Theorem an utter triviality to students, something so obvious that it could not be imagined to be false.

The process of moving from the discrete to the continuous needs careful consideration. The standard approach in calculus is to begin with "continuous" phenomena or functions (actually, "smooth" in the analytic sense), then to discretize and finally pass to a limit of some kind. The option exists to try the reverse.

Opportunities for meaningful use of ratios are frequent when one attempts to measure change. In fact, ratios form the basic mathematical tool in the formalization process. Indeed, the notion of proportion as expressing linearity, a suggestion strongly made by Karplus and Karplus (1972), gets center stage in this approach. In particular, students get a chance to deal with multiple instances of proportions, in which the context and hence units are fixed, but the numbers change, a situation that occurs all too rarely in standard approaches, but which seems to be pedagogically powerful (Kaput & West, 1993). Questions of units are likewise a frequent occurrence. They inevitably occur in substantive contexts.

While many, nay most, curricular details are yet to be determined for a coherent approach stretching through the grades, the major point is that the study of change and accumulation of quantity should begin at the early grades. It can be highly purposeful activity, certainly in the spirit of the Standards, one that contextualizes much hard thinking and even some hard computation if so desired. The research issues are abundant and will be discussed further.

Our Historical Roots: Leibniz and the Scholastics.

The kinds of activity described earlier all have the property that the underlying functions tend to treat the independent variable as an *index* for the values of the dependent variable and with a fairly strong sense of *progression,* as Leibniz used the term. In fact, these are precisely the kinds of activities that Leibniz used to get started. Looking closely at his early work, his differentials were treated exactly as if they were finite differences. Issues of limiting behavior intruded rather slowly and, in fact, rather unobtrusively.

As already noted, these activities all have rewarding geometric representations. Indeed, they take coordinate representations much further than do traditional graphing activities in schools, which are designed simply to display functions, with at best some elementary reasoning based on the resulting shape. These activities require serious thinking *with* the graphical representations and close coordination between such things as first and second differences of consecutive values in tables of data and the graphs.

Another strong parallel relates to the real beginnings of calculus with the Scholastics, who were attacking the modeling of real phenomena. The activities suggested, especially if they include rich discussion and writing, can amount to exactly what the Scholastics were trying to do—mathematize change. And given the sparseness of formal symbolic tools at an early age, the process is likely to be best handled numerically, discursively, and with rudimentary graphics. Oresme has taught us that it is not an easy matter and certainly should not be rushed. Nonetheless, students need not wait 250 years for algebra and coordinate axes to be invented.

Rather than attempt to characterize a secondary school curriculum at this point, I turn to the second major issue—how technology might be exploited to support a longitudinally coherent calculus curriculum. This approach utilizes the other major force behind the historical development of calculus, the study of motion and the associated motion imagery that was seen in Newton's version of calculus and which played a central role in the thinking of mathematicians before and since. It provides a complementary set of experiences to those described earlier that helps fill out the foundational calculus curriculum.

Following Newton, But With Dynamic, Interactive Media

While good calculator technology clearly can help with the activity previously discussed, and accumulators and difference takers can be useful, especially if tied to graphical representations (Confrey, 1993; Nemirovsky, 1991), even more dramatic impacts might be possible with more ambitious applications of technology. "The intuition that comes from driving a car, and from ordinary use of the speedometer and odometer, is a free gift to calculus teachers" (Strang, 1990a, p. 20).

We accept this gift, but would actually like to make much more of it than Strang has suggested. Our basic idea (developed over the past several years, e.g., Kaput, 1988) also builds on the approach taken by W. W. Sawyer in his classic *What Is Calculus About?* (1961):

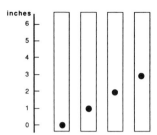

FIG. 4.8. W. W. Sawyer's (1961) "snapshots" of motion.

We are going to investigate speed, the speed of a moving object. How can we see clearly what a moving object is doing? We might make a "movie" of an object moving along a straight line. Suppose we have a camera that makes a picture every tenth of a second. Suppose successive pictures are as shown in Figure [4.8]. What is the little object doing? Every tenth of a second, it moves up 1 inch. It seems to be moving with a steady speed of 10 inches a second. (p. 11)

Thus begins Sawyer's approach to calculus. He goes on to examine, graphically, several more examples, including some with variable speed. He then suggests how an object might make a record of its own speed. Imagine it being an inked pen moving vertically, with paper sliding horizontally under it at a constant rate. He eventually goes on to discuss motion of vehicles, negative "speed," tabular recording of speed data, how to recover distance traveled from such data, what simple algebraically defined functions of speed look like, and so on, for the better part of two chapters before entering the traditional derivative computations, and then only very gradually.

MathCars: A Lived-In Motion Simulation

> I am one of the people who believe that the computer will revolutionize our subject as greatly as did Arabic numerals, the invention of algebra, and the invention of calculus itself. All these were democratizing discoveries; problems solvable only as research and by an elite suddenly became routine. The computer will do the same for our mathematics, and calculus is the place to begin. (Young, 1988, p. 173)

While from the context it is clear that Young is thinking of the computer's role in manipulating and linking traditional notations, the more leveraged purchase on the problem of learning and doing mathematics may be had from the introduction, manipulation and linkage of representations that are less formal and that capture more aspects of the world that we live in. Recall the Greek dead end that resulted from not having any way to represent and therefore think about irregular motion.

I have designed a motion world learning environment called MathCars that uses dynamic, interactive technology to extend this approach and to make it more interactive. The basic idea is to map the phenomenologically rich experience of

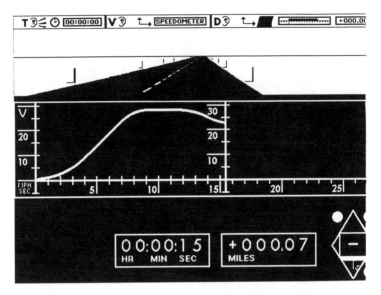

FIG. 4.9. MathCar with actively generated velocity graph.

motion in a vehicle (sights and sounds) onto coordinate graphical and other mathematical notations. The user controls the velocity of a simulated vehicle by controlling an accelerator (lower right side of Fig. 4.9), and, depending on the user's choices, coordinated representations of time, distance traveled, or velocity appear on the simulated dashboard, continuously being updated as travel proceeds. On the top of the windshield is an array of trip representation options, there are several for each of the three basic descriptors: time, velocity, and distance traveled (actually directed distance). As discussed later in the context of Fig. 4.13, any graph generated can then be studied as an object in its own right, so that the slope of a position graph might be examined at various points, and the slopes might even be plotted on another graph, and so on.

The particular scene pictured in Figs. 4.9 and 4.10 involves one car and a velocity versus time graph, together with a digital clock and an odometer. Here, the user has moved the analog speedometer to a vertical position where it slides from left to right as time progresses so that its tip leaves a trace of the vehicle's velocities at each instant traveled—the graphic here does not clearly indicate that it is a line segment whose height represents the current velocity (the result of bleaching a color graphic). One can configure the system so that the speedometer segment drags out a region or so that its tip drags out a curve, which is the case in the figures given here.

The resemblance to Oresme's initial representation is clear. But there is a deeper relation to Newton's conceptions as well—the use of time as a continuous variable. In a sense, continuity is for "free"—it is the gift referred to indirectly

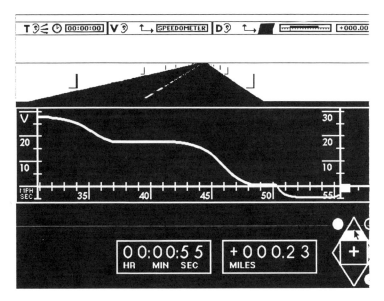

FIG. 4.10. MathCar in reverse.

by Strang in the quote earlier, and it is the fact of life exploited so effectively by Newton, discussed earlier. The curves are generated in the way Newton thought of them, with time as an implicit parameter.

In Fig. 4.10, the same trip is shown further down the road. You will notice that the driver has stopped (at approximately $T = 48$) for 2 seconds and then backed up (negative velocity), and at the instant pictured, is slowing the backward motion in order to stop, which is indicated by the pointer in the positive end of the accelerator. Of course, a perceptually more powerful indicator is provided by the windshield view, which shows objects receding at a decreasing rate.

In Fig. 4.11 many of the phenomenon-display options are shown. There are three different kinds of data that one can display given in three groups—one having to do with time grouped on the left-hand side, one with velocity in the center, and one set for position on the right side. (Recall Oresme's struggles with choices among these and how to represent them.) Within each of these are a sound cue and two or more types of visual data. On the left side there is a sound cue in the form of a beep—which one turns on by clicking on the ear icon—and analog and digital clocks. In retrospect, the trip shown in Figs. 4.9 and 4.10 had the time sound turned on, as indicated by the "marked ear" on the top-left menu. This metronome beeping sound can be coordinated with other display events, such as the depositing of dots, as indicated in Fig. 4.21.

In the center are velocity descriptors—a sound cue, given by the ear icon, a sound whose pitch varies with the velocity—up for faster, down for slower, a

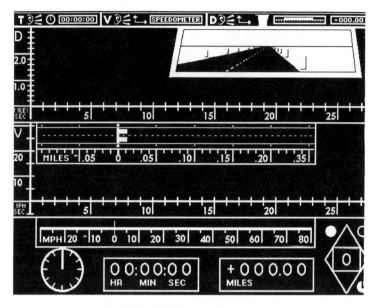

FIG. 4.11. MathCar display options.

velocity versus time graph, which was shown earlier, and an analog velocity meter, that is, a speedometer, which was also shown.

On the far right are position descriptors, in which the sound cue is an echo sound from the passage of the poles along the road—the faster one goes, the more rapid are the echoes from the posts that the driver is traveling by. There is a position versus time graph and a map. It will be noted that the map is actually a vertical view of the situation, which is basically a straight road with two lanes on it, to make provision for two MathCars on the same road. On the far right of the menu bar is an odometer than can actually be made to scroll into a table of data when linked with the digital clock (linkage not shown).

There is also a rearview mirror available whenever two cars are active. And, of course, there is an accelerator in the lower right-hand part of the screen. The lower right button turns the simulation on and off, while the upper two buttons indicate which car is being controlled—the blue or the red.

One can mix and match different ways of describing the motions that are generated to suit our purposes. Typically, one would not have more than two or three of these descriptors active at the same time. Nonetheless, these are carefully coordinated in order to link the phenomenology of the experience of motion with its formal representations. For example, the distance between the poles that one passes is coordinated with the distance between the position tick marks on the vertical axis of the position versus time graph.

In Fig. 4.12 is a different trip, in which the position versus time graph is

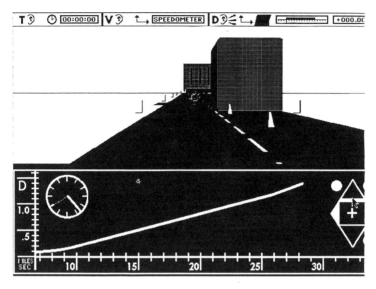

FIG. 4.12. An accident about to happen.

displayed instead of velocity, in which an analog clock is displayed, and in which the driver has used the lane changer (by clicking on the left arrow of the accelerator). Disaster looms. The driver has speeded up exactly when he should have slowed down and has not seen the truck backing up into the passing lane.

The District Attorney salvaged the record of the trip and exported the position versus time graph to an analytical environment, as shown in Fig. 4.13, in which she attempts to determine the velocity at the time of the accident. She used the slope tool to create two triangles (note the disastrous increase in velocity).

The system can support more than one active vehicle, although only one can be controlled at a time. However, the other's motion can be specified in advance. One can also be a passenger in one vehicle and control the other by remote control. This provides perceptual variation, as perspective is shifted. But, interestingly, the graphs do *not* change, providing a stability of event representation not shared by the standard "windshield view." The mathematical representation of the situation transcends the position-bound perspective of the driver.

Colors are coordinated so that all information about a vehicle is given in the color of that vehicle—this picture is bleached to accommodate the black and white medium of this book. The rearview mirror is used when one is in the lead car to get feedback on the relative velocity of the other car. In Fig. 4.14, I displayed the "map" (which includes traces) and the position versus time graph while the red car is passing at the beginning of a trip.

In Fig. 4.15, the drivers are further down the road, and the view is still from the blue car now following the red car, but preparing to pass the red car. Note that

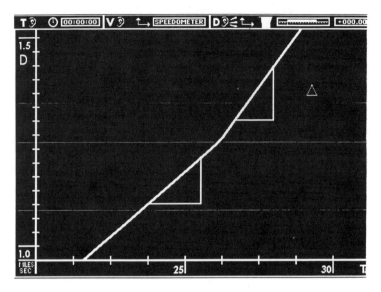

FIG. 4.13. Analyzing a previously driven graph.

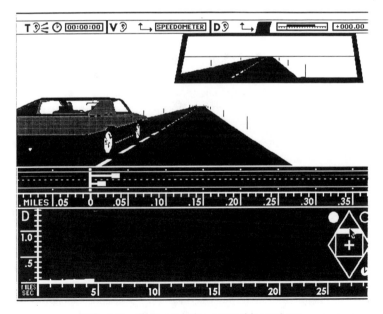

FIG. 4.14. Blue car being passed by red car.

4. DEMOCRATIZING ACCESS TO CALCULUS   143

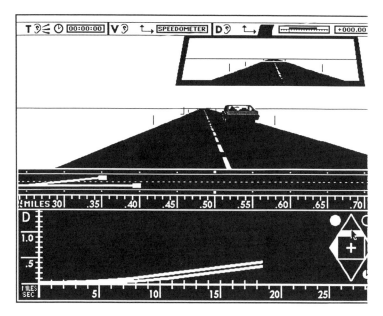

FIG. 4.15. Blue car following red car—Preparing to pass.

the blue car's position is behind that of the red car on the map, and its position graph is below that of the red car's. It is currently traveling at the same velocity as the blue car as indicated by the parallel position graphs. Of course, this sort of activity can (and certainly would) be repeated using velocity graphs.

In Fig. 4.16 the blue car is now ahead of the red car, which is visible in the rearview mirror. The position graphs have crossed, and now the blue car's graph is above the red car's graph. Furthermore, the distance between them is increasing as the red car is slowing down. Indeed, it has almost stopped, as reflected in the near-horizontal position graph.

Note that one might save a trip and then replay it, but with the option of changing the display, so that exactly the same event can be reviewed with different descriptors, for example, using the velocity displays rather than the position displays.

In Fig. 4.17 a more abstract activity is depicted, one in which the windshield view does not appear. In its discrete form, using sportscars, it might be called a *rally*. Once a more popular activity than it is today, a rally consists of a group of driving enthusiasts who begin from a particular location, where each is given instructions and perhaps a map. The instructions give places and times specifying where each driver, perhaps with a navigator, must match as closely as possible. Drivers must stay within speed limits and generally obey the laws, and their score is determined by the total variance from the target times and positions.

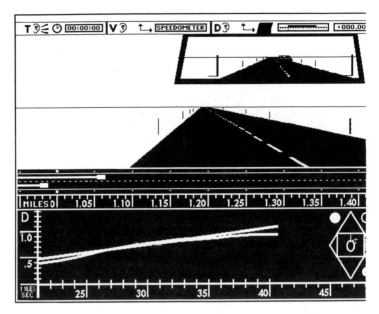

FIG. 4.16. Blue car in front—Red car in rearview mirror.

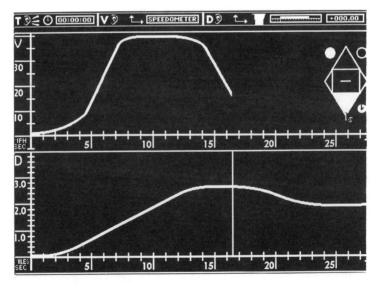

FIG. 4.17. "Enacting" the fundamental theorem of calculus.

A continuous version of this game, if one must operate strictly with velocity and time feedback, amounts to "enacting" the Fundamental Theorem of Calculus. Depicted in Fig. 4.17 (without color coding, unfortunately) is an illustration. The driver is provided a target position versus time graph (in the lower part of the screen) and must attempt to drive that graph—but with the critical proviso that the only feedback is in terms of *velocity* (where time is automatically taken care of by the time cursor, the vertical line that slides from left to right on the lower [target] graph).

The driver begins here with steep acceleration, a jack-rabbit start, and then the fairly constant velocity. She tries to match the straight position graph and then slows down, rather quickly toward 0 when she realizes that the velocity must be 0 in order for the position graph to be horizontal. In Fig. 4.18 she reverses (negative velocity) and then accelerates forward to stop again. She then moves to a forward velocity and a constant velocity at this point for a while, as she is trying to match the steadily climbing position graph, and then she decelerates. She seems to be a little bit behind the times in trying to match the position graph.

At the end of the trip she is able to check to see how well she did by superimposing the position graph she actually drove on the one she was trying to match (the "computer's" graph) via the dialog box in Fig. 4.19. But actually, in Fig. 4.19, she had previously shown the velocity graph she should have driven— it is the "computer's" velocity graph, partially obscured by the dialog box.

Using the dialog box she can now show her own position versus time graph to see what kind of a position graph she did in fact generate, which is shown in Fig.

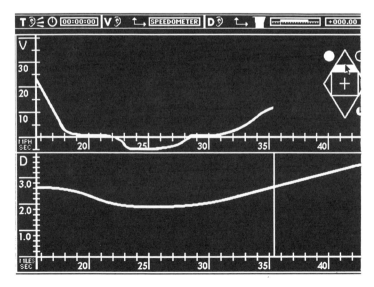

FIG. 4.18. Reversing direction.

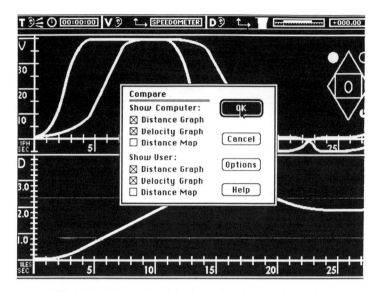

FIG. 4.19. Comparing driven graphs with target graphs.

4.20. As is apparent, it appears as if she really were a little bit behind the times. She did not quite generate the position graph, although she did seem to get a little better as she went along. She can look at her result both ways by comparing the target's velocity with her actual velocity or by comparing her actual position with the target position. She might also be asked how far her final position is from where she was "supposed" to be at the end (note the compensating error late in the trip).

This activity could have been done the other way, giving a velocity graph as a target with feedback only on the position graph, which reverses the thing. Of course, this amounts to acting out the relations between a function and its derivative given by the Fundamental Theorems of Calculus. They are not being proven in any sense. In fact, the relations specified by the theorems are instantiated in the software, in much the same way that the hierarchical structure of the number system is instantiated in the standard placeholder notation. The aim of these sorts of activities is to coordinate one quantity's rate of change with the accumulation of that quantity. Another view of this activity is that it amounts to solving in an enactive mode a graphically given differential equation. Yet another view is that the driver is "following" the vehicle whose motion is given by the target graph—but without a windshield view, using "Instrument Driving Rules."

I have ignored depictions of numerical representations of the data, but one can imagine generating multicolumned tables of data, for instance, that capture the same information. These might also be tied to algebraic representations of the data in cases in which the data are regular enough to admit a linear or quadratic

4. DEMOCRATIZING ACCESS TO CALCULUS 147

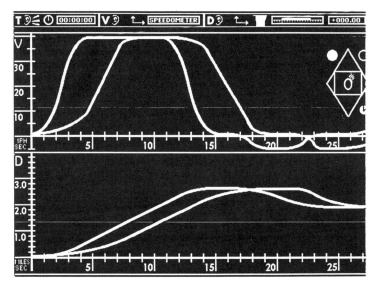

FIG. 4.20. Results of comparisons.

description. For example, one might attempt to "drive" a certain parabolic position curve by choosing one's velocity slope in advance, testing it out, and comparing it with the target (after, of course, determining that a straight-line velocity curve yields the best shaped position curve). Possibilities are abundant.

Variations On MathCars: Some Research Issues.

One direct variation on MathCars is MathBikes, in which the student is responsible for generating the motion kinesthetically. It should also be possible to plot velocity or distance against parameters other than time, for example, physiological parameters, such as heart rate or blood pressure. Then one can model, in effect, physical fitness by plotting velocity or distance against the physiological parameter. (Does it make a difference which is regarded as the independent variable?)

It is important that students experience the process of coordinating change and accumulation of change of quantity in a variety of contexts. A look at all the builders of calculus and its methods from the 17th century onward reveals that all were modelers. While MathCars deals with velocity and position, it is but a short step to an environment in which the windshield view is replaced by another appropriate quantity display, such as a fluid flow and accumulation, and in which the student controls the rate of change of fluid flow instead of acceleration. Essentially everything else is the same, reflecting the invariance of underlying mathematical structure, which is what we want students to abstract from these experiences.

Interesting research questions arise in these variations, some of which have already been addressed by Nemirovsky (1993) and Rubin (Rubin & Nemirovsky, 1991), who have examined student thinking and behavior as one changes the modeling context from motion to fluid flow. They have also begun examining the impacts of switching between computer-based models of computer-generated data and models of non-computer-generated physical data, as, for example, the data generated as one moves by hand a toy car in front of a motion detector (Mokros & Tinker, 1987) or controls air flow by manipulating a bellows. In either case, the data are displayed graphically on a computer screen. Work just beginning enables students to import such data into an analytical environment to examine it further. Greeno (1991) is likewise testing the relation between computer and physical models as contexts for students' learning and thinking. Thompson (1993) likewise is examining student learning with computer-based models of linear motion. His subjects are only fifth graders, but exhibit surprisingly potent intuitions regarding variable velocity and its relations to position and, especially, the use of mean velocities to approximate variable ones, as in the 500-year-old Merton's Mean Value Theorem. Indeed, one suspects that such a basic result may reflect an underlying normalization tendency on the part of individuals facing variation—a way of coping with variation and complexity.

Of central interest is the kind of understanding that results from dealing with graphically represented data. Rubin and Nemirovsky (1991) have identified a basic "vocabulary" of shapes that students seem to use in interpreting changing quantities (which bears strong resemblance to that used by the Scholastics). We need to determine not only the basic interpretive shapes, but how they relate to one another as representations of f and f', and how they can form the basis for an understanding of the basic functions represented algebraically. Schwartz and Yerushalmy (this volume) have built a prototype computer environment that enables one to patch together basic function shapes to model phenomena and to track the resulting algebraic description as well.

Another set of issues concerns the use of other dynamic diagrams in modeling quantitative relationships (Hall, 1988, 1990, Kaput, 1992b, 1993a, 1993b). A dramatic change results when one replaces static diagrams by dynamic, manipulable ones. This issue is discussed further in Kaput (1992b, 1993b), in which the representational framework introduced in Part I is elaborated on to account for the difference between static/inert representations and dynamic/interactive ones.

Moving from Graphic Displays to Graphic Action Representations.

I emphasized earlier the profound impact of the introduction of algebra as an action notation system in the 16th century on the nature of mathematics that was possible. I traced the explosive expansion of calculus based on the algebraic mode of mathematical reasoning in the 18th century. Of course, the introduction of coordinate graphs also proved critical. We now need to consider whether a comparable transformation of the experience of doing mathematics may be possible as a result of the new dynamic, manipulable graphics now becoming available for all manner of applications (Kaput, 1993a).

One key may be the change of graphical representations from display to action representations. Elsewhere I (1989) discussed the different meanings of the phrase "to solve an equation" as the representations of the relationships involved change. Schwartz and Yerushalmy (this volume) and Confrey (1993) have shown how the action system of algebra can now become the *recipient* of actions initiated in the coordinate graphical realm—rather than the other way around. By and large the actions on coordinate graphs are defined in terms of the algebraic structure of the functions being dealt with, which means that a rather restricted set of functions can be acted upon—those expressible in closed analytic form, that is, those "continuous" in the 18th-century sense. A similar statement can be said of Tall's *Graphic Calculus* (1986). But, of course, the progress of the 18th century forced a widening of this restriction, because ever more functions "naturally" appeared that did not fit the constraint, but which nonetheless had meaning in some context.

Now we ask, with Thomas Tucker (1988, p. 16): *Are all functions encountered in real life given by closed algebraic formulas? Are any?* Of course, the functions generated by accelerator control in MathCars—and indeed in any cars—are not given by closed algebraic formulas. This does not mean that one should not be able to analyze them, including in ways that reflect the insights of the masters.

Consider Fig. 4.21, in which it shall be assumed that the top graph (a position versus time graph) has been generated by a student in some direct way, without the use of an algebraically defined formula. Suppose, for example, that it was generated in MathCars with the "dot-dropping" option turned on, which deposits a dot on the graph for each second elapsed, which might in turn be coordinated with the time beeper.

It is assumed that the graph (as a set of ordered pairs) has been put into an analytical environment in which we have set up a second time axis and asked the question, what is the interpretation of the vertical distance between each dot? The length of the vertical line segment constructed for some of the dots (yielding a "lollipop") is, of course, the change in position over the 1-second interval between dots. If there was a table of these data at such intervals, the length of the lollipop sticks would represent the "first differences" of the consecutive position values—an idea with which both Leibniz and Newton would likely be comfortable (provided, for Newton, the graph has been continuously generated in advance).

Suppose now that we copy those vertical segments onto the new time axis as indicated in the bottom of Fig. 4.21 and ask what the interpretation of the vertical axis should be? The answer is given by the conspicuous label, namely, a distance unit per second (where, in this case, the distance unit, derived from the scale of

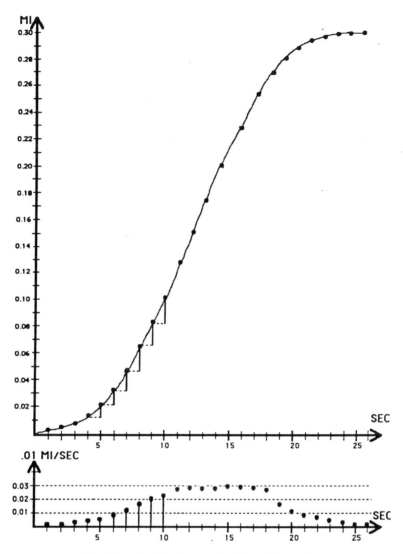

FIG. 4.21. Graphical approximation of derivatives.

the position graph, is .01 mi). After all, if the length of the vertical line segment is a change in position for each 1-second interval, then we can think of this distance as an average velocity over that 1-second interval. We further ask, what is the meaning of the entire set of dots? The answer, given glibly and quickly here, is that it is an approximation to the velocity of the vehicle on a 1-per-second sample of position change. More abstractly, it is an approximation to the derivative of the position function.

Suppose now that we decrease the beeping interval. We get more dots and a better approximation to the derivative (or velocity function). The reader will note that this process makes almost no use of algebra and does not depend on the function being given as a closed-form rule. Further, the same process could be applied to a velocity function to yield a change-in-velocity graph, namely, an acceleration approximation, that is, a second derivative. The process can also be reversed, using an area interpretation, to yield antiderivatives. (How would *you* do it?) In fact, the elements of differential equations can be approached in this manner. See Kaput (1993b) for further elaboration of these ideas.

While we are sweeping some detail under the rug, particularly issues of scale (see the discussion of a scaling curriculum in Kaput, 1993a), the reasoning involves simple ratios, which in turn require only a modest amount of symbolic representation to be dealt with efficiently.

Although the details are subject to much improvement with practice and actual instantiation in a computer environment, the key point is that we are doing calculus. In this case it is nonalgebraic calculus. But, the techniques are of wide generality (they could also be applied to relative velocity involving two vehicles or to quantities generated in any manner whatsoever).

Some Obvious Questions and Challenges.

This highly graphical approach raises questions regarding the appropriate grade level for the introduction of these ideas. My preliminary hypothesis is that middle school is probably a good place to start. But this then challenges some very fundamental assumptions about the role of algebra as a language for the expression of important ideas and the usual prerequisite relation between algebra and calculus. It challenges the prerequisite structures even more conspicuously than did the discrete activities for elementary grades discussed earlier. This approach also puts a much heavier burden on understanding graphs and manipulating graphs, understanding that is well known to require specific instruction. Other aspects of this experience do not involve graphics per se, but involve what Williams (1993), drawing on the work of Winograd and Flores (1986), refers to as "thrown-ness." The degree of immersion and style of interaction of a "lived-in simulation" such as MathCars is very different from that of a parameter-adjusting simulation, in which feedback, while prompt, is separated from one's choices.

Another series of questions concerns the extension of the described graphical approaches to more formal algebraic ones. Which kinds of functions should be introduced and how? Should they be introduced as objects in their own right or as approximations to the irregular functions generated by "real" phenomena? If the latter, then there is heavy use to be made of the Mean Value property—not an unreasonable starting point given its prominence and utility in the 13th and 14th centuries before algebra was available as an analytical tool, as already noted.

Reflections for the Longer Term Reform of Calculus

Despite our prolixity on the topics of the roots and purposes of calculus, I have not drawn conclusions regarding the *methods* of calculus and their treatment in the curriculum that would follow based on the experiences described earlier—presumably at the secondary level. It is here that much of the ongoing calculus reform discussion would seem to have its impact. However, even here, there is much that is taken for granted that should be open to question. Assuming that technology makes a large impact on the relevance and the forms of execution of particular techniques, we must ask where these techniques should now be taught? Might they best be handled in the contexts of their use, given that conceptual foundations are in place? Might we now be able to offer students opportunities to learn some version of quantitative methods in the contexts of the life and social sciences that were once reserved for the academic elite? After all, democratizing access to big ideas is only part of the battle. We must also democratize their use.

Can We Afford Not to Rethink Calculus?

Extrapolating ahead, it seems apparent that, as the incredible growth rate of mathematics and its application continues, the amount of mathematical content that will be important for productive citizens to know in the next century will continue to expand prodigiously. Much of this mathematics will involve modeling continuous change phenomena, especially dynamic and chaotic systems in computational media (Morrison, 1991). To the extent that this mathematics will be represented in the curricula of tomorrow, especially in secondary curricula, early treatment of the mathematics of change as depicted earlier is neither a curiosity nor a luxury, but an absolute necessity. In the long run, the new mathematics as well as its embodiment in school curriculum may subsume what has historically been known as *calculus,* and students will simply study the mathematics of change.

ACKNOWLEDGMENTS

The author gratefully acknowledges grant support from the Apple ACOT Program, OERI grant R117G10002, and NSF grant MDR 885617. The title "borrows" from John Mason (1985). The author also wishes to thank Ricardo Nemirovsky for many rewarding conversations on the topics of this chapter.

REFERENCES

Ayers, T., Davis, G., Dubinsky, E., & Lewin, P. (1988). Computer experiences in learning composition of functions. *Journal for Research in Mathematics Education, 19*(3), 246–259.
Baron, M. E. (1969). *The origins of the infinitesimal calculus.* London: Pergamon Press.

Barwise, J., & Moss, L. (1991). Hypersets. *The Mathematical Intelligencer, 13,* 31–41.
Bishop, E., & Bridges, D. (1985). *Constructive analysis.* New York: Springer-Verlag.
Bochner, S. (1966). *The role of mathematics in the rise of science.* Princeton, NJ: Princeton University Press.
Boyer, C. (1959). *The history of calculus and its historical development.* New York: Dover Publications.
Boyer, C., & Mertzbach, U. (1989). *A history of mathematics* (2nd ed.). New York: Wiley.
Breidenbach, D., Dubinsky, E., Hawks, J., & Nichols, D. (1991). Development of the process conception of function. *Educational Studies in Mathematics,* 247–285.
Bridger, M. (1989). *An introduction to chaotic dynamical systems* (2nd ed.). Redwood City, CA: Addison-Wesley.
Burton, D. (1985). *The history of mathematics: An introduction.* Boston, MA: Allyn and Bacon.
Cajori, F. (1929). *A history of mathematical notations* (Vol. 2: Notations mainly in higher mathematics). La Salle, IL: The Open Court Publishing Co.
Cipra, B. (1988). Recent innovations in calculus instruction. In L. Steen (Ed.), *Calculus for a new century* (MAA Notes No. 8, pp. 95–103). Washington, DC: Mathematical Association of America.
Clagett, M. (1961). *The science of mechanics.* Madison: University of Wisconsin Press.
Clagett, M. (1968). *Nicole Oresme and the medieval geometry of qualities and motions.* Madison: University of Wisconsin Press.
Confrey, J. (1993). *Function probe* [Software]. Santa Barbara, CA: Intellimation.
Cornu, B. (1983). *Apprentissage de la notion de limite: Conceptions et obstacles* [Approaches to the notion of limit: Conceptions and obstacles]. Doctoral dissertation, L'Université Scientifique et Medicale de Grenoble.
diSessa, A., Hammer, D., & Sherin, B. (1991). Inventing graphing: Meta-representational expertise in children. *Journal of Mathematical Behavior, 10,* 117–160.
Devaney, R. (1990). *Chaos, fractals, and dynamics,* Reading, MA: Addison-Wesley.
Dreyfus, T., & Vinner, S. (1983). The function concept in college students: Linearity, smoothness, and periodicity. *Focus On Learning Problems in Mathematics, 5,* 119–132.
Dubinsky, E. (1991). Reflective abstraction in advanced mathematical thinking. In D. Tall (Ed.), *Advanced mathematical thinking* (pp. 95–124). London: Reidel.
Dubinsky, E., & Schwingendorf, K. (1991). Constructing calculus concepts: Cooperation in a computer lab. In C. Leinbach et al. (Eds.), *Laboratory approaches to teaching calculus* (MAA Notes No. 20, pp. 47–70). Washington, DC: Mathematical Association of America.
Dugas, R. (1988). *A history of mechanics.* New York: Dover.
Edwards, C. (1979). *The historical development of the calculus.* New York: Springer-Verlag.
Eliot, T. S. (1962). Four quartets: Little Gidding (v). In *T. S. Eliot: The complete poems and plays, 1909–1950.* New York: Harcourt, Brace & World.
Fraser, C. (1988). The calculus as algebraic analysis: Some observations on mathematical analysis in the 18th century. *Archive for History of Exact Sciences, 38,* 317–335.
Frid, S. (in press). Three approaches to undergraduate instruction: Their nature and potential impact on students' language use and sources of conviction. In E. Dubinsky, J. Kaput, & A. Schoenfeld (Eds.), *Research in collegiate mathematics education, Vol. 1.* Providence, RI: American Mathematical Society.
Gardner, H. (1979). Developmental psychology after Piaget: An approach in terms of symbolization. *Human Development.*
Gillespie, C. (1960). *The edge of objectivity: An essay in the history of scientific ideas.* Princeton, NJ: Princeton University Press.
Grabiner, J. (1981). *The origins of Cauchy's rigorous calculus.* Cambridge, MA: MIT press.
Grabiner, J. (1983). The changing concept of change: The derivative from Fermat to Weierstrass. *Mathematics Magazine, 56,* 195–203.

Grattan-Guinness, I. (1970a). Bolzano, Cauchy and the "new analysis" of the early nineteenth century. *Archive for History of Exact Sciences, 6,* 372–400.

Grattan-Guinness, I. (1970b). *The development of the foundations of mathematical analysis from Euler to Riemann.* Cambridge, MA: MIT Press.

Greeno, J. (1983). Conceptual entities. In A. Stevens & D. Gentner (Eds.), *Mental models.* Hillsdale, NJ: Lawrence Erlbaum Associates.

Greeno, J. (1991). Number sense as situated knowing in a conceptual domain. *Journal for Research in Mathematics Education, 22,* 170–218.

Hall, R. (1988). *Qualitative diagrams: Supporting the construction of algebraic representations in problem solving* (Tech. Rep.). Irvine: University of California, Department of Information and Computer Science.

Hall, R. (1990). *Making mathematics on paper: Constructing representations of stories about related linear functions.* Unpublished doctoral dissertation. University of California, Berkeley.

Harel, G., & Kaput, J. (1991). Conceptual entities in advanced mathematical thinking: The role of notations in their formation and use. In D. Tall (Ed.), *Advanced mathematical thinking.* London: Reidel.

Kaput, J. (1979). Mathematics and learning: The roots of epistemological status. In J. Clement & J. Lochhead (Eds.), *Cognitive process instruction* (pp. 289–303). Philadelphia: The Franklin Institute Press.

Kaput, J. (1987). Toward a theory of symbol use in mathematics. In C. Janvier (Ed.), *Problems of representation in mathematics learning and problem solving* (pp. 159–196). Hillsdale, NJ: Lawrence Erlbaum Associates.

Kaput, J. (1988, April). *Applying technology in mathematics classrooms: Time to get serious, time to define our own technological destiny.* Paper prepared for the Annual Meeting of the American Educational Research Association, New Orleans.

Kaput, J. (1989). Linking representations in the symbol system of algebra. In C. Kieran & S. Wagner (Eds.), *A research agenda for the teaching and learning of algebra* (pp. 167–194). Reston, VA: National Council of Teachers of Mathematics; and Hillsdale, NJ: Lawrence Erlbaum Associates.

Kaput, J. (1992a). Notations and representations as mediators of constructive processes. In E. von Glasersfeld (Ed.), *Constructivism in mathematics education* (pp. 53–74). Dordrecht, Holland: Reidel.

Kaput, J. (1992b). Technology and mathematics education. In D. Grouws (Ed.), *Handbook on research in teaching and learning mathematics* (pp. 515–556). New York: Macmillan.

Kaput, J. (1993a). The urgent need for proleptic research in the graphical representation of quantitative relationships. In T. Carpenter, E. Fennema, & T. Romberg (Eds.), *Integrating research in the graphical representation of functions* (pp. 279–312). Hillsdale, NJ: Lawrence Erlbaum Associates.

Kaput, J. (1993b). The representational roles of technology in connecting mathematics with authentic experience. In R. Bieler, R. W. Scholz, R. Strasser, & B. Winkelman (Eds.), *Mathematics as a didactic discipline: The start of the art* (pp. 379–397). Dordrecht: Kluwer.

Kaput, J., & West, M. (1993). Factors affecting informal proportional reasoning. In G. Harel & J. Confrey (Eds.), *The development of multiplicative reasoning in the learning of mathematics* (pp. 237–289). Albany, NY: State University of New York Press.

Karplus, R., & Karplus, E. (1972). Intellectual development beyond elementary school: Ratio, a longitudinal study. *School Science and Mathematics, 72,* 735–742.

Keisler, H. (1976). *Foundations of infinitesimal calculus.* Boston: Prindle, Weber & Schmidt.

Klein, J. (1968). *Greek mathematical thought and the origin of algebra.* New York: Dover Publications. (Revised 1992)

Kline, M. (1968). Logic vs pedagogy. *American Mathematical Monthly, 77,* 264–282.

Kline, M. (1972). *Mathematical thought from ancient to modern times.* New York: Oxford University Press.

Leinhardt, G., Zaslavsky, O., & Stein, M. (1990). Functions, graphs, and graphing: Tasks, learning, and teaching. *Review of Educational Research, 60*(2), 1-64.

Luce, R. D., & Narens, L. (1987). Measurement scales on the continuum. *Science, 236,* 1527-1532.

Mahoney, M. (1980). The beginnings of algebraic thought in the seventeenth century. In S. Gankroger (Ed.), *Descartes: Philosophy, mathematics and physics.* Sussex, England: Harvester Press.

Mamona-Downs, J. (1990). *Calculus-analysis: A review of recent educational research* (Tech. Rep.). Pittsburgh, PA: Learning Research and Development Center, University of Pittsburgh.

Mason, J. (1985). *Routes to, roots of, algebra.* Milton Keynes: Open University Press.

Mokros, J., & Tinker, R. (1987). The impact of microcomputer-based labs on children's ability to interpret graphs. *Journal of Research in Science Teaching, 24,* 369-383.

Morrison, F. (1991). *The art of modeling dynamic systems.* New York: Wiley Interscience.

Nemirovsky, R. (1993). Rethinking calculus education. *Hands on, 16*(1).

Nemirovsky, R., Tierney, C., & Ogonowsky, M. (1993). *Children, additive change, and calculus* (TERC Working Paper, 2-93). Cambridge, MA: Technical Educational Resource Center.

Pea, R. (1987). Cognitive technologies for mathematics education. In A. Schoenfeld (Ed.), *Cognitive science and mathematics education* (pp. 89-122). Hillsdale, NJ: Lawrence Erlbaum Associates.

Peitgen, H., & Richter, P. (1986). *The beauty of fractals.* New York: Springer-Verlag.

Piaget, J., & Garcia, R. (1989). *Psychogenesis and the history of science* (H. Feider, trans.) New York: Columbia University Press.

Priestly, W. (1979). *Calculus: An historical approach.* New York: Springer-Verlag.

Richardson, C. (1954). *An introduction to the calculus of finite differences.* Princeton, NJ: D. Van Nostrand.

Robinson, A. (1966). *Non-standard analysis.* Amsterdam, London: North Holland.

Rubin, A., & Nemirovsky, R. (1991). Cars, computers and air pumps: Thoughts on the roles of physical and computer models in learning the central concepts of calculus. In R. G. Underhill (Ed.), *Proceedings of the 13th annual meeting, North American Chapter of the International Group for the Psychology of Mathematics Education* (Vol. 2, pp. 168-174). Blacksburg, VA.

Russell, S-J. (1991). *Investigations in number, data, and space* (National Science Foundation Elementary School Materials Development Project). Cambridge, MA: TERC. Technical Educational Resource Centers.

Salomon, G. (1979). *Interaction of media, cognition and learning.* San Francisco: Jossey-Bass.

Sawyer, W. W. (1961). *What is calculus about?* New Haven, CT: Yale University Press. (Reprinted by the Mathematical Association of America, New Mathematical Library Series, 1975).

Schoenfeld, A. (1986). On having and using geometric knowledge. In J. Hiebert (Ed.), *Conceptual and procedural knowledge: The case of mathematics* (pp. 225-264). Hillsdale, NJ: Lawrence Erlbaum Associates.

Sfard, A. (1991). On the dual nature of mathematical conceptions: Reflections on processes and objects as different sides of the same coin. *Educational Studies in Mathematics, 22*(1), 1-36.

Skemp, R. (1987). *The psychology of learning mathematics* (revised American ed.). Hillsdale, NJ: Lawrence Erlbaum Associates.

Smith, R. (1984). Sharing teaching ideas: Some sum derivations. *Mathematics Teacher, 77,* 110-112.

Steen, L. (1990). *On the shoulders of giants: New approaches to numeracy.* Washington, DC: National Academy Press.

Starr, S. (1988). Calculus is the core of a liberal education. In L. Steen (Ed.), *Calculus for a new century* (MAA Notes No. 8, pp. 35-40). Washington, DC: Mathematical Association of America.

Strang, G. (1990a). Sums and differences vs integrals and derivatives. *College Mathematics Journal, 21,* 20-27.

Strang, G. (1990b). *Calculus*. Wellesley, MA: Wellesley-Cambridge Press.
Struik, D. (1986). *A source book in mathematics 1200–1800*. Princeton, NJ: Princeton University Press.
Tall, D. (1981). Infinitesimals constructed algebraically and interpreted geometrically. *Mathematical Education for Teaching, 4,* 34–53.
Tall, D. (1986). *Graphical calculus* [print materials and software]. Barnet, England: Glentop Publishers.
Tall, D. (1987). *Building and testing a cognitive approach to the calculus using interactive computer graphics*. Doctoral thesis, University of Warwick.
Thompson, P. (1993). The development of the concept of speed and its relationship to concepts of rate. In G. Harel & J. Confrey (Eds.), *The development of multiplicative reasoning in the learning of mathematics*. Albany, NY: State University of New York Press.
Thompson, P. (in press). Students, functions, and the undergraduate curriculum. In E. Dubinsky, J. Kaput, & A. Schoenfeld (Eds.), *Research in collegiate mathematics education* (Vol. 1). Providence, RI: American Mathematical Society.
Tierney, C., & Nemirovsky, R. (1991). Young childrens' spontaneous representations of changes in population and speed. In R. G. Underhill (Ed.), *Proceedings of the 13th annual meeting, North American Chapter of the International Group for the Psychology of Mathematics Education* (Vol. 2, pp. 182–188). Blacksburg, VA.
Tucker, A. (1988). Calculus tomorrow. In L. Steen (Ed.), *Calculus for a new century* (MAA Notes No. 8, pp. 14–17). Washington, DC: Mathematical Association of America.
Van der Waerden, B. L. (1963). *Science awakening* (A. Dresden, trans.). New York: Wiley.
Vergnaud, G., & Errecalde, P. (1980). Some steps in the understanding and use of scales and axes by 10–13 year old students. In R. Karplus (Ed.), *Proceedings of the 4th International Conference for the Psychology of Mathematics Education* (pp. 285–291). Berkeley, CA.
Vinner, S. (1987). Continuous functions: Images and reasoning in college students. *Proceedings of the 11th Meeting of the PME, II,* 177–183.
Williams, S. (1990). The understanding of limit: Three perspectives. In G. Booker, P. Cobb, & T. Mendicuti (Eds.), *Proceedings of the 14th Meeting of the PME,* pp. 101–108. Mexico.
Williams, S. (1991). Models of limit held by college students. *Journal for Research in Mathematics Education, 22,* 219–236.
Williams, S. (1993). Some common themes and uncommon directions. In T. Carpenter, E. Fennema, & T. Romberg (Eds.), *Integrating research in the graphical representation of functions* (pp. 313–337). Hillsdale, NJ: Lawrence Erlbaum Associates.
Winograd, T., & Flores, F. (1986). *Understanding computers and cognition*. Norwood, NJ: Ablex.
Young, G. (1988). Present problems and future prospects. In L. Steen (Ed.), *Calculus for a new century* (MAA Notes No. 8, pp. 172–178). Washington, DC: Mathematical Association of America.
Youschkevitch, A. (1976). The concept of the function up to the middle of the 19th century. *Archive for History of Exact Sciences, 16,* 37–85.
Zia, L. (1991). Using the finite difference calculus to sum powers of integers. *College Mathematics Journal, 22,* 294–300.

Comments on James Kaput's Chapter

Ed Dubinsky
Purdue University

THE ROLE OF HISTORY IN MATHEMATICS EDUCATION

The history of mathematics in general and calculus in particular has been thoroughly studied over the past few decades. There are, for example, 20 books and papers cited in Kaput's chapter. More recently, a few researchers in mathematics education have taken the point of view that the historical development of a mathematical topic, in particular, the difficulties that were encountered and the obstacles that were overcome, can tell us something about how an individual might learn—or fail to learn—that topic. A simple and lyrical way of expressing this is that ontogeny recapitulates philogeny.

Some researchers (see, for example, Cornu, 1983; Sierpińska, 1985) have worked with the notion of epistemological obstacle as a context for making use of the lessons of history. According to these authors, in order to understand a particular idea or set of ideas, it is necessary to develop certain notions at a particular moment and then later, with considerable struggle, replace these notions with more sophisticated versions. Thus, to see a function as nothing more than a single algebraic expression is useful and even necessary at some point, but later, when a more powerful function conception is required, there is a difficulty because the old, familiar, useful idea is not easily dispensed with. This situation is referred to as an *epistemological obstacle*. According to this point of view, the study of historical difficulties and delays in development can provide us with important clues about obstacles that are likely to face students.

In spite of this interest in the relationship between learning in an individual, or, to use a Piagetian term, *psychogenesis,* and historical development in a society, there exists, as yet, no complete synthesis of these two fields of investigation. For the most part in the literature, a study will be concerned with history and perhaps make an occasional reference to educational implications, or it will take the historical development as given and try to apply what appears to be known to questions of individuals learning or psychogenesis. There are not many attempts to study historical development and psychogenesis together, to look at history from a mathematical–epistemological perspective and simultaneously to understand learning in an individual from an historical point of view.

Perhaps the first to have attempted this is Piaget who, with Garcia (Piaget & Garcia, 1989) made what Kaput refers to as "perhaps the deepest and most complete analysis of the parallels between historical and individual development" (p. 84[1]). In my opinion, the chapter by Kaput is a second, although very different, attempt at a systematic, coordinated study of historical development and psychogenesis.

Before turning to a discussion of Kaput's chapter on its own terms, it will be useful to compare the perspective of Piaget and Garcia with that of Kaput. This comparison shows how an interpretation of history can have at least as much to do with the interpreter as it does with any set of historical "facts"—if indeed such things exist. For example, both studies examine the work of Oresme. Kaput sees Oresme as involved in a "struggle for representations," whereas Piaget and Garcia see him as spending a long time (and not fully succeeding) in making what I would call encapsulations of processes—that is, the conversion of a process to a mental object to which actions can be applied. The point I wish to make is that the difference between the two interpretations is not due to anyone getting their facts wrong or seeing references missed by the other(s). The difference lies in the point of view, or theoretical perspective, on which the interpretation of the facts is based. This observation is important in its own terms, but it also has pedagogical implications: Different interpretations of the history and different theoretical perspectives lead to different conclusions about pedagogy. I return to this theme in the section entitled "The Relation Between History and Curriculum Design."

The Perspective of Piaget and Garcia

It is important to note that Piaget and Garcia are *not* mainly concerned with establishing correspondences between historical development and psychogenesis with respect to the content of mathematics and science. In addition, there is nothing in their book that explicitly addresses any pedagogical implications of these ideas. They do, however examine the parallels between the mechanisms of historical evolution of certain major ideas and the mechanisms of development of concepts within individuals.

---

[1] Page numbers without further identification refer to the Kaput chapter.

It is true that in the case of the evolution of physics from Aristotle to just before Newton they establish a "very direct" correspondence between four historical periods and four stages in psychological development (Piaget & Garcia, 1989, p. 26). They feel, however, that "it would evidently be absurd to generalize this type of parallelism of contents in the case of scientific theories proper, such as those which emerged between Newton's mechanics and Einstein's relativity" (p. 27). Similarly, they point out that "the historical evolution of geometry . . . goes far beyond anything that can be observed in the elementary stages" (p. 111).

They are quite explicit about their goal being "not to set up correspondences between historical and psychogenetic sequences in terms of content, but rather to show that the mechanisms mediating transitions from one historical period to the next are analogous to those mediating the transition from one psychogenetic stage to the next" (p. 28).

Indeed, there is a sense in which this book, which was Piaget's last publication, is a culmination and summing up of his life's work. In a very brief (pp. 26–29) listing of the mechanisms common to historical development and psychogenesis we find all the main pillars of Piaget's work in the last decades of his life: reflective and empirical abstraction, the inferential aspect of the relation between observing subject and observed object, the synthesis of differentiation and integration, causality, stages of development whose order is fixed, and, what is for Piaget perhaps the most important because of its relation to biological organization (p. 274), the intra, inter, and trans levels of thought, always appearing in that order with transition driven by equilibrium and implemented via reflective abstraction. The book attempts to show that these mechanisms of change account for both historical development of ideas and the development of concepts in an individual. Thus, interpretations of history appear as support for Piaget's general theory of genetic epistemology.

The Perspective of Kaput

Kaput's agenda is quite different. He is explicitly interested in curricular implications of historical investigations of calculus. Indeed, his chapter is divided into three parts, and he describes the second, which makes up about 60% of the whole, as a "curriculum- and pedagogy-sensitive historical overview of calculus" (p. 79). He is clear from the beginning about his goal for this study. On the first page he announces his intention to "look closely at the origins of the major ideas of the calculus for clues regarding how calculus might be regarded as a web of ideas that should be approached gradually, from elementary school onward in a coherent school mathematics curriculum" and to "look closely at dynamic graphical means for representing important calculus ideas in ways that reflect their origins." He appears to feel successful in establishing these two points, because at the beginning of the third part of the chapter he presents them as "recommendations drawn from the historical review and a survey of current conditions and possibilities" (p. 132).

It is clear from this and from the rest of the chapter that Kaput is making a cognitive application of "ontogeny recapitulates phylogeny," although he cautions against possible excesses in applying such a tempting principle (p. 83).

Another important way in which Kaput differs from Piaget and Garcia is in the role given to notational systems. Kaput considers them to be very important and devotes one (albeit the shortest) of his three parts to them. Piaget and Garcia hardly mention them at all. I believe that Kaput's views on notational systems and representations, which pervade his entire chapter, form a keystone of most of his thinking about epistemology. In the third section, I will try to show how his chapter exemplifies the central role he gives to notation and also discuss some alternative points of view.

THE RELATION BETWEEN HISTORY AND CURRICULUM DESIGN

Kaput's chapter includes a study of the history of calculus, what he calls a "longitudinally coherent calculus curriculum," and implications from this history which justify his curricular recommendations. I discuss separately his history and the recommendations he draws from it.

Kaput's History of Calculus

In his chapter, I believe that Kaput has made an important contribution to the study of the history of calculus. He begins with three root aspects of calculus. He then tries to find sources for them and trace their developments in various periods and movements of history.

The Roots.

The roots he considers are:

- Geometric issues related to computations of areas, volumes, and tangents.
- A mix of practical and theoretical interest involving the characterization and theoretical exploitation of continuous variation of physical quantities.
- Inherently theoretical concerns with the foundations of calculus.

A Piagetian Analysis of the Roots.

Before describing briefly the extent to which Kaput sees these roots in different historical periods and in the work of various individuals and movements, let me consider for a moment the roots themselves and their possible connection with the perspective of Piaget and Garcia.

Kaput proposes such a connection when he suggests that "the operant nature of the algebraic symbol system provided the means for the transition from the former stage [intraobject] to the latter [interobject]" (p. 102). I think that Piaget would disagree, and I would like to offer an alternative.

The first root is clearly intraobject (as Kaput suggests) in that it is concerned with investigation of individual objects such as areas, volumes, tangents, and so on. The key to the second root is the variation of these objects or certain qualities of them (such as size, position, etc.). Again, Kaput points out, this is already at the interlevel in that it is concerned with relations among objects or the same objects with different qualities (e.g., a body at one position compared with the same body at another position). Finally, he suggests that the transition from the first to the second is made by means of the algebraic symbol system, and this is where I think that he is wrong, at least as an interpretation of Piaget, who states in several places his contention that the means of transition is reflective abstraction (see Dubinsky, 1991a, 1991b, for specific references on this point).

More specifically, I would suggest that the main step in passing from intra to inter with reference to a particular quality of an object, such as its velocity, is the *encapsulation* of that quality which (as Kaput points out) is initially a process (e.g., change in position coordinated with change in time for the case of velocity). Only by thinking of something as an object is it possible to compare it with something else, or itself at a different time.

Similarly, I would suggest that the transition from interobject to transobject is done by means of imagining all possible processes (including their reversals and compositions) involving the objects of concern and coordinating them in a single totality or structure.

As I will try to argue, the usefulness or even possibility of a symbol system is the result of, not the means of, the transition from intraobject to interobject.

Finding the Roots in History.

Kaput looks for the roots and traces their development from the time of the ancient Greeks, through Oresme and the Scholastics in the pre-Renaissance period, the 17th century, Newton and Leibniz, Euler, and up to the present day.

Kaput explains (pp. 87–88) that the Greeks did not make much progress toward the development of calculus, indeed could not even get started, because they were unable to quantify motion. In connection with this, they did not really have a conception of a variable. To be sure, they used literals, but for them, a literal was nothing more than a place marker for an unknown constant, little different from a number.

Kaput gives great importance to Oresme and the Scholastics (pp. 88–89). He suggests that their attempts at "mathematizing genuinely experienced variation before algebra" (p. 132) is not only a root of calculus, but something we should encourage in students at a very early age. Several examples are given of attempts to represent various qualities, such as temperature, velocity, acceleration, time, and distance, as geometrical objects, such as lines and rectangles. Kaput points out several instances of children making similar attempts.

In keeping with his view of the role of notation, Kaput considers that it is the development, in the 17th century, of algebraic representations of variables via coordinate systems and via expressions that could be manipulated according to formal rules that made possible the calculus invented by Newton and Leibniz in the following century (pp. 108–110).

An important next step, taken by Newton, is the reliance on motion imagery to conceptualize continuous phenomena. In the third part of his chapter, discussed later, Kaput tries to use video technology to implement this development.

Kaput emphasizes Leibniz's contribution in inventing a powerful symbolic notation for differentiation and integration. He also mentions the role of finite difference calculus from which comes the use of discrete formulations to model continuous phenomena.

Finally in considering the period beginning with Euler and continuing to the present day, Kaput describes "the subtle shifts in both the semantics and the nature of justification as the roles of geometry and algebra shifted in the period between the 16th and 19th centuries" (p. 132) and discusses the movement to put the principles of calculus on a firm, logical foundation. For this last period, Kaput presents a relatively conventional description of the movement in the last two centuries to work out the foundations of calculus.

On Weierstrass's $\epsilon - \delta$ Definition.

Before leaving the discussion of the historical development of calculus, I would like to offer an alternative to Kaput's rather standard interpretation of Weierstrass's definition of limit. According to Kaput, Weierstrass's $\epsilon - \delta$ definition "was a static concept, replacing the motion metaphor used in Cauchy's definitions . . . no longer did values of a variable 'approach indefinitely a fixed value.' Weierstrass rendered limits atemporal concepts."

I would put it a little differently. I would like to suggest that, contrary to what most people believe, Weierstrass's definition is *not* static. The dynamics are still present in his notation, but they have been moved from "external" metaphors to "internal" mental processes. Or to put it more precisely from a constructivist point of view, a person's understanding of notation is a coordination of what is present in the notation (e.g., the marks on the paper) and what is present in the mind of the person. Kaput refers to this as "the act of 'building meaning' from the notations" (p. 81).

In the mathematical concept of limit, there are intrinsically dynamic processes. One variable is moving, in its values, along a path in its domain toward some particular value. Simultaneously, a second, dependent variable is moving in some manner within its domain. The relationship between the variables is a coordination of two processes. Up until Weierstrass, the goal was to capture the processes and their coordination in the markings of the notation. This is really not possible, at least with symbols and writing and maybe not even with videos and computers, because even videos can only represent finite processes. With Weierstrass, there is, for the first time, a realization that these processes must be constructed in the mind, which can handle temporal change and is the only tool we know that can be used to construct an infinite process. The notation is then relegated to the role of a support for the construction of the mental process.

Moreover, in this formulation, the notation (that is, the traditional $\epsilon - \delta$ notation), if viewed on its own, does appear to be static and devoid of meaning—empty formalism according to many. It is only when the notation is read together with appropriate mental processes in the mind of the reader that the concept takes on its true dynamic nature. Thus, it is no surprise that, for someone who has not or is not about to construct a certain pair of mental processes and coordinate them in a particular way, this notation is a piece of empty formalism and remains totally incomprehensible. The important point is that meaning resides in a combination of what appears in the symbolism and what is constructed in the mind of the reader of that symbolism.

The main pedagogical implication here is that symbolism itself does not convey ideas. It is necessary for instruction to be aimed at helping students make appropriate mental constructions, which they can combine with powerful representations in order to build their mathematical concepts.

Kaput's Recommendations

As a consequence of his historical analysis, Kaput draws two major conclusions that he expresses in the form of the following two recommendations (pp. 132–133):

1. Calculus needs to be studied across many years of school, from early grades onward, much as a subject like geometry should be studied. Hence, its many purposes should be examined, not merely its refined methods. But most especially, its root problems should take precedence as the organizing force for curriculum design.
2. The power of new dynamic interactive technologies should be exploited in ways that reach beyond facilitating the use of traditional symbol systems (algebraic, numeric, and graphical), and especially, in ways that allow controllable linkages between measurable events that are experienced as real by students and more formal mathematical representations of those events.

A Longitudinally Coherent Calculus Curriculum.

In his first recommendation, Kaput is really getting at an issue that is the essence of the idea of calculus in our culture. From time to time, there occurs in human endeavor, a relatively coherent set of ideas, such as calculus. It is important to make those ideas part of our culture and hand them down to succeeding generations. This is one of the purposes of our schools. The only way we know how to do this is to create academic courses. Unfortunately, these courses, by their very nature, are static, unresponsive to local needs, and tend to concentrate, somewhat unrealistically, on the end product, rather than on the process(es) by which it came about.

Kaput calls for a reexamination of the tacit consensus that calculus is what he calls a capstone course. He would like to see the ideas of calculus spread out through the total education of all students. It is difficult to disagree with this goal, but much harder to see how to bring it about. As Kaput readily admits, the present educational system is not very conducive to absorbing calculus in this way. It is a little disappointing that Kaput tells us very little (pp. 133–135) about what calculus might look like in a curriculum that followed this recommendation, and he tells us nothing about how one might actually bring about such a new curriculum.

One thing is, however, very clear about the curriculum Kaput is calling for. It should strongly reflect the root problems, which, his analysis suggests, were the driving force in the historical development of calculus. It is possible to argue with this suggestion.

Let us grant that the root problems are those given by Kaput, and that they drove the historical development of calculus. Why should they drive the development of the ideas of calculus for students today? Perhaps we don't need to aim cannons to fly over the walls of medieval castles. Perhaps we don't (or shouldn't) need to map out the trajectories of intercontinental missiles. Perhaps we have a greater need to analyze DNA molecules or understand the behavior of the market. In the latter case, we might wish to reverse the traditional use of calculus in which the discrete is an approximation of the continuous and investigate fluctuation of prices in which the continuous is a model for the discrete.

In other words, the root problems come from the needs of society. If those needs change (or should change), what are the implications for calculus curricula? Kaput does not deal with this question. He assumes tacitly that the historical root problems form the only alternative.

A Motion World Learning Environment.

Most of the last part of Kaput's chapter (pp. 136–151) is related to his second recommendation and consists of a description of a video system MathCars that uses dynamic, interactive video technology to simulate driving a car. The basic idea of the system is "to map the phenomenologically rich experience of motion in a vehicle (sights and sounds) onto coordinate graphical and other mathematical notations" (pp. 137–138). The user views a video screen, which presents the motion from various perspectives. He or she controls an accelerator and can coordinate representations of time, distance traveled, or velocity. These representations, in the form of graphs, can be studied as objects in their own right.

The descriptions and pictures presented are very interesting (and are even more exciting when viewed on video as was the case at the conference where this chapter was presented). It will be even more interesting to see how this system can be used in the kind of curriculum Kaput is proposing and to hear about results that are obtained. One hopes this is on Kaput's agenda for the not too distant future.

Justification

Perhaps the most serious weakness in Kaput's chapter lies in the connection between his historical analysis, which, as I have indicated, is quite profound, and his recommendations, which certainly must be on the table as the revamping of mathematics education proceeds. I do not think that Kaput has really made a strong case for his contention that the recommendations follow from, or are even justified by, the historical analysis. His case is restricted to a relatively small number of uncoordinated examples in which historical difficulties are mirrored by students. I feel, however, that Kaput does not argue persuasively either that his conclusions are justified by his historical analysis, or that these same conclusions cannot be obtained from other analyses.

We have already indicated earlier one argument that accepts Kaput's root problems but suggests an alternative conclusion. Here is another. It is possible to conclude from the historical analysis, especially the many centuries it took for certain ideas to emerge, that the concepts of calculus are very difficult and therefore should not be introduced earlier, but later! Instead, let the period from kindergarten up to say the second year of college be spent in preparing students for calculus. In my opinion, most of that time should be spent with the various aspects of functions, but that is another discussion.

Even if one accepts Kaput's conclusions, it is not clear that the historical analysis is needed to justify them. Indeed, Kaput himself presents (pp. 134–135) several arguments not based on his historical analysis that are aimed at justifying his proposed curriculum. Moreover, it is somewhat jolting, in this age of technological explosions in education, to imagine that Kaput's second recommendation needs any justification. It is as if someone were standing in the middle of a hurricane trying to argue that the wind should be blowing.

NOTATION AND REPRESENTATION

Kaput begins his discussion of notation with the following statement on page 80 (I have added the statements in brackets, making use of Kaput's subsequent qualifiers for the specific case of mathematics):

I take the point of view that we organize the flow of [mathematical] experience jointly using two structures, one mental—the structures of mind [mathematical knowledge]—and the other material—the material artifacts [mathematical notations related to the knowledge], including spoken and written language, produced and used in accordance with our cultural inheritance in one or more physical media.

No one can argue with this statement, in particular the joint (and, presumably, balanced) role of the two structures of conceptual understanding and formal symbolism. Nevertheless, a "chicken and egg" question is inevitably raised. Does one of these two structures dominate the other? Is it the case that one is made possible by the other which cannot otherwise progress?

One quote of Cajori that is given by Kaput (p. 78) does suggest such an asymmetry: "Without a well-developed notation the differential and integral calculus could not perform its great function in modern mathematics." Others have tilted matters in the opposite directions as did Tolstoy when he suggested that there is often difficulty in learning a new word, not because of its sound, but because of the concept to which the word refers. He concluded as follows: "There is a word available nearly always when the concept has matured" (Tolstoy, 1903).

There certainly is an issue here, and I will first argue that, although in his other works Kaput may discuss the construction of concepts, in this chapter at least, he really does appear to emphasize the idea that notation must precede, if not dominate, conceptual understanding. I will then try to present an alternative point of view, which seems more reasonable to me.

The Role of Notation in Kaput's Chapter

First I will try to show that Kaput's chapter presents a point of view about the relation of conception to notation that is quite unbalanced. Then I will discuss his idea of action notation, which is a very important notion. Although I will later present an analysis of this idea that is different from Kaput's, I certainly agree with his description of its role in the development of mathematics.

Predominance of Notation Over Conception.

I base my contention that Kaput's chapter presents an unbalanced view on the fact that I find, in the chapter, a large number of specific places in which he expresses this dominance. I list a few examples and leave it to the reader of this volume to see if I am right in suggesting that there are very few examples in the other direction:

- Action notation "had profound impact on the nature of mathematics" (p. 82).

- "I mention the terminology to give a sense of the difficulty experienced by those who are attempting to develop a coherent mathematical theory of variation before a systematic language for the expression of a theory was available" (p. 89).
- "The Scholastic philosophers were striving to express their ideas in words and geometrical diagrams and were not so successful as we who realize, and can make use of, the economy of thought which mathematical notation affords" (p. 98).
- In reference to Descartes, Kaput writes "his symbolism for exponents . . . led to a new conception of number and variable that opened up a whole new world of possibilities" (p. 103).
- Although he opens his section on Leibniz with the balanced statement about "the development of notation in concert with the development of concepts," he titles his section "Leibniz: The Power of Notations" (p. 113). Nowhere does Kaput discuss, much less use as a title, the power of concepts to engender notations!
- He refers to a pattern of "syntactically driven operations with symbols" leading to conceptual extensions, in particular, to the construction of conceptual entities (p. 118).
- In considering functions, Kaput suggests the following causal relation, "Euler's general definition was not applied even by him, and then for a century hence, because there was no way to represent—and hence meaningfully study, compute, or apply—such general objects" (p. 127).

I am not suggesting that I think Kaput is wrong in all of these statements. I think that several of the points have some truth to them. I present them because, as a whole, they (and the absence of very many balancing points) suggest to me that Kaput really does believe that you can develop a notation, and then this can lead to understanding a concept, or that understanding a concept must wait for the development of appropriate notation. I will try to show how at least some of the points can be interpreted reasonably by assuming that it is the other way around.

Action Notation.

Kaput makes a useful contribution to analyzing notational systems when he divides them into two kinds. A *display system* serves "primarily to display information for the user to read or respond to," whereas an *action system* provides "systematic means for the user to act on it physically" (p. 101). Examples of the latter, which are clearly the more important, abound: algebraic expressions, use of 0 in writing integers, dy/dx, and so on.

No one could disagree with the very strong case that Kaput makes (pp. 101–110) for the critical importance of action notation as an indispensable tool for making progress in mathematics. It is so valuable that it (rightly) maintains its preeminent position among mathematical devices, in spite of the fact that there are two major drawbacks involved with using action notation.

One of these is pointed out by Kaput (p. 132) in terms of Skemp's notion of *instrumental* understandings as opposed to *relational* understandings. The very power of many examples of action notation creates the possibility (unfortunately too often realized in our schools and colleges) that students will learn to perform mechanical operations without much understanding of what is behind them. This problem is so serious that we actually have a society that is convinced that the manipulation of symbols is the essence of mathematics, and that mathematical brilliance consists of multiplying or factoring very large integers. (The recent film, "Little Man Tate," whose director and star is a Princeton graduate, is a case in point. It depicts a young boy who is purported to be a mathematical prodigy. With one exception, everything that this child does consists of mentally performing arithmetic calculations with large numbers very quickly.)

The second drawback is that as the concepts get more complicated, and as the portion of the population that must be skilled in mathematics grows, it becomes more and more difficult to get students to be successful, even with the manipulation of symbols—especially when it is not completely rote. When unusual letters are used in familiar situations, things fall apart. I have seen graduate students become disconcerted when asked to explain what is meant by $(x + y)(F) = x(F) + y(F)$. And how many of us who teach calculus have seen students insist that the derivative f' of the function f given by $f(x) = a^x$ is given by $f'(x) = xa^{x-1}$? Thus, we have two problems in student performance with notation. One is that it can be hard for them to develop skills, and second they can become wedded to very specific choices in notation, unable to function with even the slightest variations.

An Alternative to Kaput's View on Notation

First I point out some questions that arise and are not, in my opinion, dealt with sufficiently in Kaput's use of notation. Then I present a different point of view, use it to reformulate some of Kaput's specific comments, and indicate very briefly how it might be applied pedagogically.

Some Difficulties with Kaput's Idea of Notation Predominance.

Kaput's historical analysis does not tell us anything about where notation comes from. We learn that there are instances in which long waits were necessary (centuries even) before appropriate notational schemes were devised (p. 88). But there is nothing about what causes the delay, or what sorts of developments had to take place in order for effective notation to arise.

Kaput refers to the "real achievement . . . of the masters who built the concepts initially and wrote them in language that embodied the organizing syntax that we consumers can now use with confidence" (p. 131). Is he telling us that progress must wait for geniuses? And what about our students, or at least those whose reaction to various notational schemes is something other than "use with confidence"?

There is another, related difficulty. Because the historical analysis doesn't tell us about where notational schemes come from or how they arise, it is not very helpful in pointing to pedagogical solutions to students' notational and conceptual difficulties.

Now I am fully aware that Kaput does, in this chapter and in many other works, propose approaches, using computers and other forms of technology to help students develop representations of mathematical concepts. What I am suggesting is that although this historical analysis convinces us that we should not be surprised at certain difficulties of students, it tells us very little, in principle, about how to deal with those difficulties.

Predominance of Conception Over Notation.

Therefore, I would propose an alternative point of view, in which construction of concepts comes before development of notation. Kaput points out (p. 118) that several people (including this author) have worked with the idea of mental construction of objects. The construction of mathematical concepts includes a very important step that takes place after the construction of a mental process. It is the *encapsulation* of that concept into an object. What is suggested by the research of myself and others is that first one encapsulates a process into an object. This is very difficult. It is what takes time and for which one must devise special instructional treatments, in which actions are applied to processes. Once it has been accomplished, however, a notational scheme can be developed and connected to the concept by the relatively simple act of associating a syntactically governed set of symbols with a mental object that an individual has already constructed.

That is, to paraphrase Tolstoy's comment, I would say that a notational scheme can be learned and used effectively by an individual, with understanding, once the concepts to which this scheme refers have been constructed, by that individual, as mental objects or entities.

As I indicated earlier, it is this act of encapsulation that is the key mechanism in the transition from the intraobject level to the interobject level in the sense of Piaget and Garcia.

Some Alternative Explanations.

Let me consider a couple of examples. Kaput writes about the struggles of the Scholastics to find reasonable ways to represent acceleration, and the fact that they did not succeed. From the theoretical perspective just introduced, one could argue that before it is possible to construct the concept of acceleration, one must first develop an understanding of velocity as a process of comparing distances at different times. Then this process is encapsulated to a conceptual entity so that velocity becomes a mental object. Only then is it possible for velocity to have different values at different times, so that one can go on to the idea of acceleration.

Or consider the long wait before Euler's general definition of function became usable (p. 127). I would suggest that the wait was not, as Kaput suggests, for ways of representing functions as objects, but for people to develop the ability to encapsulate function processes as objects. Indeed, if Kaput is right, why do we have so much trouble getting students to work with functions as objects (the derivative of a function is a function, the solution of a differential equation is a function, the composition of two functions is a function, etc.)? We have an excellent notation to show them, but it doesn't help very much. They still must struggle to make encapsulations.

Pedagogical Applications.

At the very least, we can use this alternate formulation to propose, implement, and evaluate instructional methods for helping students develop various concepts that require the construction of mental objects. We can do this using computer systems in which students can write programs that accept functions as inputs and produce functions as objects, that provide the opportunity to perform actions on certain concepts, which are to be the referents of notational symbols. (See Ayers, Davis, Dubinsky, & Lewin, 1988, and Breidenbach, Dubinsky, Hawks, & Nichols, 1991, for examples that indicate some success in this endeavor.)

These initial indications of success are a far cry from establishing the validity of the theoretical perspective I am proposing, or its preferability over Kaput's point of view. It does suggest, however, that it might be a viable alternative.

REFERENCES

Ayers, T., Davis, G., Dubinsky, E., & Lewin, P. (1988). Computer experiences in learning composition of functions. *Journal for Research in Mathematics Education, 19*(3), 246–259.

Breidenbach, D., Dubinsky, E., Hawks, J., & Nichols, D. (1991). Development of the process-conception of function. *Educational Studies in Mathematics,* 247–285.

Cornu, B. (1983). *Apprentissage de la notion de limite: Conceptions et obstacles* [Approaches to the notion of limit: Conceptions and obstacles]. Doctoral dissertation, L'Université Scientifique et Medicale de Grenoble, Grenoble.

Dubinsky, E. (1991a). Constructive aspects of reflective abstraction in advanced mathematical thinking. In L. P. Steffe (Ed.), *Epistemological foundations of mathematical experience.* New York: Springer-Verlag.

Dubinsky, E. (1991b). Reflective abstraction in advanced mathematical thinking. In D. Tall (Ed.), *Advanced mathematical thinking* (pp. 95–123). Dordrecht: Reidel.

Piaget, J., & Garcia, R. (1989). *Psychogenèse et histoire des sciences*. Paris: Flammarion.

Sierpińska, A. (1985). La notion d'obstacle epistemologique dans l'ensignement des mathematiques. *Proceedings of the 37th CIEAEM meeting*. Leiden.

Tolstoy, L. (1903). *Pedagogicheskie statli* [Pedagogical writings]. Moscow: Kushnerev.

Comments on James Kaput's Chapter

Jere Confrey
Erick Smith
Cornell University

In "Democratizing Access to Calculus: New Routes to Old Roots," James Kaput raises a number of serious challenges to the ways we currently introduce and develop the ideas of calculus. Unlike numerous other critiques, he offers an historical perspective on the subject and argues for its potential to influence the cognitive construction of the new calculus curriculum. The importance of this endeavor lies in its serious consideration of epistemological issues about what should be taught and how it should be taught.

Kaput sets his goals out at the beginning of the chapter. He seeks to challenge the idea that calculus should be the "capstone" course, its ideas held up until the student has mastered the algebraic techniques necessary for facile manipulation of equations of functions, and he seeks to question the introduction of the subject solely through algebra. We are in complete agreement with his challenges on these two issues. However, we push the interpretation he has made regarding the implications of history and propose a related set of claims. We discuss these claims in light of our own work on exponential functions and multiplication relations and finally tie these to our multirepresentational software, Function Probe© (Confrey, 1991a). To do this, we must first make an argument about the conduct and role of historical work. Historical work is not the documentation of facts. This becomes particularly evident in mathematics, in which records focus more on validation arguments than on conjectures and refutations, necessitating careful reconstructive efforts. Modern lenses can be obstacles to that process. In the history of mathematics, unless these modern lenses are identified and explicitly challenged, an historical analysis can lead toward: (a) a eurocentric view of mathematics, often with gender and race biases (Joseph, 1991); (b) a formalist view, in which symbolic representation is most highly valued (Unguru, 1976);

and (c) a separation between mathematics and other disciplines that was not necessarily characteristic of the times. We do not presume that a reconstruction can be effected with no lens, however, we do wish to emphasize that one's choice of lens must be subject to examination concurrently with the claims one wishes to make. One must be as explicit as possible about what perspective one wishes to take, because what one chooses to say will be largely a product of it.

For example, the process of "rational reconstruction" was superbly demonstrated in Lakatos's *Proofs and Refutations* (1976). His position was clear: "The history of mathematics and the logic of mathematics discovery cannot be developed without the criticism and ultimate rejection of formalism" (p. 4). He described the goal of his work:

> Its modest aim is to elaborate the point that informal, quasi-empirical, mathematics does not grow through the monotonous increase of the indubitably established theorems but through the incessant improvement of guesses by speculation and criticism, by the logic of proofs and refutations. . . .
>
> The dialogue form should reflect the dialectic of the story; it is meant to contain a sort of *rationally reconstructed or 'distilled' history. The real history will chime in the footnotes, most of which are to be taken, therefore, as an organic part of the essay.* (1976, p. 5; emphasis in original)

Lakatos did not simply study history and discover in it a basis for a challenging formalism. Rather, his reconstructive process was influenced by his conceptions of mathematics. He was a student of Popper and influenced by Hegel. His approach to history reflects certain commitments from these mentors. Popper argued for the essential role of refutation in directing the development of a discipline. Hegel argued for the dialectic of thesis, antithesis, and synthesis. The historical work of Lakatos reflects and modifies these intellectual perspectives. The editors of Lakatos's work suggest he would have moved further away from a Hegelian view of mathematics if he had lived longer, leaving less of the remnants of a view of mathematical problems as existing relatively independently of human invention and imagination (1976, p. 146).

Lakatos did historical research with the goal of reinterpreting progress in mathematics. Kaput presents historical analysis for different purposes, which have a significant impact on his presentation of history. We try to show, in fact, that the two purposes he proposes are somewhat contradictory, and that the novelty and significance of his claims would be enhanced if he were to dismiss one of them, exploring the other more vigorously. Kaput identifies these purposes as "to give a better understanding of the present" and "to help expose opportunities for major alternatives" to the current pedagogy (p. 79). These two goals seem to be in conflict, if one assumes, as he does, that the present state of calculus instruction needs reform. The conflict arises because a conventional or standard history is not likely to fundamentally challenge current pedagogy, whereas a nonstandard history, when contrasted with standard history, will challenge rather than explain current practices. We see Kaput as struggling with the tension between these two views and proposing an uneasy resolution, which includes a measure of novelty and a measure of convention. In our own work on the history of mathematics, we seek to demonstrate that a nonstandard approach to history, based in an epistemology of multiple representations, resolves this tension and allows one to argue even more convincingly for reform.

We would suggest that an "epistemology of multiple representations" (Confrey, 1991b) is actually more in line with Kaput's presentation of the use of symbol systems and his exploration of the use of computer technology. Kaput claims that knowledge evolves cyclically in relation to our conceptions (mental states) and notations; we have numerous pairings of these conception-notations. The cognitive operations which allow us to move between and among these pairings constitute knowledge as well as the conceptions and understanding of the individual notations. His emphasis here validates the importance of examining the belief systems of the person engaged in a mathematical pursuit and recognizing the impact of available notations in shaping the conceptions and vice versa.

The tension in the Kaput work is exacerbated by the selection and interpretation of the Piaget and Garcia framework. That framework proposes three stages in the development of an idea: (a) the intraoperational, where one "performs actions within the objects, with attention to the objects themselves"; (b) the interoperational with a focus on the "relationships and transformations between objects and invariances across objects"; and (c) the transoperational involving the building of "a higher level structure that embodies these relations as its elements and one attends to the properties of the structure" (p. 84). Kaput reveals further his interpretation of Piaget and Garcia as he discusses Grabiner's four-stage description of the development of the derivative and summarizes these as "use, discovery, development, and formal definition" (p. 85). Notice that in making this parallel argument, Kaput has identified structure with formalization.

Grabiner's description is exciting in many ways, for it argues for a reversal of so much of the curricular presentation. In it, use precedes the articulation of the idea as an independent entity, development and elaboration allow ideas to be connected, and the formal definition is given as a culminating activity. A calculus curriculum with this as its basis would differ considerably from current work. Students would have the opportunity to see the mathematical ideas embedded in problems and situations first, which would create the need for the idea. They would then gain a sense of the concept by differentiating it from other ideas and developing the idea in its own right. The formalization through definition would come about last. This would challenge current practice, but not as fundamentally as we and, we suspect, Kaput, would like.

Equating Piaget and Garcia with Grabiner's work has certain risks. Assuming the Piaget and Garcia work is not a radical departure from Piaget's other work, questions can be raised about equating the transoperational stage with formalization through definition—Grabiner's final stage. The three-stage model in Piaget and Garcia bears close resemblance to scheme theory (Confrey, 1991b; Piaget, 1971; Steffe, in press).[1] In our interpretation of scheme theory, the knower constructs knowledge adaptively through a three-stage process involving a problematic, an action and a reflection. Piaget posits that after operating successfully to resolve a perturbation, the actor separates the patterns of action from the immediate experiential content, uses them for prediction and explanation, that is, becomes aware of the schemes, and a level of reflective abstraction is reached. Reflective abstraction requires the conscious recognition of the scheme as an effective pattern of action. However, abstraction used in this sense is not identical with abstraction as it appears in formal definition. For Piaget, the idea remains embedded in its contextual conditions for its emergence, and tied to its operations. The idea is abstract in that it is stored as a routine for future action and can be examined as an idea in its own right.

Abstraction in mathematics has very different, but perhaps related, connotations. For one, it is typically intertwined with an algebraic presentation. For instance, algebra is considered more abstract than a graph or a table. This view of abstraction has to do with the placement of an object within a formal structure with rules governing the system. Lakatos, for instance, describes the formalist school as one which tends to identify mathematics with its formal axiomatic abstraction. Abstraction is then tied to formalism. The algebraic notation relied on to express that formalization tends to be heralded for its decontextualization. Thus, we would extend Kaput's claim that algebra is largely responsible for holding calculus in its present position to include abstraction, formalization, and decontextualization. Thus, what we add to Kaput's analysis is that to successfully challenge the related concepts, the idea of abstraction itself must be reexamined and reconstructed.

For example, a primary goal in Leibniz's treatment of calculus was to make it mechanical: "It is unworthy of excellent men to lose hours like slaves in the labor of calculation which could safely be relegated to anyone else if machines were used" (quoted in Edwards, 1979, p. 232).[2] Leibniz also saw that his proposed notational systems that he had intended as simplification also had the unfortunate consequence of obscuring the roots of his work, which rested largely in the use of tables. As he saw the trend toward others learning only the algebraic algorithms, he recognized not only the gains from his work but also the losses. Near the end of his life he stated: "It is most useful that the origins of memorable inventions be

---

[1] It also bears resemblance to stage theory, with its final stage in formal operations. If one chooses stage theory for the genesis of the model, then one would probably disagree with us on Piaget's position on abstraction. We would nonetheless argue that it is scheme theory with its local application to the development of concepts, not stage theory, which is longitudinally applied over long periods of time and based in maturation as well as in experience, that ought to guide educational activity.

[2] We point out the sexism implicit in this quote as large numbers of the human calculators were women and would fall into the category of "anyone else" in this quote.

known, especially if they were conceived not by accident but by an effort of mediation. . . . One of the noblest inventions of our time has been a new kind of mathematical analysis known as the differential calculus; but while its substance has been adequately explained, its source and original motivation have not been made public" (quoted in Edwards, 1979, p. 234). We take this as an argument in support of multiple representations.

One can see an interesting trend in the Kaput article; Kaput applies Piaget and Garcia to the development of history in a way that does not differentiate the mathematical treatment of abstraction from reflective abstraction. As a result, Kaput's historical presentation ultimately endorses rather than challenges the dominance of algebra as the culmination of mathematical activity. That is, we see Kaput tending to describe the development of algebraic[3] and formal approaches as (a) logically necessary, (b) signaling cognitive advances, and (c) entailing no loss over alternative methods. By presenting such a view of abstraction and identifying it with formalization as Lakatos (1976) described it, Kaput weakens his reform argument. Many readers will assume that the changes he proposes will apply only to the early grades, and that the outcome of the calculus sequence will thereby remain largely intact. In our opinion, this is clearly not what Kaput intends, as evidenced by such statements as "Algebra and knowledge of important classes of functions as algebraically defined objects have acted as an insurmountable barrier to calculus access for all but the intellectual elite" and "I am . . . concerned . . . with rethinking the entire enterprise of calculus as a school or university subject" (p. 79).

Another way to understand how a problem with formalization surfaces in the Kaput chapter is in viewing the three roots of calculus and how the Piaget and Garcia framework has an unintended impact on them. We believe that the segmentation of calculus into (a) geometry and the calculation of areas and volumes, (b) continuous variation of physical quantities, and (c) Zeno's paradoxes and the development of limits provides a powerful curricular base. However, if these three roots are unintentionally connected sequentially to the three stages of Piaget and Garcia, then the endorsement of formalization in mathematics occurs again. That would imply that geometry is related to actions with objects, that variation marks attention to transformation and variation, and that the higher structure is "inherently theoretical" (p. 83) and corresponds to the "formal theory of limits in the 19th century that now logically frames the other two strands" (p. 83). We try to demonstrate how this interpretation can be seen in the Kaput discussion of history. We offer instead a presentation that legitimizes all three strands as equally explanatory. We suggest that whereas the Grabiner model of historical development, similar to discussions of tool or object distinction of Artigue (1992), may be useful in analyzing development over short historical episodes, it is not applicable as a model of long-term trends such as from Greek mathematics to the

---

[3]Taking Kaput's lead, we note our restriction of our use of the word "algebra" in this chapter to the formal symbol systems that dominate the secondary curriculum.

present. Also we believe that "object" status in a Piagetian sense is not equivalent to formal definition in the Grabiner sense.

Kaput highlights several historical topics, raising persistent and significant conceptual issues, which help to raise the questions about an appropriate frame for conducting historical investigation. These include a distinction between the mathematics of variation and the mathematics of form, a related distinction between action and display notations, representational issues particularly in pre-Cartesian "graphing," the relationship between deductive proof and creative freedom, and more traditional calculus issues related to continuity versus infinitesimals. We endorse his identification of legitimate issues that need more complete historical explorations.

We would claim that his historical writing supports a view of history as an inevitable progression, what Confrey has called "progressive absolutism" (Confrey, 1980). Whereas absolutism views history as an accumulation of facts, progressive absolutism, like conceptual change approaches, recognizes that viewpoints are rejected and replaced over time. However, progressive absolutism also assumes that, on a global level, the discipline is steadily advancing toward mathematical or scientific truth, a position denied in the conceptual change tradition. By relying on the framework of Piaget and Garcia, Kaput endorses a view that epistemological progress can be defined as movement "upwards in abstraction" (p. 84). Thus, increasing abstraction becomes the universal standard for mathematical progress. Typically this has two unfortunate tendencies. First, it tends to treat abstraction as an ahistorical concept, that is, it assumes that we can interpret historical mathematical events in terms of some timeless concept of abstraction. Second, it encourages the creation of an historical record in which only those events that are viewed as part of the story of increasing abstraction are considered important, while events that do not fit into this framework are often considered superfluous or wrong. As a result, abstraction is not critically examined by Kaput, and one sees evidence that "more abstract" is associated with (a) distillation to a single all-encompassing essence, (b) disassociation from context, and (3) formalization into definition and theorem proof, typically in algebraic notation.

One way to avoid progressive absolutism is to take a view of history that recognizes that while each new perspective appears progressive in relation to the previous one, this perception of progress is a local and not necessarily a global characterization. Constructivism, in as much as the theory argues that knowledge evolve in relation to its viability, must postulate selection in terms of local progress from the perspective of the knower. The question is whether an assumption of global (over significant time periods) progress is warranted. We argue that it is not. Using an analogy to a section of rope, one can describe another section as immediately in front or in back of it, but when the rope is coiled on the ground, the ordering among the sections globally can change dependent on one's orientation. Surely history is at least as multidimensional as a coiled rope.

We take such a view of history, and it leads us to examine the views of any historical period potentially to be equally as valid as our modern perspective. Historical reconstruction becomes a documentation of alternative viewpoints and serves to assist us in recognizing and acknowledging the legitimacy of student invention. Locally the changes from one perspective to the next are fascinating, and they create a deeper insight into the genesis of the ideas. No overall progression or genesis is assumed. As we look backwards, it is often easy to see what we now understand but which those before us seemingly did not. Because it often seems that, the closer in time they are to us, there is less that they did not understand, we "see" progress. However, what is much more difficult to see is what they did understand that we do not and perhaps cannot, because we cannot enter the historical and social world in which they lived. However, if we do not seriously attempt to understand the viability of the worlds of those who came before us and to seek out their alternative perspectives, we are too easily led to a self-reinforcing progressive version of history.

Furthermore, we offer a different view of the epistemology of mathematics. For us, epistemological progress lies in the coordination of multiple representations. That is, to know a piece of mathematics is to act on it in different representational forms and to coordinate and contrast those forms in order to resolve problematic situations. Although we are convinced that abstraction needs to be reintroduced and revised in such a way as to reject the necessity of decontextualization and the dominance of algebraic formalization, we are not yet certain of what its characteristics in "an epistemology of multiple representations" look like. We consider the treatment of "reflective abstraction" in Piaget as a possible alternative.

Nonetheless, an epistemology of multiple representations also creates a very different lens through which to view the historical events presented in the Kaput chapter. For example, we would question the implicit message which seems to weave through the historical presentation that the movement from geometric to algebraic understanding parallels a movement from more concrete to more abstract. We further question whether this historical development is an accurate historical interpretation or a necessary path for mathematics to follow to advance to modern levels. Also, although Kaput partially discounts the idea that "ontogeny recapitulates phylogeny," his version of the historical development of calculus could be seen as supporting such a view. We would argue that it is not simply that the "blind watchmaker" is "inefficient and irregular" (p. 83), but that it is impossible to predict what path and what products he or she will create. Looking backwards, the paths seem obvious with a certain touch of inevitability. Looking forward, they simply do not exist. Thus, to advocate a version of "ontogeny recapitulates phylogeny" seems to require a somewhat Lamarkian view of cognitive development. A distinction which emphasizes a progressive-absolutist versus a multirepresentational framework would be to suggest that a progressive view suggests a temporal version of ontogeny recapitulates phylogeny (temporal ontogeny recapitulates temporal phylogeny). In contrast we might describe a multiple representation view as a reflective version—holistic ontogeny recapitulates holistic phylogeny—we use the problems and representations of history as a way to reflect on student thinking and vice versa.

We discuss several particular concerns we have with the analysis and provide one particular historic example to illustrate how this version of history can be seen as undermining our understanding of mathematical concepts.

HISTORIANS OF MATHEMATICS AND THE HISTORY OF MATHEMATICS

Many mathematics historians have written their story of mathematics as if it took place primarily within a Greek–European framework and portrayed the history as a fairly straightforward logical progression. One version of this approach has been to see Greek mathematics as an undeveloped version of modern mathematics and thus to see arithmetized magnitude, algebra, and even real analysis as what the Greeks were "really" doing. These views of Greek mathematics have been strongly criticized in the work of Unguru (1976) and Fowler (1987). Although Kaput does not seem to "arithmetize" Greek mathematics, there is a tendency to portray it as an undeveloped version of modern mathematics, a concern we return to later. More generally, however, we take the work of historians such as Unguru and Fowler as warnings to look with caution at the implicit interpretations in historical works. One concern we have with Kaput is his uncritical inclusion of many historical interpretations. A prominent example is a quotation from Edwards (1979) suggesting that had mathematics not followed the particular historic path presented in textbooks from Greek deductive systems to modern calculus, it might now be a "dead and forgotten science" (p. 87). To make sense of such a claim seems to require a view of mathematics as embedded in the formal systems of modern textbooks. Although the point of the quote for Kaput was to emphasize certain limitations of Greek mathematical thought, including it without comment serves to perpetuate this view of mathematics, a view that would seemingly go against Kaput's own views and current dominant views in both mathematics philosophy and education.

FROM GEOMETRY TO ALGEBRA: CONCRETE TO ABSTRACT?

There is little question that when looking at the history of Greek and European mathematics into the 19th centuries, one sees a shift from an emphasis on geometry to algebra and numeric analysis. Kaput tends to describe the historical change in terms of a change from concrete to abstract thinking. For example, he describes analytic geometry in contrast to classical geometry as allowing "statements to be made about general classes of figures, rather than particular figures" (p. 84). Such a statement seems to neglect the fact that Euclid's work allowed many statements to be made about classes of geometric figures and implies that "general classes of figures" must be defined by algebraic criteria (based on degrees of equations, etc.), in which case Kaput's claim becomes a tautology. Likewise, he describes algebra as a "forward 'propellant,'" providing us with a "generality-elevating" role for geometry (p. 85).

An alternative view would be to see the changes from visual to algebraic representation as a swing of a pendulum, each swing of which contributes valid and complementary insights. In Kaput, Descartes is attributed with providing us with "operational freedom" (p. 103) by relieving us of the necessity of coordinating the geometric dimension with the results of calculations. Kaput does state, however, that Descartes "maintained a geometric referent for his quantities." In a multiple-representational view of history, Descartes plays a pivotal role of proposing a system to coordinate algebra and geometry. In so doing, he demonstrates how the construction and manipulation of geometric figures to create loci of points can be concurrently described in terms of the relationships among variable quantities. As we pointed out in a previous paper (Smith, Dennis, & Confrey, 1992), for Descartes the construction and orientation of the coordinate axes was an important part of the constructive process. The importance of the constructive process is now lost from view as the "Cartesian plane" has become a set of orthogonal axes for displaying algebraic expressions of functions. According to a multiple-representational viewpoint, the elimination of the geometric perspective in algebra can be viewed as a loss of conceptual diversity while at the same time it acknowledges the local progress associated with the development of algebra.

SYMBOLS AND MATHEMATICAL NOTATIONS

A central and unifying theme in the historical analysis is the role of symbols and mathematical notations. Kaput makes two observations on which we offer comment. The first concerns Kaput's description of the change from display to action notations, which he located primarily during the 16–17th centuries. The second concerns the "extraordinary stability of the standard approach to calculus across the past two centuries" (p. 78) combined with "the storage and accumulation of human knowledge across generations [which] has been made possible by the invention of writing" (p. 131).

There is a definite need to explore the roles of various representations in our construction of mathematical concepts, and the relationship between algebraic and geometric representations is central. Thus, Kaput's articulation of a possible relationship between changing notations and the development of the mathematics of variation and calculus is noteworthy. Kaput has been one of the few mathematics educators to examine notation in advanced mathematics in terms of its impact on our thinking and has, for years, imagined how we might better incorporate movement imagery into our notational systems. However, we have a concern about the distinction between describing the two representations as display and action notations. The question is where to place the distinction between action and display—in the notation or in the intentions of the mathematizer. To describe a notation as intrinsically display- or action-oriented is at odds with a constructivist understanding of mathematics. We build our understanding of representations through our constructive actions and reflections on those actions, and it is in these activities, not the form of the representations, that a distinction between display and action can be made. There is a definite need to explore the roles of various representations in our construction of mathematical concepts, and the relationship between algebraic and geometric representations will play a key role in this analysis. Thus, Kaput's articulation of a possible relationship between changing notations and the development of the mathematics of variation and of calculus is noteworthy, but the action/display distinction is, perhaps, too great a simplification.

One problem is that Kaput tries to locate the distinction in the representational system itself as "intrinsically" action or display. Our work with students belies such a global characterization of the notation and suggests that it is the user who determines such characterizations, as they engage purposefully, operationally, and reflectively in mathematical activity. In particular, it would seem that the dominant curricular practice of viewing graphs as the end result of mathematical activity, something one does as an alternative display of results after algebraically solving a problem, leads us to characterize graphs and geometric representations as "display" representations. Because we have participated in a world in which algebra has been made the place for taking actions, there is a real danger of not being able to see actions in a geometric (or graphical) context. For Descartes, however, it seems that the importance of his constructions lay at least as much in the dynamic action of creation as in the display of the result. The construction of the hyperbola, for example was created by a linkage construction which involved sliding a triangle along a straight line. The hyperbola was the locus of points of intersection of the extension of one side of the triangle and a line, passing through the opposite vertex and pivoting through a fixed point in the plane. Neither the algebra nor the geometry of Descartes can be reduced to a categorization of solely display. Likewise, in our curriculum we would suggest that creating opportunities for students to build representations of their mathematical ideas in various representational forms will provide them with the opportunity to take actions on those representations that reflect the way in which they constructed them. In traditional instruction in which the construction of a graph is often nothing more than an algorithmic plotting of points from an algebraic equation, we should expect that students may be unlikely to see the possibility of action within the graph.

We do agree with Kaput, however, that computer environments offer the potential for students to increase their ways of creating representations and of acting on them. However, we would not accept the idea that computer representations can be inherently action-oriented without attempting to model the "problematic" (Confrey, 1991b) as constructed by the students during the activities in which they create their representations. We believe, for example, that Kaput's MathCars has great potential to be used as a tool for actively creating representations, but that this may not happen unless both the mathematics and the pedagogy of traditional classrooms is seriously questioned. In our software, Function Probe, for example, one can act directly on the graph through mouse actions. Using prototypic functions, a student can transform the functions through translation, reflection, and stretching, while keeping track of the coefficient of the action in a register and history window. In many problem situations, a solution can be achieved through such visualizations. Thus, figuring out how to code these actions in terms of their numeric impact on the point values or as algebraic expressions need not be the initial or primary activity (Borba, 1993). However, the distinction between action and display in this environment is a product of the activity of the learner. Each of the representations may initially be "action" representations as the student undertakes a goal-oriented activity but becomes "display" after the actions are taken, operationalized, and internalized through reflective abstraction or as one engages in communication about completed work.

ALGEBRAIC NOTATION: POSSIBILITIES AND LIMITATIONS

Kaput's recognition of the role of changing notational systems in the development of calculus merits careful consideration, especially with regards to algebraic notation. Algebraic notation is easily manipulable, a characteristic that played a significant role in its formulation. The introduction of the notation helped locally to solve certain outstanding anomalies, and at the same time it constrained the problems that would be investigated. Any notation acts as a tool, shaping and molding the character of the material to which it is applied. However, Kaput seems to create a narrow view of the history when he makes such claims that Leibniz, Euler, and Lagrange created "unjustified algebraic leaps." This clears a view of algebra as discontinuous from previous work and promotes the view that the algebraic formalism can embed the elusive abstract idea without compromising or contaminating its generality. We suspect that such views mystify mathematical development, allowing it to be portrayed as the product of the isolated minds of geniuses, mostly white males, who miraculously produce results. Though we do not wish to rule out the possibility of genius, we believe that more careful documentation and analysis may offer more insight into the rational basis of their work.

TABLE 1.

x	$y = 3x + 2$
10.00	32.00
100.00	302.00
1000.00	3002.00

In our own work, we are examining closely the work of Wallis during the post-Descartes, pre-Leibniz period on (a) using tables effectively to raise questions of interpolation and intercalculation, and (b) relating such tables to the creation of infinite series. Leibniz also was known for his capable and constant use of tables, for example, constructing the "chain" rule through the use of finite values for Δx. Effective and insightful use of tables is a talent that has been largely ignored in modern mathematics and so decentering to understand that its use in the development of calculus can appear very difficult. Understanding the role of infinite series in the development of the function concept is also neglected (Dennis, Confrey, & Smith, 1993). We suspect that such work may help to demonstrate both the power and limitations of algebraic expression and fill in and reinterpret some of the transitional periods documented as mystery in standard texts.

We believe that there is a fertile territory here that is largely overlooked in creating new approaches to calculus. In an "epistemology of multiple representations," one assumes the adequacy and inadequacy of every representational form, asking what does it do for me, and how does it both expand and limit my perspective? In our use of computer technology and the software Function Probe, we have witnessed remarkable uses of the table. These uses have convinced us that tables are not simply transitional tools, on the way to the graph or to the algebra, but genuine mathematical objects in their own right. For example, in a Guess My Rule exercise for linear functions, students input x-values, and the computer supplies the corresponding y-value. Students figure out the value of trying multiples of 10 for the x-value (see Table 1, the equation would not appear in the second column when working with students). One quickly sees the students' strategy: to separate the multiplier from the adder by turning the result into an extended string. This use of large numbers is not an isolated tactic. We have seen students apply it to find the coefficients for higher degree polynomials, again to separate the magnitude of the changes.

ONTOGENY RECAPITULATES PHYLOGENY—REVISITED

The problem with a progressive-absolutist view of history is that it tends to emphasize what mathematicians of the past could not do when viewed from a modern perspective, instead of seeking to understand the breadth and variation of the understandings they did have within the cultural and historical period in which they lived. Educationally there is a danger in that when we examine the mathematics of students, we will emphasize what they cannot do instead of what they can do. We think that both portrayals are historically and pedagogically ill advised. A basic commitment that we have as constructivists is that students and historical figures make sense within their experiential worlds. Thus, we question, for example, Kaput's characterization of some of Oresme's work as "decidedly confused" (p. 79), for the confusion most likely lies in our own inability to imagine the world as Oresme saw it, rather than as a sense of confusion on his part. Likewise, to claim that the Greeks "had no concept of acceleration" (p. 87), simply because they did not include it within their formal studies of astronomy or mathematics, is a strange statement for it seems to claim that the experiential world of the Greeks did not include a sense of changing velocity. Likewise, to describe students as "without a concept of acceleration" would seem to be a claim that they have never ridden in an automobile, unless "without a concept" means without an abstraction—a symbolic or notational disentanglement. However, this is again evidence of an unexamined acceptance of abstraction as the basis for mathematical understanding. We suggest that we would be better off by trying harder to describe both what the Greeks and modern students do understand in relation to acceleration and to use these multiple understandings in the building of curricula. The second example provided by Kaput, in which he describes how students relate quantity to the length of time segments more easily than to the position of points and draws a parallel to Oresme's representational works, is more in the spirit of the approach we advocate. One can begin to imagine how this knowledge of what students do know, and are able to use, might be enhanced by creating opportunities for them to explore within alternative representational forms, an approach also advocated by Kaput (p. 101).

AN EXAMPLE: RATIOS AND THE DEVELOPMENT OF EXPONENTIAL FUNCTIONS

So as to avoid being overly general in our critique, we seek to include an example from our own historical work on exponential functions. In it, we try to demonstrate the validity of alternative perspectives. A primary concern we have with Kaput's presentation is that his focus on presenting calculus in a standard framework of historical development leads him to focus on what was lacking in earlier mathematical works, rather than allowing us insight into the valid alternative perspectives of earlier times. His approach does allow us to see history develop progressively toward our own perspective, but as it stands, it seems to minimize the value of alternative perspectives in relation to our own understanding. An example directly related to our own work is the treatment of ratio and exponential functions.

Kaput's reference to Greek understanding of ratio seems to focus on what it lacked. We also believe it is important to focus on what it contained and what it can contribute to our modern understanding. Like other historians, he characterizes the Greek ratio concept as narrow, because it is restricted to "ratios of 'like' measures," which he attributes to "the underlying semantic linkage to a static geometric model for quantities and operations on them" (p. 103). What such a description misses is the richness of a ratio concept derived from geometric models. We have previously described one way of understanding the ratio of Greeks as built up from a primitive or intuitive understanding of similarity (Confrey & Smith, 1989; Smith & Confrey, 1989a, 1989b).[4] From this perspective, similarity does not depend on first understanding angle, quantified magnitude, and equivalent fractions, which are prerequisites to the formal definitions typically presented in textbooks. Instead, similarity is seen as basic to our construction of depth perception and thus of object constancy as objects become closer and farther away from us. Proportion then becomes what we identify, in relation to magnitude, as "the same" across similar figures. To say that two similar figures are proportional is a statement about the equivalence in the relationship between corresponding magnitudes within each figure. "Ratio" is the particular name we give to the relationship between two specific magnitudes across proportions (or across similar figures). Thus, to understand ratio is to be able to identify and objectify what is the same across equal proportions.

Although this concept of ratio does appear to be limited to relationships within the same dimension, we see it as enriching our own concept of ratio for several reasons:

1. It does not require the numeric quantification of magnitude.
2. Ratio is not derived from magnitude, but is created through an understanding of the relationship between magnitudes. In other words, it is based on a relationship between magnitudes but not on magnitude itself.
3. It is closely related to one way of describing what makes two constructive actions the same. For example, starting with a line segment and carrying out a set of constructive actions will create some geometric figure. Carrying out the same constructive actions on a different line segment will create a similar figure, and thus proportional relationships and equal ratios. Thus, ratio is closely connected to an understanding of making in the same way.[5]

---

[4]We do not think that it is possible to "know" the concepts of others, especially those separated by time and culture. Thus, we do not offer this as the Greek understanding of ratio, but rather as one possible interpretation that seems to be compatible with the historical record. Our goal was to use their records to enhance our own understanding of ratio.

[5]Of course, not all like constructions create similar figures. For example, if the construction includes an external referent to a fixed magnitude, one would not expect all figures made from the construction to be similar.

4. From this perspective we can generalize to relationships outside single geometric dimensions. We see ratio as what is the same across equal proportions, allowing equal proportions to be built up in a variety of ways. We would argue that numeric quantification is one of these ways of constructing equal proportions.

We have also described Oresme's treatise which starts with the assumption that ratio is inherently different from quantity and magnitude (Smith & Confrey, in press). He developed what we would now call an isomorphic relationship between ratios and magnitude, in which the primitive relationship among magnitudes is addition, and the primitive relationship among ratios is multiplication. Using terminology similar to Euclid's, he then claimed that just as any given magnitude can be divided into "parts," the same is true of ratios. For Euclid, a *part* is equivalent to what we would now call a unit fraction. Thus, just as a magnitude can be divided into 2-parts (halves), 3-parts (thirds), or n-parts (n^{ths}), the same is true of ratios. However, Oresme pointed out that creating a part of a ratio must be true for ratio operations. One creates a 2-part of a given magnitude with the intention that two 2-parts when added together will be equivalent to the whole. Thus one creates a 2-part of a ratio with the intention that two 2-parts when multiplied together will be equivalent to the whole (ratio). Thus one-half (one 2-part) of the ratio 3:1 is $\sqrt{3}:1$ or, in modern notation, $3^{1/2}:1$. Likewise, three-fifths, or three 5-parts of the ratio 3:1 would be $3^{3/5}:1$. (Oresme also suggested that there should be "irrational" parts such as $3^{\sqrt{2}}:1$.) By focusing on the operations through which we construct ratio and the distinction from magnitude, Oresme described an alternative perspective that has been subsumed into our abstracted world of real numbers and continuous exponents.

COVARIATION

In addition, an historic form of representing the relationship between variables which was common from the time of Aristotle until the modern development of algebra is closely related to what we have described as a *covariational* representation of function rather than the common *correspondence* representation (Confrey & Smith, 1991; Rizzuti, 1991). If we take an algebraic correspondence representation of a linear function, we might write, for example, $y = 2x + 5$. One historic way of writing such a relationship in an algebraic covariational form would be:

$$(y_4 - y_3) = (y_2 - y_1) \leftrightarrow (x_4 - x_3) = (x_2 - x_1) \quad (1)$$

A more generalized form would be (Smith & Confrey, in press):

$$(x_4 - x_3):(x_2 - x_1) = (y_4 - y_3):(y_2 - y_1) \quad (2)$$

From our modern perspective, we might tend to be critical of these forms. Because they specify neither the starting value nor the slope they do not allow us to move directly from a given x-value to a corresponding y-value. This seems to violate modern definitions of function and our emphasis on numeric quantification of magnitude. However, such representations beautifully capture the operational relationship between the two variables by explicitly placing both variables in an additive world. Equation 2, for example, can be seen as equivalent to a statement of constant slope, but emphasizes the operational relationships within dimensions rather than between dimensions. This form is also closely related to the construction of differential equations as it reflects the relative change between two variables.[6] The power of this form becomes even more evident when we relate it to what we would now call exponential functions. For example the correspondence equation, $y = 2^x$, would be written in a covariational form as:

$$(y_4 / y_3) = (y_2 / y_1) \leftrightarrow (x_4 - x_3) = (x_2 - x_1) \qquad (3)$$

In this case, the form indicates a relationship between a multiplicative world and an additive world.[7]

Looking at Oresme's understanding of ratio together with this covariational understanding of relationships can lead to alternative ways of thinking about Oresme's representations of motion as described by Kaput. One can see, for example, in the quote attributed to Bacon (p. 91) as well as in the quote attributed to Oresme (p. 92), the kind of covariational expression described in equation 2. Oresme's statement describing how the relationship between velocities is related to the relationship between times under constant acceleration seems to emphasize the operational relations between velocities compared to the operational relationships between times, a very different approach from our own concern with matching specific times with specific velocities.

Oresme's approach to ratio would lead one to question whether he ever found it necessary to think of a "particular unit length reference" (p. 94) for the quality being represented. As described earlier, his concept of ratio may have been, like the Greeks, developed as an understanding of relationship between magnitudes, which would not require that the magnitudes themselves be assigned a numeric measure. One sees throughout the quotations from this period a focus on how change in one dimension is related to change in another, with ratio being the kind of unit used for describing these changes. When looking at his representations of

---

[6]This is even more relevant in the covariational understanding that our students have constructed, particularly when co-filling columns in the table window. In such situations the covariational relationships have been expressed with more numeric specificity. In the form written earlier, one might, for example, write: $(x \rightarrow x + a) \leftrightarrow (y \rightarrow y + 2a)$.

[7]Thomas of Bradwardine created a more generalized form of this exponential equation similar to the generality of equation 2, however, his argument cannot be included here. See Smith and Confrey (in press) for a description of his work.

velocity and time (Figures 4.3–4.5), this interpretation allows some insight into his focus on the shapes of the figures themselves and his relative lack of interest in a consistency in orientation (left-to-right versus right-to-left) (p. 92). It suggests that Oresme's interest may have been more related to how he could use ratio and similarity to describe characteristics of a temporal event rather than using magnitude and position to describe particular instances of a relationship between variables.

Whereas the earlier description suggests that traditional historic accounts, including that of Kaput, may undervalue the work of Oresme and others of that period by, in effect, neglecting aspects of their perspective, which seem to differ from modern mathematics, we also believe that there is a tendency to overvalue the work of later mathematicians who were responsible for developing many modern representational forms, particularly algebraic forms. This is not said to take anything away from the brilliance of the individual accomplishments of these mathematicians, rather as a suggestion that when viewing their work in a historical context, we must balance the beauty of their abstractions against the resulting loss of conceptual diversity, which has plagued mathematics ever since.

As a particular example, we might look at Euler's development of the natural log and exponential as described on pages 120–121 in Kaput. To understand the issue we wish to pursue, it is important to understand two ways we might view an exponential function of the form $y = ca^x$. In formal mathematics, y, c, a, and x are treated simply as real numbers, that is there are no qualitative distinctions in their treatment. However, we have argued that the basis for the way we construct a concept of exponential is through the recognition or objectification of a multiplicative action (or a ratio action). When in situations in which we repeat this action, we have called the action a "multiplicative unit" (Confrey & Smith, in press). The unit is the repeated action. From this perspective, which is closely related to the ratio world of both Oresme and the Greeks, the parts of the equation are looked at quite differently: y and c are quantities (or magnitudes), a is a ratio (or multiplicative unit), and x is a count.[8] From this perspective, Euler's derivation is more confusing. We need look only at the first few steps to realize a certain discomfort with the procedure. He initially claims that a^ϵ can be written as: $(1 + k\epsilon)$ where ϵ is an arbitrarily small number, and k depends on a. Next he lets N be arbitrarily large such that $N = x/\epsilon$. He can then write:

$$a^x = a^{N\epsilon} = a^{\epsilon N} = (1 + k\epsilon)^N = (1 + kx/N)^N.$$

Because N is "infinitely" large, by setting $k = 1$ and $x = 1$, Euler was able to create e.[9] In this procedure, the ratio a has disappeared, and the index of counting, x, has been moved from its position in the exponent to become a component of the base. To argue that this procedure arbitrarily and erroneously mixes ratios

---

[8] By "count" we are not limiting x to integers, but using the more general sense of counting ratios, as in Oresme, where one can "count" any rational (or real) number of ratios.

(a) and counts (x) might be interpreted as a return to the "semantic linkage" of the Greeks, which Kaput suggests played a role in unnecessarily restricting mathematical progress for centuries. We do not wish to undervalue the work of Euler. However, to dismiss "semantic linkage" as restricting mathematical progress, while implying that progress is linked to abstraction (as in Piaget and Garcia) needs explicit examination and critique. We would strongly argue that building up mathematical conceptions is a process in which actions, semantics, and representations are interconnected. In the particular case of exponential functions, the standard practice of using base e to model exponential situations, which has its origins in the work of Euler, is justified in calculus as the base for which the function is its own derivative. In addition, the extent of its use was certainly extended in the precomputer age by the desirability of using standardized bases to facilitate the standardization of logarithmic tables. Thus, we must acknowledge the importance of e in these respects. However, we believe that it is also important to point out how, for most of us, the use of e effectively restricts any way of relating the form of the equation, $y = ce^{kx}$, to our understanding of the repeated action of multiplicative units. As an example, we might imagine a situation in which some population is growing by 3% a year. Thus, we imagine that the initial population, P_o, is increased to $(1.03)P_o$ by the end of a year and to $(1.03)(1.03)P_o$ after 2 years. From this we begin to see the repeated action that creates the multiplicative unit, 1.03. However, it is not uncommon to see the described situation modeled as $P_t = P_o e^{.03t}$, in which t is time in years. One can see formally in Euler's derivation how a part of the ratio or multiplicative unit gets moved up to the exponent position, but this does little to help us understand why this equation works. We have to somehow make sense of two seeming anomalies, first that $1.03 \neq e^{.03}$, and second how we can understand that which we have built as a ratio, .03, appearing in the exponent or "counter" position. Thus, although e has played an important role in the development of formal mathematics and has added important breadth to our understanding of mathematics, we question whether we should define it as "progress" in the sense that it is an abstraction that can replace that which has come before it. Because there has been a strong tendency to do so in both the mathematics curriculum and in standard scientific notations, we have equated progress with the introduction of a concept that has at some point acted as an epistemological barrier for almost all of us and whose place is still misunderstood in much of the curricular and scientific literature. Because its typical curricular treatment tends to cut off any means to relate the representation, e, to mental actions of repeated multiplication, it also encourages cookbook memorization in too many secondary and college classrooms.

Because we have placed a significant effort into the development of our own understanding of ratio, proportion, and exponential and logarithmic function, we have become particularly aware of the neglect that these areas have received both in the curriculum and in formal mathematics, despite the rich historical traditions related to their development. Our own historical investigations have played an essential role, particularly in the development of the splitting concept (Confrey, in press), multiplicative units, rate, and ratio (Confrey & Smith, in press). We have also gathered significant evidence of the strong role these concepts can play in the activities of young students (Confrey, in press), high school students (Confrey, Smith, Piliero, & Rizzuti, 1991), and college students (Confrey, 1991b) when opportunities are available. For those reasons we may be particularly sensitive to standard portrayals of history that downplay not only the basis of these concepts but other alternative ways of seeing and understanding. It is often difficult to describe exactly why standard histories can be so troubling, for they are often mutually supportive. The problem often lies in the lens chosen for historical inquiry. Kaput states, for example, that: "It is also important to note that not only did Oresme *fail* to provide a modern Cartesian graph" (p. 100; emphasis ours). Kaput is factually correct and does relate this point to a useful discussion of the difference between Cartesian graphs and alternative graphical representations. Our concern is with the use of the word "fail," which tends to emphasize a lack of success. Such portrayals lend credence to a progressive-absolutist view, particularly a view that Oresme really was somewhat trying to do the same thing that his successors were trying to do, that is, to develop modern mathematics and thus that it is reasonable to characterize his work in terms of what he failed to do. To do so, however, too often masks the recognition that Oresme and others had great success in representing the world as they saw it and in solving the problems that were important to them. We believe that attempting to understand their successes from their perspective rather than their failures from our perspectives should be a primary goal of historical research.

CONCLUSIONS

In summary, our critique of Kaput is largely based in what appears to be an unexamined acceptance of the meaning and significance of abstraction to the mathematical enterprise. We believe that essentially he would agree with the need to radically reconsider the conceptual basis for calculus, and we believe that he has identified key areas for examination. However, we are concerned with his framework for conducting his historical investigations and for how it might impact negatively progress toward his own goals for reform in education. Thus, we advocate: (a) a rejection of a progressive-absolutist view of history; (b) an introduction of a multiple-representational epistemology for assessing mathematical change; (c) a more in-depth characterization of the mathematical enterprise situated in cultural and historical period, concentrating on gain and loss in light of outstanding problems; and (d) a more radical redefinition of the current curriculum in which the computer provides a tool for implementing new perspectives of mathematics. We are convinced that an examination of the meaning of abstraction will be an essential part of this process and are confident that attributing to abstraction a privileged status for algebraic notation is unwise. Whether a new view of abstraction more closely tied into human activities of construction, or connected to the act of coordinating multiple representations can emerge, or whether abstraction will become a limited idea (associated closely with formalism), we do not presume to know.

We believe that rethinking calculus is of fundamental importance. In Kaput's chapter, the emphasis is on the use of history to provide alternatives to the standard curriculum, and we need to cooperate in the undertaking of such an endeavor. In fact, we suggest that the pedagogical enterprise might just provide some provocation to historians to reconsider their own epistemological bases for historical investigation. Coupled with historical work, we also advocate careful and systematic investigation of students' thinking. In the reform efforts on calculus in the United States, this work has been neglected. This state of affairs needs remediation and will be undertaken with much more promise if the investigators are familiar with these historical issues.

REFERENCES

Artigue, M. (1992, August). The importance and limits of epistemological work in didactics. In W. Geeslin & K. Graham (Eds.), *Proceedings of the Sixteenth Psychology of Mathematics Education Conference* (Vol. 2, pp. 195–216). Durham, NH.

Borba, M. (1993). *Students' understanding of transformation of functions using multi-representational software*. Unpublished doctoral dissertation, Cornell University, Ithaca, NY.

Confrey, J. (1980). *Conceptual change, number concepts and the introduction to calculus*. Unpublished doctoral dissertation, Cornell University, Ithaca, NY.

Confrey, J. (1991a). *Function probe* [software]. Santa Barbara, CA: Intellimation Library for the Macintosh.

Confrey, J. (1991b). The concept of exponential functions: A student's perspective. In L. Steffe (Ed.), *Epistemological foundations of mathematical experience* (pp. 124–159). New York: Springer-Verlag.

Confrey, J. (in press). Splitting, similarity, and rate of change: New approaches to multiplication and exponential functions. In G. Harel & J. Confrey (Eds.), *The development of multiplicative reasoning in the learning of mathematics*. Albany, NY: State University of New York Press.

Confrey, J., & Smith, E. (1989). Alternative representations of ratio: The Greek concept of anthyphairesis and modern decimal notation. In Don Emil Herget (Ed.), *Proceedings of the First History and Philosophy of Science in Science Teaching* (pp. 71–82). Tallahassee, FL: Science Education & Department of Philosophy, Florida State University.

Confrey, J., & Smith, E. (1991). A framework for functions: Prototypes, multiple representations, and transformations. In R. Underhill & C. Brown (Eds.), *Proceedings of the Thirteenth Annual Meeting of Psychology of Mathematics Education-NA* (pp. 57–63). Blacksburg, VA.

Confrey, J., & Smith, E. (in press). Exponential functions, rates of change, and the multiplicative unit. *Educational studies in mathematics*. Dordrecht, The Netherlands: Kluwer/Academic Press.

Confrey, J., Smith, E., Piliero, S., & Rizzuti, J. (1991). *The use of contextual problems and multi-representational software to teach the concept of functions*. Final Project Report to the National Science Foundation (MDR-8652160) and Apple Computer, Inc. Cornell University.

Dennis, D., Confrey, J., & Smith, E. (1993, April). *The creation of continuous exponents: A study of the methods and epistemology of Alhazen, Wallis, and Newton.* Paper presented at the annual meeting of the American Education and Research Association, Atlanta, GA.

Edwards, C. H. (1979). *The historical development of calculus.* New York: Springer-Verlag.

Fowler, D. H. (1987). *The mathematics of Plato's academy.* Oxford: Clarendon Press.

Joseph, G. G. (1991). *The crest of the peacock.* London: I. B. Tauris.

Lakatos, I. (1976). *Proofs and refutations: The logic of mathematical discovery* (J. Worrall & E. Zahar, Eds.). Cambridge: Cambridge University Press.

Piaget, J. (1971). *Biology and knowledge.* Chicago: University of Chicago Press.

Rizzuti, J. (1991). *High school students' use of multiple representations in the conceptualization of linear and exponential functions.* Unpublished doctoral dissertation, Cornell University, Ithaca, NY.

Smith, E., & Confrey, J. (1989a, March). *Ratio as construction: Ratio and proportion in the mathematics of ancient Greece.* Paper presented at the annual meeting of the American Education Research Association. San Francisco.

Smith, E., & Confrey, J. (1989b, July). *Ratio in a world without fractions: A study of the ancient Greek world of number.* Paper presented at the Thirteenth International Meeting of Psychology of Mathematics Education. Paris, France.

Smith, E., & Confrey, J. (in press). Multiplicative structures and the development of logarithms: What was lost by the invention of function. In G. Harel & J. Confrey (Eds.), *The development of multiplicative reasoning in the learning of mathematics.* Albany, NY: State University of New York Press.

Smith, E., Dennis, D., & Confrey, J. (1992). Rethinking Functions, Cartesian Constructions. In S. Hills (Ed.), *The history and philosophy of science in science education, Proceedings of the Second International Conference on the History and Philosophy of Science and Science Education* (Vol. 2, p. 449–466). Kingston, Ontario: The Mathematics, Science, Technology and Teacher Education Group; Queens University.

Steffe, L. (in press). Children's multiplying and dividing schemes. In G. Harel & J. Confrey (Eds.), *The development of multiplicative reasoning in the learning of mathematics.* Albany, NY: State University of New York Press.

Unguru, S. (1976). On the need to rewrite the history of Greek mathematics. *Archive for History of Exact Sciences, 15*(2), 67–114.

5 Making Calculus Students Think With Research Projects

Marcus Cohen
Arthur Knoebel
Douglas S. Kurtz
David J. Pengelley
New Mexico State University

> **Arthur Knoebel,** *New Mexico State University*
> **Leon Henkin,** *University of California, Berkeley*
> **Jere Confrey,** *Cornell University*
> **Michael Ranney,** *University of California, Berkeley*
> **Ed Dubinsky,** *Purdue University*
> **Susanna S. Epp,** *DePaul University*
> **John Addison,** *University of California, Berkeley*

Arthur Knoebel. I would like to tell you about a remarkable and surprising discovery that my colleagues and I have made in teaching calculus. It is about getting students to solve problems that are much harder than normally expected. The students are given a longer time than usual and are expected to write their solutions up well. The results are very heartening.

Let me begin with a philosophical comment. Mathematicians are fond of chiding engineers for not being concerned about the foundations of the mathematics that they use and in particular for not paying attention to questions of convergence and the like. In presenting our work on teaching, I find the roles reversed. I feel that I am in the role of the engineer now and could be chided for not paying more attention to the underpinnings. Like the engineer, I say, "What we do works, but we don't know entirely why." Our pragmatic approach leaves us open to criticism from the theoretically inclined among you, and we welcome it.

First I will describe our program, then talk about evaluation, and finally raise a number of issues that may be of interest to you and hopefully will provoke some discussion.

Here is a little background. New Mexico State University is a medium-sized public institution with mathematics courses ranging from remedial to doctoral levels. The origin of our program goes back $2\frac{1}{2}$ years. A couple of young professors in our department, Marcus Cohen and David Pengelley, happened to live across the street from each other. One warm spring evening they were bemoaning the sad state of calculus instruction and how demoralized they were by their students' lack of interest and poor performance. After some conversation Marcus said, "Why don't we give our students something more to do? We think they're not doing enough so let's give them something special." Out of this grew the idea of projects. These two started using them for a semester or so, when the National Science Foundation (NSF) put out a request for proposals, for reforms in calculus. So, they chatted with people around the department and interested three more of us, Edward Gaughan, Arthur Knoebel, and Douglas Kurtz. The five of us together submitted a proposal to NSF.

We submitted a 1-year planning grant proposal. NSF's reaction was surprising. They said, "Well, really you're already in the planning stage because you already started this." The result, after some negotiation and rewriting, was that we ended up with a 3-year development grant. By the end of the summer of 1990, it will be essentially over. I think the NSF was very clever in their strategy, because they got a planning grant out of us for nothing.

The idea of student research projects is to pose problems much harder than typical homework problems and give the students one or two weeks to do them. We provide access to help when students need it. The help is usually limited to "Yes," "No," "You're on the right track," or if they're not on the right track, giving them an indication of where to go. This is in contrast to the typical situation in which the student says, "I can't do this homework problem," and the instructor responds, "Here's how you do it," explaining it as quickly as possible in order to get on to something else. The object is to give the students an idea of mathematical research, to have them learn the agony and ecstasy of intellectual discovery. We want to give students a taste of doing mathematics as mathematicians do it. After all, after 12 years of study, shouldn't students have some idea of what professionals in the field do?

Let me describe some of our projects. They include some of the original ones, and their success is what heartened the first two people before we got the grant. Figure 5.1 is a project on the harmonic series. The harmonic series diverges, but if you consider the decimal representation of the denominators and leave out those that contain the digit 7 anywhere, then the series converges. It is a remarkable result, and unexpected. And it is something you normally wouldn't hand students before the junior year. Yet, our students were able to do it (with a good bit of help). Along a similar line is a project (Fig. 5.2) showing that one can rearrange the terms in the alternating harmonic series so that one gets a series that converges to any predetermined sum whatsoever. Again, this problem is something that you wouldn't expect to give students at this level. With help they manage to do this. Of course, they have to spend some time, quite a bit of time, some of them claim, but they do it.

5. MAKING CALCULUS STUDENTS THINK

If $a_1, a_2, a_3, \ldots$ are the positive integers whose decimal representations do not contain the digit 7, show that

$$\sum_{n=1}^{\infty} \frac{1}{a_n}$$

converges and its sum is less than 90.

Here are some suggestions: Break up the sequence of positive integers into carefully chosen blocks (not necessarily all of the same size), count the number of terms in each block, and see what fraction of them is left after removing those with an offending 7. Then bound the size of the sum of the reciprocals in each block, and try to compare this result to a convergent series.

FIG. 5.1. A harmonic series with missing digits.

The great virtue that we see in projects is the variety of challenges that can be posed. They range from the briefly stated to detailed floor plans; some are open-ended, some not. They may bristle with formulas or contain not a single mathematical symbol. Here are some examples. Figure 5.3 is a problem about a greenhouse, in which there are no formulas, not even any numbers stated. The students find this baffling at first. They are floored by it but eventually see that they have to define some variables and work from there. The way they go about this is amazing. What do they think they have to do? One student went to his parents' home and measured it to figure out how to knock out the wall that is specified in the problem. Fortunately, he didn't go as far as knocking it out.

Projects may use computers or calculators. Projects may be pure, applied, or what we like to call pseudo-applied. An example of the latter is one we have called *Houdini's Escape* (Fig. 5.4). The students are given a formula for a flask, a very large flask, in which Houdini is going to be put. He is to be shackled, but he can unshackle himself in 10 minutes. He is also to stand on a little pedestal,

The alternating harmonic series

$$\sum_{n=1}^{\infty} \frac{(-1)^{n+1}}{n}$$

converges to ln 2. Now suppose s is any given number.

You are going to prove the amazing result that the alternating harmonic series can be rearranged (that is, its terms can be written in a different order) so that the resulting series converges and has s as its sum.

You will need to understand the precise definitions of the limit of a sequence and the sum of a series.

HINT: Start by trying to reorder the terms to alternately overshoot and undershoot s with various new partial sums.

FIG. 5.2. Rearranging an infinite series.

Your parents are going to knock out the bottom of the entire length of the south wall of their house, and turn it into a greenhouse by replacing some bottom portion of the wall by a huge sloped piece of glass (which is expensive). They have already decided they are going to spend a certain fixed amount. The triangular ends of the greenhouse will be made of various materials they already have lying around.

The floor space in the greenhouse is only considered usable if they can both stand up in it, so part of it will be unusable, but they don't know how much. Of course this depends on how they configure the greenhouse. They want to choose the dimensions of the greenhouse to get the most usable floor space in it, but they are at a real loss to know what the dimensions should be and how much usable space they will get. Fortunately they know you are taking calculus. Amaze them!

FIG. 5.3. A greenhouse extension.

and the question is how high should that pedestal be, given the shape of the flask, so that Houdini will be able to get out of there in 10 minutes given that the flask starts empty and fills up with water. Can you time it so that he will get free of the shackles just at the time his head is going under water? They like that kind of project. It's ridiculous and unrealistic, but it gives them something to visualize.

Projects may be 1 week or 2 weeks long, worked by individuals or groups. We started out using only individual projects; now we're doing more and more group projects. There are quite a few advantages to that. The students learn from one another. Working in groups gives them a more realistic approximation to how they will work after graduation. Finally, instructors spend less time helping students and grading papers.

How do we create these projects? Well, ideas for them are everywhere. You obtain some out of books, but we tend to eschew amplifying the problems in standard textbooks. We like to create our own. Sometimes students will raise a question in class that goes beyond what is expected, and this will lead to an idea for a project. You can get ideas while walking across campus or at home. For example (Fig. 5.5), if you go to a pizzeria and order a pizza, you typically see sliced olives on it. They are little disks with the centers removed where the pit was. This led to an idea for a project. If you take a sphere, and you have two blades that are always at a fixed distance, and you put the sphere symmetrically through them and slice off a couple of caps, you'll have some circular caps. If you further remove a cylinder from your olive, which is exactly the size of that circular disk, then you'll end up with a pitted olive. What is amazing is that the volume left in your pitted olive is independent of the original size of the olive, assuming it's big enough so that there is something to slice off. We had students prove that, and then we asked them, "Olives aren't spheres. They're more like spheroids. If you start looking at various shapes other than spheres, does that still hold?" This in turn led to a study of the Jacobian in order to perform the integrations so that one could decide for which shapes of olives this would be true. One student actually reported a breakthrough after having dinner at a pizzeria.

Another instance illustrating sources of ideas for these projects is in Jere

5. MAKING CALCULUS STUDENTS THINK 197

Harry Houdini was a famous escape artist. In this project we relive a trick of his that challenged his mathematical prowess, as well as his skill and bravery. It will challenge these qualities in you as well.

Houdini had his feet shackled to the top of a concrete block which was placed on the bottom of a giant laboratory flask. The cross-sectional radius of the flask, measured in feet, was given as a function of height z from the ground by the formula

$$r(z) = \frac{10}{\sqrt{z}},$$

with the bottom of the flask at $z = 1$ foot. The flask was then filled with water at a steady rate of 22π cubic feet per minute. Houdini's job was to escape the shackles before he was drowned by the rising water in the flask.

Now Houdini knew it would take him exactly ten minutes to escape the shackles. For dramatic impact, he wanted to time his escape so it was completed precisely at the moment the water level reached the top of his head. Houdini was exactly six feet tall. In the design of the apparatus, he was allowed to specify only one thing: the height of the concrete block he stood on.

(a) Your first task is to find out how high this block should be. Express the volume of water in the flask as a function of the height of the liquid above ground level. What is the volume when the water level reaches the top of Houdini's head? (Neglect Houdini's volume and the volume of the block.) What is the height of the block?

(b) Let $h(t)$ be the height of the water above ground level at time t. In order to check the progress of his escape moment by moment, Houdini derives the equation for the rate of change dh/dt as a function of $h(t)$ itself. Derive this equation. How fast is the water level changing when the flask first starts to fill? How fast is it changing when the water just reaches the top of his head? Express $h(t)$ as a function of time.

(c) Houdini would like to be able to perform this trick with any flask. Help him plan his next trick by generalizing the derivation of part (b). Consider a flask with cross-sectional radius $r(z)$ (an arbitrary function of z) and a constant inflow rate $dV(t)/dt = A$. Find dh/dt as a function of $h(t)$.

EXTRA CREDIT: How would you modify your calculations to take into account Houdini's volume, given Houdini's cross-sectional area as a function of height?

FIG. 5.4. Houdini's escape.

Confrey's talk at this conference. She was talking about power and exponentiation, and she had a little picture (Fig. 5.6) of spirals that got me thinking as follows. I think she had something like the solid and dashed lines that look like an interesting new coordinate system. How could we specify that coordinate system? If I understood correctly, the spirals that she was dealing with there would be of the form $r = \theta$. I let the constant of proportionality equal 1 so, if α is 0, this first equation corresponds to the heavy line, and if you vary α, you get the other solid lines. If you change the sign of θ to the negative sign, you'll get the

In order to stay in school and take more calculus, you have to take a summer job. You have been hired by Jacobi's Pizzeria as a consulting engineer to help them test and calibrate a new automated olive pitter, which they are considering buying. The advertising flyer for the Pits Company says:

> Stage one of our auto-pitter has two cutting blades spaced 6 mm apart. Olives of various shapes and sizes (but always greater than or equal to 6 mm in every direction) are fed between the horizontal blades, cutting off congruent caps. In stage two a cylinder is punched out of each olive by a combination of pressure and vacuum, taking the pit away. The bottom and top of this cylinder are the ellipses left where the symmetrically placed caps were first lopped off. The auto-pitter has been designed to yield a pitted olive of constant volume independent of the diameter of the olive.

Now Mama Jacobi is very skeptical of this last statement. Before she even orders the auto-pitter, she has hired you to model mathematically its effect on olives of various shapes. In order to use calculus, you assume the olives are all ellipsoids. You go to Papa Jacobi (who in his youth did some work of his own in calculus!), who shows you a method involving coordinate transformations. He then shows you how to apply his method of change of variables in a double integral to make part (a) below even easier by transforming it to polar coordinates in the xy-plane.

Read about the Jacobian and use Papa Jacobi's method for double integrals on parts (a) through (c). For example, for (b) transform the integral to polar coordinates; for (c), in order to use the technique of (b), transform the elliptical cross section to a circle.

(a) Your first set of model olives are perfect spheres of diameter at least 6 mm. Show that, surprisingly enough, the company's claim about volume is accurate! Find the volume of the pitted olives.

(b) Your next set of model olives are spheroids—they have an elliptical vertical cross section of at least 6 mm. in length and a circular horizontal cross section. They are fed between the blades, giving off caps with circular edges. Is the company's claim still valid?

(c) Next, you feed your machine an ellipsoidal olive, the skin of which has the equation,

$$x^2/a^2 + y^2/b^2 + z^2/c^2 = 1.$$

Find the volume of the pitted olive. Is it constant, i. e., independent of the parameters a, b, and c?

You are now preparing a report of your results. You find that you must do the following:

(d) Explain Papa Jacobi's method. Find an appropriate theorem on changing variables in a double integral by means of the Jacobian, and explain how this theorem applies to the changes of variables in the previous parts of this project. Does the theorem you cited cover the entire region of integration?

(e) Prove the theorem of part (d). Let the equations, $x = x(u,v)$ and $y = y(u,v)$, be an arbitrary change of variables in a double integral. Use the particular changes of variables you employed in the earlier parts as particular examples to illustrate each step in your explanation of the proof of Papa Jacobi's method.

Mama and Papa Jacobi now offer you a permanent job, promising you won't have to do calculus anymore. Do you take it, or go back to school?

FIG. 5.5. Jacobi's Pizzeria.

5. MAKING CALCULUS STUDENTS THINK 199

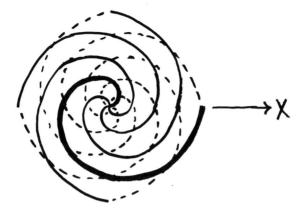

FIG. 5.6. Spiral coordinates.

dashed lines by varying β. Now you have a new coordinate system. You can write it like so:

$$\begin{cases} r = \theta + \alpha, \\ r = -\theta + \beta; \end{cases}$$

and solve for α and β, or you can solve for r and θ. You find that the Jacobian is 2.

How would a project come out like this? Well, I have not worked out the details yet, but there are various possibilities. One might be to give students a shape, inside a couple of solid lines and a couple of dashed ones, and ask them to find its area. Maybe one doesn't tell them about the coordinate system but lets them reconstruct it in order to do the integration easily by a change of variables. Or, give students a function over the region and ask them to integrate it. They would have to use a transformation of integrals and use the Jacobian and so forth. There are lots of possibilities. Sometimes we try to anticipate too much what students might or might not be able to do. Maybe the idea should be to get them started so that they end up drawing this picture and proving something about it.

At first, we thought we would quickly run out of ideas for projects. But our experience demonstrates the opposite: There appears to be no end in sight for new ideas. We are presently compiling a book of completed and polished projects. (See Cohen et al., 1991.)

Leon Henkin. What is the density of projects in the time span of the course?

Arthur Knoebel. We have two or three project assignment periods during each semester, and each time projects are assigned we give a class anywhere from one to five different projects. We use anywhere from about two to nine in a particular course for a semester. For a student it would be two or three. David Arterburn, at the New Mexico Institute of Mining and Technology, is doing something quite extreme. He has each student do five projects per semester but no examinations. He gets pretty well worn out by this, so I don't recommend it, but it's possible to do.

Our initial reason for assigning more than one project at a time to a class was to avoid collaboration between students. However, most of us have come to feel this is not so important. We find pedagogical value in students talking to one another, as long as each student writes up his or her own ideas. Moreover, we schedule individual postproject interviews, which insure that each student's work is his or her own. These interviews also help students do better on future projects. And problems of collaboration have become less important as we have moved more to group projects.

Time is of the essence. One of the issues that emerged quite early is that projects can eat up a lot of time, both teachers' and students'. We have concentrated on cutting down the teacher time. A partial solution we have found is a laboratory that is open when the projects are assigned. Students can go there for help. The laboratory is typically run by teaching assistants who are also teaching calculus sections with projects. As part of the NSF grant they have a reduced teaching load so that they can do this. One problem arose: One laboratory for all projects sections of all calculus courses was too unwieldy, with students having to wait in line, and each teaching assistant having to be familiar with too many different projects. Our solution now is to run a separate laboratory for each course. Next semester we're going to institutionalize project courses so that we can continue without the NSF grant, relying on university funding.

It is noteworthy that students also have trouble budgeting their time. A student once hooked on a project often won't let go. Frequently we witness a bulldog tenacity, going beyond all reason, causing a student to shortchange other courses. Logically, a C student should stop work when about 75% of the project is completed, but illogically the student often keeps on going.

That is a brief description of the program. There is a lot more to it than I've described. With over 25 people having a variety of roles I can only highlight its main features. In addition to the program at New Mexico State University there are people at three other schools using projects in selected sections as part of this NSF grant: Adrienne Dare at Western New Mexico University, David Arterburn at New Mexico Institute of Mining and Technology, Richard Metzler at University of New Mexico, and Carl Hall at University of Texas at El Paso. Also we have two evaluators, Marsha Conley and Carol Stuessy, to help gather statistics and interpret them.

Jere Confrey. How do you work with the teaching assistants so that they are capable of working in this lab, in the ways that you hope? Did you find that they already are capable, or did they need to be assisted in being able to handle the projects?

Arthur Knoebel. When we started we were very careful in our choice of teaching assistants. We chose students who were typically working on their thesis so that they have some flexibility in their time. We also picked the very best. We found that we didn't have to train them to any extent. Then last fall we included a more typical teaching assistant (T.A.), and we found out that training is needed. This T.A. did not know what we expected, and we were surprised. We learned that we need to train teaching assistants in ways that we had never trained them before. We train them intensively by working more closely than is usually the case. The ones who go through this obtain a kind of experience that the usual T.A. doesn't get. This is an unexpected dividend. It's also changed our attitude. Typically there is one faculty member designated as coordinator for all the sections of a course, who is supposed to make sure things run smoothly and make sure that the teaching assistants are doing what they should be. It is the kind of supervisory role that is not relished. We find now that we're talking to teaching assistants informally about the problems or the good things that are happening in the course. We think that this is a good thing. A teaching assistantship has become a true apprenticeship.

Also surprising is the amount of positive and detailed discussion taking place among the five principal investigators. This is in marked contrast to the normal scene in the faculty lounge where teachers typically bitch among themselves about their terrible classes and trade horror stories about how bad their students are. Now we brag to each other about the best project write-ups our students have turned in.

Let me present some of our findings. I will first give you our impressions, which in some ways are what really motivate us.

The main thing that you should realize from Fig. 5.7 is that we're asking students to do an integral, which is beyond what they can normally handle. There are steps given that lead them on, give them an idea, and introduce them to some of the tricks. One student who did very well on her project also submitted some art work with it. I've never seen this handed in before in a math course. Figure 5.8 shows the pictures this student drew for the front and back covers of her project, indicating how she felt before and after doing the project. Note the objects in their "before" and "after" states. Other students write story lines along with what they turn in. They know it is not going to count any more for their credit, but they just love to do it. We found out that students can do much more than we ever expected.

Something that has occurred over our experience with assigning these projects is what I call *project inflation*. Projects have been getting harder and harder. We find students can do more than we ever expected. They learn how to learn. They really learn how to tackle something that is going to take deliberate effort and more than a few minutes to complete. At first we had students slacking off and waiting until the last night to start on their projects. The word got around that that just doesn't work. Now we have students coming in on the second day after a project is handed out. They have a total of two weeks, but they come in early on and they've seriously started on the projects. This is part of their learning how to learn. Students find out what mathematics is. You can argue that this is not at the

Often, two integrands can look similar while one is easy to integrate and the other very hard. Consider, for example, the integrals:

$$\int_0^{\pi/2} \frac{\sin x}{\sin x + \cos x} dx, \quad \int_0^{\pi/2} \frac{\sin^2 x}{\sin^2 x + \cos^2 x} dx, \text{ and } \int_0^{\pi/2} \frac{\sin^n x}{\sin^n x + \cos^n x} dx.$$

The second one is easy to integrate while the other two are not. Still you are going to evaluate

$$\int_0^{\pi/2} \frac{\sin^n x}{\sin^n x + \cos^n x} dx$$

where n is any positive integer. To do this you will need one trick and some ingenuity. We will supply the trick.

Before you work on the entire problem, evaluate the second integral ($n = 2$) without doing anything "hard." Next, read about integrating rational functions of $\sin x$ and $\cos x$ and use that idea to evaluate the first integral ($n = 1$).

We could use this approach for larger values of n, but imagine how hard it would be. We need to find an easier way.

Now, here is the trick. Let f be a continuous function over the closed interval $[0, a]$. Then,

$$\int_0^a f(x)dx = \int_0^a f(a-x)dx.$$

Prove that this equality is true.

Use this result to evaluate

$$\int_0^{\pi/2} \frac{\sin^n x}{\sin^n x + \cos^n x} dx$$

for any value of n. Remember, you will need to use some ingenuity here also. Make sure that your answer agrees with your previous results for $n = 1$ and $n = 2$.

FIG. 5.7. A special technique for integration.

forefront of mathematical research, but I think we're having students go through the same kinds of things we do when we do mathematical research, but on a smaller scale. We replace incremental learning by jumps. In other words, dramatic inspiration is replacing piecemeal progress. The students find out that if they work hard on this, they may get stumped, but if they keep working, finally they will have some insight.

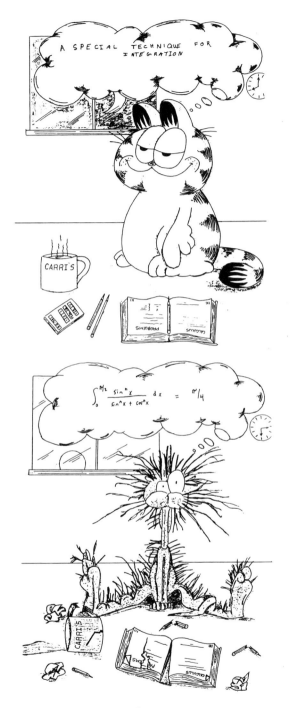

FIG. 5.8.

Consider the two functions:

$$f(x) = 1 - x \text{ and } g(x) = \frac{1}{x}.$$

We can compose them in two ways:

$$f(g(x)) \text{ and } g(f(x)).$$

We can go further and compose these functions with themselves, and also with the old ones, in a number of ways. Keep composing these functions with new ones as they are generated and figure out simplified formulae for them in terms of the variable x. (Don't forget to compose functions with themselves, like $f(f(x))$.) You might think that more and more new functions will be generated. Surprisingly only a finite number of new ones get generated by composition, even though there may be many different ways of composing f and g to get the same function. Remember that two very different-looking formulae may represent the same function.

(a) How many distinct functions are there, including f and g themselves?

(b) List them.

(c) How is each one composed from f and g?

(d) How do you know that these are all there are?

(e) For what real numbers are all these functions simultaneously defined?

FIG. 5.9. Composing functions.

We are introducing students to epistemological issues, asking, "what do you know and how do you know it?" As an example of this, Fig. 5.9 is a project that was given in a calculus course for biology and social science majors. It may not look terribly hard, but it was enough work for them to do. The answer here is 6. You're getting the symmetric group on three letters. We got answers from 4 to 37. One of the problems that the students ran into was that they did not know how to tell when two functions were equal if they are given by radically different expressions. They also didn't know how to tell when the functions were different. Perhaps more importantly, they did not realize that these were two very different tasks. The students are confronted with these issues. Unlike an exam in which they take a stand very quickly, get through the exam, and then throw their paper away as soon as they see their grade, they think about these things. Often they resolve them. What about teachers? Well, we love it too. My own exclamation after using these projects in the first semester was, "It's amazing how much one's estimate of students goes up after using projects."

Regarding precise evaluation, we have some instruments. We monitor grades of students in project versus nonproject sections, both contemporaneously and longitudinally. Overall, students in project sections are doing as well or better than those in nonproject sections, measured either by final grades or grades on a common final. This is important, because some students feel they are at a disadvantage taking fewer exams in a project section. Even more importantly, our statistics support the idea that a student is not put at a disadvantage in moving from a project section to a nonproject section the next semester.

We also obtain student evaluations of the program. When students drop or change sections they are required to fill out a form saying why they are doing so. This is to see if students are specifically dropping or changing sections, because they want to avoid projects. Some of them are. There are others who specifically enroll in a section, because they know it is a project section. It works both ways. Also, at the end of the semester we hand out an evaluation form in all sections of calculus, project or not. It asks a variety of questions. The replies we get are something like this: "I like the projects." "They teach me a lot." "I've never learned so much mathematics before, but it takes too much time. Therefore I'm not going to take a project section again." In summary, many students love the projects, while others want to avoid them. One reason students like projects is because they have time to think and work out details deliberately and correctly. They prefer projects to exams, where snap decisions have to be made. (See Conley et al., 1992.)

Leon Henkin. You said that some of the students avoid projects. You also said earlier that your estimation of students goes up after using projects. Would you put those two things together? That is, how does your estimation of students' abilities vary when they drop out of project courses? You might have decided that those people are no good.

Arthur Knoebel. It is always a disappointment if somebody says, "I'm not majoring in mathematics. I want to put my efforts into engineering courses." It's a disappointment, but if they have done well on their projects, I feel that we've accomplished what we wanted in that sense. When we institutionalize project courses in the next academic year, we are specifically going to give the project sections 4 hours of credit versus the traditional 3 hours of credit. This is to give students credit for the extra work. We won't ask them to do something for nothing. That may take care of the problem of students not wanting to enroll in project sections, because it takes extra time.[1]

Michael Ranney. On the question of expectations, do you have a sense that a certain type of student, a priori, might gain more from the projects course than others? Are the better students enjoying the projects, or is there a subset of poorer students that find the projects invigorating?

Arthur Knoebel. Good students always excel and get a lot out of the projects. However, poor test takers may also do well on projects. Sometimes you

---

[1] Logistically, having both 3- and 4-credit options proved difficult. Moreover, since this talk was given, we now try to have all sections incorporate project-like activities, and thus, while all sections have been enhanced, they remain 3 credits.

pick up students who really do have an examination problem, and who latch onto the projects and do better. Thus, they may move from a C− to a B grade. There are also D students whose problem is not in taking exams but in a lack of motivation, and projects help some of these students succeed by capturing their interest. However, there is a subset of students at the lower end of the spectrum who find projects baffling. We are aware of the problem but have not paid it the attention it deserves. I think we ought to. In fact, I think there's a split right about at the C level. We're seeing what we call "dividing of the waters." The students at C+ and above are getting better grades in the courses, and the C− students are doing worse. As just pointed out this is not uniformly true, and projects provide a way of rescuing some of these traditionally poor students.

Ed Dubinsky. I'd like to suggest that the bifurcation is a fairly prevalent phenomenon that's occurring in a lot of "reform" courses. One possible explanation is that you're just getting a more accurate reading of student performance. That is, in our standard way of evaluating students, there are ways a D or F student can earn a C, and you eliminate some of these.

Susanna Epp. I have a couple of questions about logistics. In terms of motivating students to take the projects seriously, how much is their grade on the projects weighted in terms of their grade for the course? How do you grade the projects of students who do not succeed in fully solving the problems? I'm curious about how you monitor how much help the teaching assistants in fact give the students and how you train them. What kind of help do they give?

Arthur Knoebel. We leave these issues up to the individual instructors. Project grades count from 25% to 50% of the course grades. We find that it needs to be a substantial part of the grade for students to do well. On partial credit, sometimes they simply do several parts correctly, not always the first few. There are projects where they can jump in the middle. We like the kind of project where if they get stumped in the first part they can go on to the next part. In this case it's just a question of saying, "Ok, you did half the project." What about the person who tries to do all the parts and sort of does something but is not sure? These are difficult to grade.

Monitoring of the amount of help students are given is somewhat difficult because of the number of people involved. Typically in a course there will be four sections using projects, taught by two faculty and two teaching assistants. We try to keep an ear out for whether the teaching assistants are giving too little or too much help. This semester I was working with a teaching assistant who was really too tight with his advice, and I said something to him. Now he is just the opposite. So, you don't get rid of all the problems working with teaching assistants. The faculty also vary. We try to talk among ourselves and come to some consensus. When creating a project, we keep in mind the kind of help students will need. Typically, a person will work up an idea for a project, then get together with another instructor, and they will bat it back and forth and go over what they think would be a typical solution. The teaching assistant is brought in somewhere along the line. He or she is expected to know how the problem could be solved in at least one way. At this point various issues come up concerning whether students could really solve the project using just what is given and what would be a reasonable amount of help. We prepare before each laboratory session regarding how much help we're going to give. Each instructor does not give the same amount, because we have different styles. We had to learn how to help students with projects in the lab and to get away from saying, "Well, here you do this, this, and this, and you solve the problem." We don't want to do that. We want to say, "You don't know how to do that? Why don't you go off and try this." Then a student comes back and says, "Ok. That worked, but now I'm stuck again."

John Addison. What about different students working on the same project? Are they encouraged to collaborate or ordered not to collaborate?

Arthur Knoebel. On the individual projects they're only supposed to look in books or to come to one of the instructors or the laboratory for help. On the group projects they should collaborate among themselves, within a group.

Susanna Epp. I found that when I've worked with projects, I've had students who just can't get anywhere on it unless I give them a big hint. Then I have students that need only a tiny nudge. It's hard to gauge.

Arthur Knoebel. It is hard to gauge. We train ourselves and our teaching assistants to give varying amounts of help according to when the project is due. One could have an emergency mode where, if a student is really stuck on one section and it's essential that it be completed before he or she can go on, one could bargain: "We'll tell you this, but you won't get credit for this part."

There is a question of whether students can discover things on their own or whether they need guidance. You have to give students some idea of where to go, but you don't want to give them too much of a detailed floor plan. Should we be telling students how to discover mathematics and harness their unconscious, or just let them loose and see if they can do it alone by giving them a project with minimal directions? Given the organizer of this conference, it seems silly to ask that question. But we are a pragmatic group. We haven't incorporated everything that's been done by other people. I tend to think we should be looking more in the direction of George Pólya (1981) and Alan Schoenfeld (1985).

To illustrate this dilemma, let me give you an example of a recent minimalist project (Fig. 5.10). In this type of project, students are given minimal information with no directions and are expected to discover the appropriate mathematics in their work. You have a sugar bowl shaped like a paraboloid, and there is an ant

An ant at the bottom of an empty sugar bowl eats the last few remaining grains. It is now too bloated to climb at a vertical angle as ants usually can; the steepest it can climb is at an angle to the horizontal with a tangent equal to 1. The sugar bowl is shaped like the paraboloid,

$$z = x^2 + y^2 \ (0 \leq z \leq 4),$$

where the coordinates are in centimeters.

(a) Find the path the ant takes to get to the top of the sugar bowl, assuming it climbs as steeply as possible. Use polar coordinates (r, θ) in the xy-plane, and think of the ant's path as parameterized by r; then find a relation between the differentials $d\theta$ and dr, and integrate this relation to get $\theta(r)$.

(b) What is the length of the ant's path from the bottom to the rim? To answer this, first discover a formula for arc length involving dz, dr, and $d\theta$ in three dimensions.

(c) Draw a graph of the sugar bowl and the path the ant takes to get out.
HINT: You may want to start with the projection of the path in the $r\theta$-plane.

FIG. 5.10. An ant in the sugar bowl.

in the bottom. Suppose the steepest the ant can climb is a slope of 1. What is the shortest path it can take to get out?

One final issue has already been alluded to: Should projects be more fully integrated into the course? There are two sides to this. The pedagogical instinct in most of us says to integrate to the hilt. Carefully fine-tune the course so that each topic leads naturally into the next assignment and each assignment flows smoothly into the next topic. This implies in our case that a textbook be written around the project idea with only those projects that meld directly into the topics of the text. But this may well be self-defeating, because only those who closely share the philosophy and pedagogical aims of the authors will use their text. My own feeling is that it is better to provide instructors with a wealth of interchangeable tools, which they can intermix as they see fit.

REFERENCES

Cohen, M., Gaughan, E. D., Knoebel, A., Kurtz, D. S., & Pengelley, D. (1991). *Student research projects in calculus*. Washington, D.C.: Mathematical Association of America.

Conley, M. R., Steussey, C. L., Cohen, M. S., Gaughan, E. D., Knoebel, R. A., Kurtz, D. S., & Pengelley, D. J. (1992). Student perceptions of projects in learning calculus. *International Journal of Mathematics, Science, and Technology Education*, 23(2), 175–192.

Pólya, G. (1981). *Mathematical discovery*. New York: Wiley.

Schoenfeld, A. (1985). *Mathematical problem solving*. Orlando, FL: Academic Press.

A Discussion of Cohen, Knoebel, Kurtz, and Pengelley's Chapter

Barbara Y. White
Ronald G. Douglas

>   **Barbara Y. White,** *University of California, Berkeley*
>   **Ronald G. Douglas,** *State University of New York at Stony Brook*
>   **Arthur Knoebel,** *New Mexico State University*
>   **Leon Henkin,** *University of California, Berkeley*
>   **Jere Confrey,** *Cornell University*
>   **Andrea diSessa,** *University of California, Berkeley*
>   **Alan H. Schoenfeld,** *University of California, Berkeley*
>   **Marcia Linn,** *University of California, Berkeley*

Barbara White. I certainly agree with Art that projects are a potentially motivating way to help students learn about the nature of mathematics. However, I think that there are a number of serious issues about project use that need to be addressed, many of which Art has alluded to. Among them are:

- What is a project—how does it differ from the problems at the end of the textbook chapter, and what are the kinds of activities that you want to include in a project-based curriculum?
- More broadly, what are the pedagogical roles that projects can play? Could you conduct an entire mathematics curriculum based solely around having students do projects?
- How do you design and sequence projects, both to facilitate students' learning and to make the teaching load manageable for the instructors?
- Last but not least, how do you assess project-based learning?

I'm going to take on each one of these issues in the context of what Art has been saying.

The first issue is: What is a project? All of the examples of projects that Art presented look much too much like standard, respectable mathematics for my taste. They all look like harder versions of the problems at the end of the textbook, or like Putnam problems. I would like to see a broadening of the range of projects. When I say broaden the range, I don't mean to include "disreputable" or "indiscrete" mathematics. I mean, consider other kinds of projects that might serve a useful pedagogical role. For example, you could have the students do teaching projects such as: "Design a series of activities for eighth-grade students to help them understand the concept of a limit." You could also get the students to do metacognitive projects, in which they would work together in groups to write reports on issues such as: "Why should we learn about integration and differentiation in this age when we have computers to do it for us?" They could also do metaconceptual projects, in which they would focus on issues such as: "What is the distinction between pure and applied mathematics?"

Arthur Knoebel. I agree. Let me note that my selection of projects has not been entirely representative. The ones I've discussed are projects that were finished and polished. We have done some new projects with calculators and computers, and I'll discuss those if anybody asks me. Projects can also be historically based. You mentioned some other things such as designing a learning module for students at a lower level. I think that would be great too. We had never thought of that. There are a lot of ways we could extend the idea of project.

Since Barbara has sniffed out what might appear to be shortcomings, I should say a few words about the makeup of our group and how we work together (or not together). Three of us are pure mathematicians, one applied, and the remaining one, me, a hybrid. Pure mathematicians are quite certain what is good mathematics, even though they may not be able to put it clearly into words. Certainly, proofs are an integral part, and math without proofs is not math for them. Most applied mathematicians would agree, although rigorous proofs would play a less important role for them. Some of the group would ask, where is the mathematical content in some of Barbara's suggestions? They want to see honest mathematics from students, not just numbers or graphs. The conservative consensus of our group is obviously dictating respectability. I personally would prefer some indiscretion or even disreputability. That is what most of our students will face in their disciplines and I think that they should be trained by us to deal intelligently with the mathematical aspects of it. Originally half a dozen consultants in other departments were recruited to give us ideas for projects. Although they generated quite a few ideas, only a few of those have been polished into viable projects, since most of us turned away from applied directions. Also, we have a cardinal rule that each project must be seriously solved by at least two members of the group. Given the unpredictability of what can happen in a project, this is a very sound idea. Unfortunately, it also has the side effect of creating a common core of conformity. I would like to see us broaden the range of projects to truly applied areas.

Barbara White. The second issue is: What are the possible pedagogical roles that projects can play? I entirely agree with Art that projects have the potential to help students learn about the form, development, and application of mathematical knowledge. I think that when you are developing projects, you need to consider their pedagogical utility from each of these perspectives. That is, you need to consider what they will require of students and teachers in terms of (1) understanding the subject matter, (2) understanding the process of creating new mathematical knowledge, and (3) understanding how to apply mathematics. This is important both in determining if your projects are covering the range of things that you want covered, and also in helping you to define the difficulty of particular projects.

Such considerations lead to this issue: How do you design good projects? I think that it is much more difficult to design good projects than it is to design good problems at the end of the chapter. A project requires students to engage in a long and complex process. You are introducing them not just to learning a piece of mathematics, but also to learning about constructing and applying mathematical knowledge. I was very disturbed by Art's statement that he found a bell-shaped curve, where good students do very well and poor students do badly at projects. I am equally disturbed by attempts to explain this result by arguing: "We get a bell shaped curve because that's the way students really are distributed and we're just measuring reality better." I would much rather see instructional designers take responsibility for the result, and say to themselves: "There's something wrong with the projects we've designed, or the sequence of projects we've given students, or the instructional supports." We should not blame the students. Instead, we should blame the instructional designers and insist that they work harder on designing good projects and making available to students a wide range of pedagogical tools. For example they could make available various kinds of computer tools for supports—such as those that Judah Schwartz and Jim Kaput have talked about at this meeting, or tools that employ AI models of mathematical knowledge and problem solving expertise to help the students (and to reduce the load of the teaching assistants).

The third issue—how do you sequence projects to optimize learning?—raises a number of considerations: (1) what is the role of structured versus open-ended projects, (2) how should one sequence applied versus theoretical projects, and (3) should projects address depth of mathematical understanding rather than breadth? One could start, for example, with well structured projects and gradually progress to more open-ended projects that require more of both the students and the teachers. One could also start with applied projects, like Art's greenhouse problem, where the focus is on modeling and application. Then one could progress to projects that involve manipulating the mathematics itself—in other words, where students are dealing with the models and the formalisms in their own right. Similarly, one could start by concentrating on one application area, such as creating models of force and motion, in order to motivate students, for instance, to derive an aspect of the calculus for themselves. Then one could progress to a number of different application areas in order to help students see the power and the range of that particular piece of mathematics. Such decisions about the sequencing of projects need to be made so that the progression of projects facilitates learning, motivation, and also reduces tutoring needs.

Arthur Knoebel. I don't have a great deal to say about other uses of projects because our efforts to this point have been limited to trying to use projects successfully in the courses we offer. If we could do that we felt we were successful. I realize we should look beyond what we have done and so I agree we should look at these things. In fact, we are already becoming more sophisticated pedagogically.

When asked how to fit projects into the curriculum, there are two extreme viewpoints in our group. I don't mean that they're necessarily opposed, because I share both of those viewpoints. One is that you try to choose a project that's right where the students are working in the curriculum. That's the most obvious thing to do. For example, the project about the ant in the sugar bowl was assigned just when they were learning about the gradient. There's even an idea that you should assign a project involving concepts which the students are about to see in their textbook, to prepare them to work with those concepts. Some instructors are even going beyond that; several projects handed out simultaneously are centered around a common theme. In fact, they are so created that when the projects are turned in an instructor can lecture on their content, tie them together, and prove further results based on them. That works rather well, except that it's an extremely difficult scheduling problem to get all the sections in a particular course synchronized. That's almost impossible to do. The other viewpoint, which is championed by Doug Kurtz, is that it really doesn't matter what the subject matter of the project is as long as the students have the necessary prerequisites. The idea is to teach them how to do mathematics. Therefore, if they come into a calculus course for the first time, you can have them do something involving high school algebra and nothing about calculus. They will learn what the projects are, and in that way, mathematics. A typical example is *composing functions* (see Fig. 5.9).

Barbara White. That leads nicely to the fourth issue, how do you assess project-based learning? If you want to make claims that students are learning how to learn, and learning about the nature of mathematical knowledge, one thing is clear: The standardized tests that we have to date do not provide measures of such meta-knowledge. This raises the need for alternative forms of assessment. Some things have been tried elsewhere, and I don't know if Art's group has tried them. There are portfolio-based approaches to assessment where the students are asked to create a portfolio of their projects to illustrate various features of their learning. For example, one feature looked for in assessing an item in a portfolio might be the evolution of a student's understanding of a particular piece of mathematics, or alternatively, the student's range of expertise with respect to a particular aspect of mathematics. Another alternative form of assessment is exhibitions where the students have to demonstrate their project-based expertise to their teachers and fellow students. Both of these forms of assessment provide a means to assess not only the students' understanding of the subject matter, but also the students' meta-knowledge relevant to their learning and problem-solving processes.

I personally think that project-based learning is a very exciting and productive way to go. I can see its potential to motivate students and also to introduce students to what it means to be a mathematician (or a scientist or a historian). I would like to see all learning of all subjects be project-based. However, before we advocate such a thing, I think we need to give serious consideration to some of the issues that I've raised here. In that context, let me present a rather disturbing anecdote.

A number of years ago Paul Horwitz and I spent two months in a sixth grade classroom working extensively with a teacher. We had the children learning about force and motion in the context of a series of interactive computer microworlds. The children were discovering things like vector addition, and they were introduced to the idea of a limit as a way of modeling continuous forces like gravity and friction as a lot of little impulses closely spaced in time. At the end of all of this we were disgustingly pleased with ourselves because we had some tangible evidence that the children had learned some significant physics and mathematics, *and* the teacher claimed that this was going to revolutionize the way that she would teach in the future. Then we had occasion a little while later to observe the same classroom again to see what was going on. To our horror the day that we went, the teacher had the children doing a project that she called Mr. Potato Head where the children were designing things with potatoes. That would have been fine if they had been doing things like building bridges and talking about the structure of bridges. Instead, they were all making cute little animals and faces. When I asked the teacher what she thought this had to do with science and scientific inquiry, she looked at me in a rather surprised fashion and said, "Creativity." So, you certainly do need to consider the above issues deeply for project-based learning to be effective.

I have to conclude by saying that I really admire Art and his colleagues for taking on a college-based calculus course in a project-oriented fashion. It's one thing to do middle school science where hardly anyone cares what you do—Mr. Potato Head is as good as Newtonian Mechanics—but it's another thing entirely to tackle calculus, especially when it's a prerequisite for so many other things.

Arthur Knoebel. On the assessment issue, one of the things that we do is post-project interviews. After projects have been graded the instructor will ask students in for a 10-minute interview. This is in addition to the contact we have with the students while they're doing the projects. For that reason we feel we have a good idea of what's going on. I don't always call in all the students but I tend to call in the students who have done poorly to try to find out why. I find that very difficult in some ways because some of them have negative attitudes towards both mathematics and mathematicians. It's difficult to get them to talk, and to try to really uncover what the difficulty was. However in general the interviews are rewarding, both for students and instructors; because of what the students tell us, we're enthusiastic. Many students say they really like the projects.

Self-assessment by students should be emphasized. Students typically treat their exams with contempt when they are handed back; after a cursory glance they conclude, for example, that it was a tactical error on their part not to have studied particular sections of the text more thoroughly, and then they toss the graded exam aside. By way of contrast, graded projects are gone over much more carefully. With post-project interviews it is possible to pinpoint precisely what a student should do in the future to do better. Some of us are finding out that it is extremely instructive for both students and teachers to have students keep logs and journals of their progress, noting their successes as well as their wrong turns.

Ron Douglas. I wanted to say some general things about what I like about this approach and then perhaps either raise some questions or mention some limitations. Next month (January 1990) it'll be four years since the Tulane conference occurred which I organized. Some of the people who were there are also at this conference. I organized the Tulane conference because I thought that the way calculus was taught at many places left a lot to be desired. In fact that's perhaps putting it mildly. I was smart enough, I think, to recognize that I could not have gotten my colleagues interested in that question, so we combined how it was being taught with what was being taught, in other words, methods with content.

Most of the calculus innovation that's going on around the country emphasizes content. That's no surprise. It's much easier to decide that you're teaching the wrong thing as opposed to deciding that you're not doing it the right way. On the other hand, this project, what's being done at New Mexico State, focuses largely on how the course is being taught. One can easily argue that they are not really grappling with that in a fundamental way because most of the rest of the course is still the conventional calculus course. Of course, as was pointed out by Judah, when you want to innovate or change something, it doesn't make any difference how good your ideas are if you can't get people to try them.

What we have here is a method of changing the way the course, or at least a part of it, is taught. It is in a way that is acceptable to most mathematicians, who are the people who teach mathematics courses. Solving problems—working on problems like the ones that were suggested here—is something that mathematicians will accept. And when the people teaching calculus accept it, they may not see it is just a start. I doubt very much that will they see that it is the proverbial camel's nose under the tent. Once you start looking at this and changing what you're doing in part of the class, it's hard to simply confine it.

We talked about what happens with the teaching assistants. It's very good. I think it also addresses another issue. Leonard Gilman raised the question, "Are the good results due to the fact that the teachers and the students are putting more time into the course or is it due to what they are doing?" From my point of view anything that we're going to do that's going to improve calculus is going to require more time. One of the big issues is, how do you get the people teaching the course *and* the students to put more time into it? So, I don't care about the answer to Gilman's question. In other words, one of the good points about this project is that people are putting more time in to calculus.

I think all of those things are good. I'm delighted that the results, so far at least, for a lot of the students and a lot of the people teaching the course, have been favorable, and that all are enthusiastic. I am also, of course, disturbed that there are students that are being left behind. On the other hand, many of the conventional calculus courses leave almost everybody behind. So, I have to assume that there is improvement here. That is certainly not the case everywhere.

One of the other things I like about the project is that it is something that doesn't require machinery. I'm not going to tell you that the computer, calculator, and everything else aren't useful, indeed they are. The fact that here is something that you can try and it doesn't require that you immediately invest tens of thousands of dollars in computer labs, and all the other things that are necessary for that, is good. As I said, once you start changing the way calculus is being taught, it's very hard to simply box it off and simply say, "Well we've gone this far, but we're not going to go any further." So I think that from the point of view of trying to change calculus at other places it's good, and I think it's also attractive from the point of view of the people teaching mathematics.

Now, the limitations are very much along the line of what's already been said. Namely, I think it would be a mistake to think that this is enough. As a way of getting started, I think it's very good. I don't think the people at New Mexico State see it as the end. I think they see themselves in a process which, although it hasn't just begun, certainly has a long way to go. One of the things I was convinced of at the Tulane conference is that what we teach in calculus probably should change. There are changes that need to be made. This project doesn't address that except in a most indirect way. I think that it would be bad if one got the idea that one could keep calculus static. In other words, that you could keep whatever it is you're doing but that you're going to put a band-aid on the course.

I think one has to look beyond this project. I think one has to look both at the way the course is being taught and at what is being taught. I think this project provides a good beginning. The first thing that is needed in order to change calculus is to have the faculty at a university or college interested in calculus. I started an article which I wrote for *UME Trends* by saying that something new was happening: People are actually talking about teaching in common rooms around the United States. Now, perhaps that's an exaggeration, but it's not an exaggeration to say that up until recently people were not talking about teaching.

If you're not talking about it and you're not interested in teaching at a given college or university, you're certainly not going to do anything. One person or even a small group of people cloistered in an office or a computer lab or someplace else is not going to effectively change instruction. I'm talking about calculus because we're talking about calculus here. I talked about calculus four years ago because I wanted to focus on something. Actually, my interest is in the whole undergraduate curriculum and indeed on K–12. Lastly, I think there is one thing that one has to be very careful about and that is the issue of oversell. In the presentation that was made here, there were no false claims, no one was selling snake oil. This isn't a cure-all and it isn't advertised as a cure-all. But, I think one should understand that there are, at this stage, limited but extremely important consequences that come out of this project.

As far as evaluation is concerned, my concern is not the kind of evaluation that's been talked about so far. My concern is very pragmatic. I'm interested in what happens to the students after they take this course. For example, do they take more math courses? Do they stay in a math-related curriculum? Do they go on to be majors in economics? Do we get more math majors? Are more going to go to graduate school in mathematics? In other words, I want to know the answers to very pragmatic kinds of questions which are just as difficult to answer as the kinds of questions which have been discussed. I've raised these issues when I've spoken to people at New Mexico State because these are fundamental questions for everybody who's changing calculus. That's because there is a relationship between the need for more math majors and what we do in calculus. I haven't seen statistics to prove it, and I'd be very interested in such, but I think there is almost a complete turnover of the students coming in, wanting to major in math, and the students who go out that do major in math. If you accept that, it means that we basically recruit our math majors in calculus. The question I usually throw out to most people is, "If you took the calculus course that your department presently offers, would you have become a math major?" So, that's perhaps a harder test for calculus reform.

Arthur Knoebel. Let me just react to what you said. It's very complimentary. I'll accept the compliment. On the long term effect, we have seen some indication that the number of our math majors is increasing. We can't tell over the short period of time if that's significant or not, but it seems to be the case. There are students who really are getting turned onto mathematics who are not necessarily in mathematics. For me, as somebody who's half engineer and half mathematician, whether somebody studies math as a math major or as an engineer is immaterial. I do know there are engineers who find out in their senior year that mathematics really is interesting, that there's something going on there. "I really think like a mathematician more than an engineer, but gee, it's too late to change." I think you're quite right, that if we can change the way we teach calculus, so that they can become enthusiastic about mathematics at an earlier stage, then they'll realize that they really are mathematicians and will go into the right field.

Ron Douglas. I'm also interested in more students not just becoming math majors but staying in the sciences, staying in economics, staying in engineering and so forth, and indeed attracting more students in to the sciences.

Leon Henkin. The question of poor students and good students has come up. I want to say that when Art began describing the project system, it reminded me of a program that was devised some years ago at Berkeley, especially for poor students. The Professional Development Program (PDP) aimed at helping minority students who were doing miserably in calculus to do well. One of the issues discussed here reminded me of that. It wasn't projects, but it was getting students to work much harder problems than the ones in the textbook. That was a major component of PDP—and PDP was successful.

Abstracting across both programs, there are two aspects that perhaps explain what makes the use of projects or hard problems successful. One has to with what we were talking about after Alan's talk. That is, we want students' experience of mathematics to be something other than using recipes to get numerical answers. That's involved in the hard problems. What makes a hard problem hard is that you cannot take the topic that was just talked about in this section of the book and plug in numbers and get the answers. The second thing had to do with emotions. We hardly ever talk about them, but they are a tremendous factor in the learning process. It's not just the ideas, it's how you feel about them. And how you feel about yourself. I think that when you get people to do hard things, they feel good about themselves. They see that they have the power to do much more than they have been doing up until now. I think that's certainly true, even more true for poor students. We should challenge them, not make it easy for them. I think we can bootstrap students to higher and higher levels, for I think that all students are underachieving tremendously compared to their potential.

Jere Confrey. I think it might be interesting to have students make up projects. It would be interesting to see what they come up with once they get intensely involved in that. Have students make up projects with certain specifications, some interesting projects that use certain things, and make your friends use certain things.

Arthur Knoebel. We've done a little bit of that. There's a way you can design a contest. I'll give an example you can see applies to lots of concepts. The idea is that students design their own series, and then the other students are supposed to find out whether the series converge or diverge. Of course, the student who designs the series has to be able to prove that the series behaves the way he or she claims it does. Then you set up a system to grade students on how many of the series they can figure out and how many they are stumped on.

Andy diSessa. Along the same lines, I've run project-based courses at the freshman level. I get this image of the way projects go. With ordinary problems, kids have to figure out how to fit themselves into this particular box. There's no place to go. With projects they can rattle around a little bit and do some strange things that have them really becoming more mathematician-like to determine where they're going to go. Still there's a place to go, which is something like the right answer. If they don't get there then it's a failed project in a certain sense.

If you give students more freedom along some dimensions, you find that they actually do very different things. For instance, they'll invent their own project. They'll say, "That that was ok, but I really want to do this other thing." Or, once they're done with your project, that springboards them into some other new kind of thing. I'm wondering if you had any experience with kids coming back with something new, new results or new problems of their own. Did their efforts spin off into that kind of territory? I'm guessing that with the scale of projects that you ask kids to do, and the limited amount of time that they have to work on them, that they can't get into that kind of territory.

Arthur Knoebel. No. For the most part they don't. Some kids start asking questions that I don't think they would normally ask about material. But no, they don't seem to get off doing their own projects. However, we are devising more and more extra credit challenges to our projects, and these are often very open-ended, designed to encourage the kind of exploration you suggest.

Andy diSessa. We did one project per term, so the students had a lot more time.

Arthur Knoebel. We're under constraints in that we're expected to cover certain topics. This certainly makes it difficult to get into open-ended things. Also, let's say you're teaching two sections of calculus and you have ninety or one hundred students, and you have them go off and do something open-ended and each one turns in a ten-page paper doing something different. How do you grade all these, particularly if you don't know all the answers yourself?

Andy diSessa. I actually don't think coverage is that much of a problem, because you can do things in the ordinary way and have projects on the side. Grading might be more of a problem.

Alan Schoenfeld. I want to make two brief comments about grading. The first is to agree that it's much, much harder to give precise and accurate grades for projects than it is for tests. The second is to say that it really doesn't matter. That's true in two ways. First, I'd argue that we put much too much emphasis on grades to begin with. In my opinion, to worry about trying to grade projects in the same ways we grade tests—breaking them down into two points for this, one point for that, is to invest our energy in the wrong directions. Second and more importantly, issues about grading minutiae become moot when you have the students doing honest-to-goodness mathematics. When you have real projects that provide kids the opportunity to do some real mathematical explorations, they get a surprisingly accurate sense of what they accomplished and what they didn't accomplish. You can give a ballpark grade and they're perfectly happy with it.

My problem-solving course is about as open-ended as you can get, and my grading standards are pretty ill-specified. Yet, I never have any complaints about the grading. Typically, I go through a problem and say, "Well this one's worth about six out of ten." The students almost always agree. The fact of the matter is, they know how much work they've put into a problem, and how successful they've been. It's no longer an issue of "I almost solved this equation, and if I had I'd have gotten an extra two points." When you're dealing with issues of substance, grading at the fine-grained level turns out not to be much of an issue.

Earlier the issue was raised of what do you do when you have ninety students who produce a wide range of projects. Sitting behind me, Marcia Linn said, "You become an English professor." I think that's true. Conceived this way, the enterprise is different. Note that English departments have composition sections where poor teaching assistants get to do most of the grading! It is labor intensive, but that's the name of the game. If we're interested in mathematical literacy, then we have to take on some additional burdens. As Ron said quite directly, what we do is dismal—but it has been cheap. It's going to take more out of our hides if we want to do it right.

Arthur Knoebel. Most of us have done subjective grading and feel comfortable with that. I've never tried it because I don't think I would feel comfortable with it. I will say the response of students to being graded on projects is much more positive, even typically when they don't do well, than it is on exams. You just don't get bargaining for a point here and a point there. Even when I break down the scoring and say "You got two points here, and three points here," they say, "Ok, I understand where I fell down."

Marcia Linn. One thing I found interesting in working with the English department on campus is that they have moved to what they call a holistic grading scheme. They don't give points on individual things, but they have demonstrated good inter-rater reliability by setting up a narrative description of a scoring rubric, delineating their expectations for work that rates an A, a B, etc. They also say that they don't get a lot of flack from students regarding such holistic grading. In fact, they find it far better than when you say, "I took off a point for grammar here, two points for a conceptual issue here." So such grading seems an interesting and workable idea.

The question I had concerned the kind of reasoning processes that students need to use, in order to work through longer problems such as the ones you're asking them to solve. I've been working in programming, having students write very long programs in introductory computer programming courses. Typically they have to work in teams to accomplish that. One of the things that we found was that there's a limit to students' ability to break down complex problems. We went to writing case studies, which are in the form of an expert commentary on the solution to an analogous problem. That's been quite successful in helping students understand the sort of thinking skills that are required to work through this kind of a complex problem. I'm curious in your course how those skills get inculcated into the group. I think that it might make a larger portion of the students feel comfortable with these projects if they had a model for the kind of reasoning that was necessary.

Arthur Knoebel. We do some of that. We took a correct solution and then made mistakes in it and then we showed how these were corrected. That's as far as we've gone.

Marcia Linn. I'm talking about something more extensive. What we do is model how you might think about a solution. "How do I think about this problem? Well, I have these things to deal with. Now, what exactly can I do with them? How do they fit together? . . ."

Arthur Knoebel. I think we should start discussing strategies with students. I think that would help.

6 A Theory and Practice of Learning College Mathematics

Ed Dubinsky
Purdue University

The metastructure of this chapter is what might be called *binary synthesis*. Theory and practice, what you know and how you know it, what you do and why you do it, research and development, epistemology and pedagogy, how you learn and what you learn—there are many dichotomies and in each case I am interested, not in choosing between them or finding compromises, but rather in building syntheses between two apparently disparate notions.

In mathematics education there must be a theory of how people learn and it must be constructed in conjunction with the practice of regular classroom activity. This is the first dichotomy. Theory and practice must be developed more or less simultaneously with each taking from, and contributing to, the other. In this chapter, I elaborate on this point by considering two additional dichotomies: first, research and development, and second, beliefs and choices.

Consider, for example, research and development. Research has to do with studies of how people learn (or don't learn) mathematics, what actually goes on inside the mind of an individual when he or she tries to understand a bit of mathematics. Development is about projects to change a curriculum—its content and/or delivery, the dynamics of the classroom, the societal pressures for, and implications of, change (or lack thereof).

Henry Pollak has raised the following issue: What is the appropriate balance of research and development in the field of mathematics education. At this particular point in time, which should be primary? In the first section of this chapter, I argue that this is the wrong question. Instead of balancing and choosing we should be synthesizing.

There is another way to think about it, which I also consider in the first section. Which is more important, the content of a course and how people learn that content, or the instructional methods that we use in the classroom? Within the profession there are many differences about content and which teaching methods one should use. We also have competing views about just how the learning process works. Which of these questions are the most important, and how do we decide them? Again I will argue that none of these issues are isolated. They are all interrelated, and our beliefs about one thing tend to determine our choices for another.

In the remaining two sections of the chapter I return to theory and practice. I try to make the generalities in the first section more specific. In the section on Theoretical Perspective, I describe the beginnings of a theory that I am trying to construct and how I go about doing that. In the section on practice I briefly mention some examples of instructional treatments derived from this theory and indicate some of the results of their implementations.

In getting specific it becomes necessary to make some restrictions and also to deal with details. My restrictions have to do with the content and the students. I am interested here in postsecondary-level mathematics as opposed to the pre-college or graduate levels, and my concern is with the "average" student as opposed to the student with exceptional talent in mathematics or the student with special difficulties in learning mathematics. Actually, all three groups of students—exceptional, average, and weak—are important and should be considered very seriously in mathematics education research. I do not suggest, however, that my approach is automatically applicable to all three. It was designed with the middle group in mind, and it is in that context that I will discuss it.

The details have to do with the use of computers, and I tend to suppress them in this report. It turns out that having students make constructions on a computer (program, write code, modify programs, etc.) can be used to implement the approach described here. I use a programming language (ISETL) whose syntax is very close to the language of mathematics and leads to very few programming distractions. It also has the advantage that most people familiar with mathematics can understand it at the level required in this chapter without any specific explanations of what the symbols mean. Hence, I omit any such explanations.

Finally, there is the question of results. In several studies our group has produced evidence suggesting that the approach described in this chapter is practical and can lead to significant improvements in student learning. The specifics are laid out in a number of papers, and there is no point in repeating them here. These results are mentioned in the last section of the chapter on conclusions, and references to the literature are provided.

TWO DICHOTOMIES AND THEIR SYNTHESES

I consider two kinds of dichotomies: research and development, and beliefs and choices.

Research and Development

As indicated earlier, I will argue that research and development should not be considered as separate or even complementary, but rather as different parts of a single whole. Here are three arguments based on practicality, urgency, and theory.

Practicality.

In the present state of the art it is impractical to work on research or development alone, because each is too weak to stand on its own.

We all have the idealistic notion (from our myths about the physical sciences) that one can perform small experiments in isolated situations and generalize from them to things that will "work" in the classroom. I think that for now and the foreseeable future neither our understanding of the learning process nor the sophistication of our experimental designs is advanced enough to warrant such a vision.

Conversely, the inadequacy of going into a classroom and trying new methods and working with new content in the absence of any theoretical understanding is probably one of the major reasons that educational experiments do not work, or at least not when tried by people other than their originators. One might go even further and suggest that such "nonprincipled" development is how our present sorry state of affairs, with traditional methodology and content and students mainly not learning mathematics, came about.

Moreover, research and development *do* inform each other, and it is often most efficient to let them do so early in the process.

Urgency.

Mathematics education at the college level *is* in a sorry state. Students are turning away from mathematics, and those who do stay do not seem to learn very much. Our students do very poorly on national and international assessments. Our school teachers seem almost to be afraid of the subject. Industry complains and does its own teaching to make up for employees' mathematical deficiencies.

We cannot wait for clean paradigms in which to work. We have to plunge right in and find out something about specific difficulties students have. What are they? How do they occur? Can they be avoided or must they be allowed to develop and then be overcome? How can we do it? These are questions that mathematicians are beginning to ask education researchers.

The answers to these questions must be good answers. That is, they must be plausible, they must be implementable on a large scale, and the results of their implementation must represent significant improvement in student learning. This daunting, but essential task is a challenge for mathematics education. It can not be met by research or development alone—it needs both.

Theory.

Finally, there is a theoretical argument. I propose in this chapter a theoretical perspective that guides our work. One of the lynchpins of that perspective is that the nature of knowledge is not separable from its acquisition. Mathematical knowledge is not a *thing* that you have, but an *activity* that you (might) engage in. This may be what Dewey was trying to tell us when he suggested that we "learn by doing" (Archambault, 1964) and what Piaget was suggesting when he said that "to know an object is to act on it. To know is to modify, to transform the object. . ." (Piaget, 1964, p. 176).

Beliefs and Choices

Beliefs and choices present another kind of dichotomy. In the case of research and development I argued that it is a mistake to treat them separately, and that mathematics educators should form syntheses. Here my point is that there exists, of necessity, an intimate relation between beliefs and choices, and that it behooves us to be aware of it.

We all make choices about what we do in a classroom. Whether we are aware of it or not, these choices are to some extent determined by the beliefs that are part of our conception of how people learn mathematics and what mathematics is.

How People Learn.

The following are four possible beliefs that one might hold about how people learn:

- *Spontaneously.* If you believe that students learn mathematics individually and spontaneously by looking at diagrams or listening to a speaker, and that little can be done to help directly, then your answer to the prime question of what we can do to help students learn might be to present material to them in verbal, written, or pictorial form, and expect them to learn it on their own.
- *Inductively.* If you believe that students learn inductively by working with many examples, extracting common features and important ideas from these experiences, and organizing that information in their minds, then your answer to the prime question might be to have your students spend a very high proportion of their time with examples.
- *Constructively.* If you believe that students learn by making mental constructions to deal with mathematical phenomena, then your answer to the question might involve a study of just what these constructions are, how they can be made, and what can be done to induce students to make them.
- *Pragmatically.* If you believe that students learn mathematics as a response to problems in other fields, then your answer might involve introducing students to many applications.

Content.

Correspondingly, there are four categories of beliefs about the nature of mathematics:

- *Knowledge.* If you believe that mathematics is a body of knowledge that has been discovered by our society (over several hundred years), and that we must pass it on to future generations by transferring it from our minds to the minds of our students, then you might present the mathematics to the students who must somehow ingest it.
- *Techniques.* If you believe that mathematics is a set of techniques for solving standard problems, then you might have your students spend most of their time practicing these techniques on large collections of problems.
- *Thought.* If you believe that mathematics is a set of ideas that individual and collective thought has created, then your teaching goal might be to help students construct these ideas on their own with, however, a great deal of guidance that will allow them to "stand on the shoulders of giants."
- *Applications.* If you believe that the essence of mathematics is its power to describe, explain, and predict phenomena in the physical world, then your teaching might be about topics in the physical sciences with emphasis on the role of mathematics.

Teaching.

It is possible to draw some inferences about what we believe by looking at what we do when we are teaching. I think it is fair to say that the overwhelming majority of mathematics teaching is based on the (perhaps implicit) belief that mathematics is learned spontaneously and inductively, and that mathematics is some combination of knowledge and techniques. My feeling, after 30 years of teaching and talking about it with other people, is that most mathematicians *say* that they believe people learn from examples (inductively), and that mathematics is a body of knowledge we have discovered. However, they *teach* as if they believe that students learn spontaneously by listening and watching and as if what they are supposed to learn is a collection of techniques for solving standard problems.

I think that all of this is wrong. I think that at least part of the blame for the failure of mathematics teaching and learning in this country (and perhaps elsewhere) must be placed on the teaching choices we have made. These choices are not wrong because they fail to follow logically from reasonable assumptions (beliefs). They *do* follow and the assumptions *are* reasonable (as I have learned from a multitude of commons room discussions). But reasonable is not enough. If we want to explain the disastrous results of our courses, the high attrition and failure rate, the low levels of understanding students bring to the science, management and engineering courses that rely on mathematics, and the apparent turning away from mathematics by the brightest and best of our students; if we want to significantly improve learning, I think that it is time to question the original assumptions, the beliefs about learning and content, and to ask if other beliefs and instructional treatments they imply might be more appropriate and effective in reaching the ultimate goal.

Indeed, we must do more than question. Perhaps the issue of content can be resolved by conferences, panels, special sessions, and ad hoc, isolated projects. The question of how people learn cannot. For this, research is needed. I don't mean the sort of mindless statistical analyses that overflow the journals in some fields. I mean hard thinking about theoretical notions and teaching experiments that simultaneously follow from these notions and that are designed to enhance and sharpen the theory. In the following pages I present such a theoretical perspective and describe some classroom experimentation that is based on it.

THEORETICAL PERSPECTIVE

A Paradigm

It is a mistake to think that a theory can be presented without background. Aside from its explanatory power, an important point to consider in evaluating a theory is its source. What is the research paradigm which has led (or is leading) to it? By paradigm I mean a set of research methods that reflect and elaborate a theoretical perspective.

In the case of the particular theoretical perspective presented in this chapter, the paradigm can be represented as a scheme such as in Fig. 6.1. The basic form of this paradigm is a circle. By that I mean that the path represented is to be traversed, over and over, on higher and higher planes of thought.

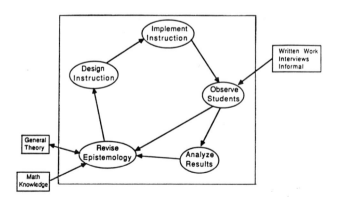

FIG. 6.1. Research paradigm.

It also means that a discussion of it, which must be linear, has to break in at some more or less arbitrary point. Let me begin with the design of instruction. Somehow or other, one decides to do something with a class that relates to the mathematical concept that is to be learned. That something can be a traditional explanation, an example, or a practice approach. Or, one can use small group problem solving or writing, or any other method. Computers can be involved as well.

This instructional approach is implemented, and the students are observed. Observations can take many forms, from informal awareness of what seems to be going on in class to formal examinations. Talking with and listening to students can be important. Interviews of some sort are probably essential.

Next, all of the data that come from the observations are organized and analyzed. The analysis depends on the data. It can consist of organizing responses into categories, counting the number of answers of different types, rating performance, or any kind of reduction to a manageable amount of information that represents at least some aspects of the entire collection of data.

The results of the analysis are then considered in light of the theoretical perspective—that is, the researchers' present understanding of what it means to understand this particular topic. Serious consideration is given to a general theory on which the particular investigation is based and also to the researcher's personal understanding of the mathematics in question. All of this is coordinated with the data and the theory as understood at the moment. If necessary, the theory is revised. It may also happen that the result tends to confirm aspects of the theory. As time goes on, not only the analysis of the particular concept under study may be revised, but aspects of the theoretical perspective are also occasionally reconsidered.

One expected outcome is that, ultimately, the theoretical perspective becomes a full-fledged theory that explains what it means to learn various topics in mathematics and suggests instructional treatments to help students succeed in that learning.

The next step in the specific investigation is that the researcher's new understanding of what it means to learn the particular topic being considered is used to redesign the instructional treatment, and the entire activity is repeated, perhaps at a later time with a different class. The iterations continue in an attempt to converge on a better understanding of the student's construction of this topic and how instruction can help her or him make that construction. In particular, it is also expected that the effect of the instructional treatment on student learning improves as the paradigm is iterated. Ultimately, this is the real test of the theory and the paradigm.

The Nature of Mathematical Knowledge

The paradigm just described is very general. I try to explicate some aspects of it in the remainder of the chapter. One aspect is the theoretical perspective I am using. I begin with a brief statement that I believe encompasses much of what is the nature of mathematical knowledge and its acquisition: A person's mathematical knowledge is her or his tendency to respond to certain kinds of perceived problem situations by constructing, reconstructing, and organizing mental processes and objects to use in dealing with the situations.

Actually, this is a very general statement that one might try to apply to any kind of knowledge. It becomes specific to mathematics when one begins to elaborate the particular constructions and reconstructions that are made and the processes and objects that result from them. I give some examples of this later in the discussions of mathematical induction, quantification, and functions, but first, there are some comments to make about the general statement.

There are several points that need to be elaborated. First, there is the idea of a *tendency to respond.* It refers to the fact (observed by many researchers and, I am sure, familiar to all teachers) that a person may respond in different ways at different times and in different places. At one moment, an answer convinces you that the student really does understand what is going on. At perhaps the next moment, and with what seems to you to be the same question, the student displays what can only be described as "blissful ignorance." A little later, the answer can be better. What do we say about the student in toto? Does he or she "know the material" or not? What grade do we give?

My reaction to this is that we just simply must accept the fact that the existence of a certain kind of knowledge in a person does not imply that he or she will exhibit that knowledge all the time. The implications of this conclusion for testing are, of course, very disturbing.

Next, there is the *perceived problem situation.* A respondent will answer the question he or she thought was asked or perhaps would like to have been asked. In any case there is no guarantee that this perception of the question is the same as that of the person who posed it.

There is also the idea of *constructing* or *reconstructing.* The point here is that a person's knowledge is not static. It is rather the case that each understanding is put together each time it appears to be necessary. Some times this *re-presentation* results in something quite similar to what has been used before. In other cases, because of the nature of the problem situation, it may be necessary to adjust it and introduce variations. This is a reconstruction, and it is the main way in which knowledge grows.

Finally, there is the question of what is constructed. Up to this point what I have presented is not very different from the constructivism of Piaget (as described, for example, in Beth & Piaget, 1966) called *radical constructivism* by von Glasersfeld (1985).

Applying this perspective to mathematics (or any other subject) consists of (at the appropriate step in the paradigm) determining the nature of the specific processes and objects that are constructed and how they are organized when one studies mathematics. The results of this can sometimes be expressed in a quite compact form called a *genetic decomposition* of the concept. As will be seen in the next section and in the remainder of this chapter, it is the nature of these objects and processes that gives the theory a mathematical flavor—if that is the area to which it is being applied.

Processes and Objects

A general discussion of mathematical objects and processes is now provided. This will be related to specific mathematical topics in the rest of the chapter.

Figure 6.2 shows the various ways in which objects and processes are constructed. These means of construction are called *reflective abstractions*. They are discussed in some detail by Dubinsky (1991).

Again there is a circle that is repeatedly traversed, which must be broken into somewhere, so I begin with mathematical objects, such as numbers, geometric figures, sets, and so on. An *action* is any repeatable physical or mental manipulation that transforms objects to obtain objects. An action requires a definite recipe, which the subject is aware of following. The steps of the algorithm and the objects at all phases of the action must be present, either physically or in the imagination of the subject.

When the need for an explicit algorithm becomes less necessary and when the total action can take place entirely in the mind of the subject, or just be imagined as taking place, without necessarily running through all of the specific steps, the action has been *interiorized* to become a *process*. At this point it is possible for the subject to use the process to obtain new processes, for example, by reversing it or coordinating it with other processes.

Finally, when it becomes possible for a process to be transformed by some action, then it has been *encapsulated* to become an *object*. According to this theoretical view, although there are several ways to construct processes (interiorize actions, reverse or combine processes) there is only one way to make a mathematical object—by encapsulating a process. The importance of this lies in

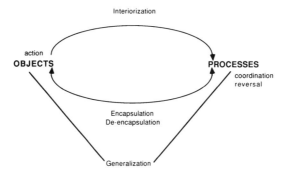

FIG. 6.2. Reflective abstractions.

the fact that in many mathematical situations it is important to be able to go from an object back to a process. One of the tenets of the theory is that this can only be done by *deencapsulating* the object, that is, to go back to the process that was encapsulated in order to construct the object in the first place.

The presence of generalization in Fig. 6.2 is just an indication of the fact that in some cases, even when the situation is new, little or no additional construction is necessary. Existing objects and processes can be used to deal with the new situation. The only learning that takes place here is that the tools that one already has can be used to handle a new situation.

PRACTICE

I now turn to specific examples of mathematical topics. I consider mathematical induction, quantification, and functions. I give indications of some of the difficulties students have with these topics, a genetic decomposition (when one is available), instructional treatments that have been used in implementing the research paradigm, and some indication of results that have been obtained. There is not, unfortunately, room here to provide very much material on the observations of students, although those data form a major source for what we do discuss. For more details on all of these, the reader can consult the papers in the bibliography.

Mathematical Induction

Most people who teach induction have their students discover and then verify closed-form expressions for finite sums. It is not clear how much understanding is necessarily involved when the content is restricted to such a narrow class of problems. It is possible to learn how do them just by imitating the behavior of the teacher. Students encounter considerable difficulty if they are asked to go beyond such problems and work on questions like the ones listed in Fig. 6.3.

I also use a number of problems (not exemplified here) in which the students are asked to both discover and prove the results. A particular kind of problem which has application to computer science is the discovery and proof of "loop invariants" in programs. The research on this topic has led to a reasonable genetic decomposition of mathematical induction that has proved useful in designing instruction. This is shown in Fig. 6.4.

Let us consider some of the ways in which the general theory is illustrated in this example. Fig. 6.4 refers to the act of interiorizing the process of a function which assigns to each integer a proposition. Of course, in order to do this, there must have been earlier an encapsulation that made propositions objects. Consider, for example, the implication. This is really a process (modus ponens) that

Prove the following statements by induction. In each case, if it is not true for all values, show it for all suffiently large values.

1. $11^{n+2} + 12^{2n+1}$ is divisible by 133.

2. Any integer composed of 3^n identical digits is divisible by 3.

3. If the prime integer p divides the product of a finite set of integers, then it must divide one of them.

4. For n a positive integer and $x > -1$ a real number,
$$(1+x)^n \geq 1 + nx.$$

5.
$$\sum_{i=1}^{2n} \frac{i}{2} = (n + \frac{1}{4})^2.$$

6. Let A be a set of character strings with the property that $A^2 = A$. Denote by A^* the union of all sets A, A^2, A^3, ... (A^2 is the set of all strings obtained by concatenating two strings, each from A, and similarly for any power A^n, $n = 1, 2, 3, \ldots$). Give a verbal description of the set A^* and show that A^* is equal to A.

7. $2^{n-1}(3^n + 4^n) > 7^n$.

FIG. 6.3. Inductive problems.

moves from the precedent to the antecedent. That process must be encapsulated to an object before one can understand the notion of an implication-valued function. I suggest that this might explain why students (who have not made such an encapsulation) have so much difficulty understanding the underlying notion: that in mathematical induction, one's direct work is not to prove the proposition but to establish the implication. At least it can be said that students who do seem to have made this construction appear to have less difficulty with the mathematical issue.

Much more must be done in the instructional treatment of mathematical induction. However, I narrow down our discussion and look at Figs. 6.5 and 6.6 to see how the computer is used to help students build these concepts.

Figure 6.5 shows two bits of computer code that students are asked to write in connection with the casino chips situation. The first is a procedure, and the second builds a (finite) sequence. Both represent the function that assigns to each positive integer the value true or false depending on whether or not the value n can be made with chips of denominations 3 and 5. Students who actually write and use this code seem, from discussions with them, to almost automatically construct an appropriate function in their minds by interiorizing a process.

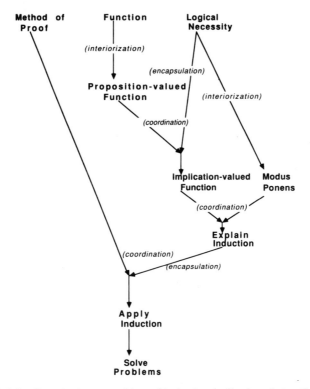

FIG. 6.4. Genetic decomposition of induction (reflective abstractions in parentheses).

Suppose that a casino has chips of denominations 3 (dollars, francs, zlotys, . . .) and 5. Which quantities can be obtained using only these chips?

1. A **func** representing the proposition-valued function that decides if a given number of dollars can be obtained using only $5 and $3 chips.

```
Chips := func(n);
            if is_integer(n) and n > 0 then
                return (exists x,y in [1..n] | 5*x + 3*y = n);
            end;
        end;
```

2. A **func** that produces a tuple containing the first k values of the function represented by **Chips**.

```
Chip_tup := func(k);
               return [Chips(i) : i in [1..k]];
            end;
```

FIG. 6.5. Proposition-valued functions of the positive integers.

Here is a **func** that takes a proposition-valued function P of the positive integers and returns the corresponding implication-valued function.

```
impl_fn := func(P);
            return func(n);
                    return P(n) impl P(n+1);
                   end;
           end;;
```

At this point it is reasonable to ask the student about the value of

impl_fn(P)(7).

FIG. 6.6. The corresponding implication-valued function.

Fig. 6.6 shows a procedure, written by the students, that converts this proposition-valued function to the corresponding implication-valued function. Again, students who perform this task seem to be able to understand that their goal in induction is to "prove the implication."

Results of Dubinsky (1986, 1989) suggest that after using this treatment, class averages on problems like the ones discussed earlier will be in the 55%–65% range or 65%–75% (depending on the strength of the students). They measure essentially full success and do not reflect "partial credit." Perhaps even more important, one can get such results when some of the problems are completely different from what the students have practiced with. It is claimed (Dubinsky, 1989) that transfer is occurring.

Quantification

One of the least considered and most important (for understanding) mathematical topics is quantification or the predicate calculus. There was an informal experiment in Grenoble a few years ago with 17-year-old students of average ability studying mathematics in the Lycée. We asked 26 students the two questions shown in Fig. 6.7.

For each of the following statements, explain why it is true or why it is false.

For every positive number a, there is a positive number x which is less than a.

There is a positive number y such that for every positive number b it is the case that y is less than b.

FIG. 6.7. Grenoble questions.

The first was answered correctly by 21 of them, but only 4 got the second correct. In interviews they stuck to their answers, and some who got the first right and the second wrong insisted that the second was a true statement, because it was "the same as the first." When the interviewer pointed out the difference in language, they claimed to have simply ignored it. We conjecture that the language was ignored, because the concept was not understood.

When a similar group was given the same questions in reverse order, again the majority got the first right and the second wrong. These results have been replicated informally on several occasions.

It is very easy to display this kind of difficulty, which underlies a great deal of the trouble students have with more advanced topics in mathematics. Concepts such as limits and continuity, one-to-one and onto, or linear dependence and independence all rely heavily, not on explicit knowledge of the mechanics of existential and universal quantification, but on the ability to work with concepts that use them.

Figure 6.8 shows a genetic decomposition of quantification. It is actually less complicated than might appear. First there is a single-level quantification. The subject begins with a proposition-valued function defined on a set and constructs the action of forming the conjunction or disjunction of all the propositions. This is interiorized to the process of iterating over all of the propositions with a "control" in the form of an existential or universal quantifier. This gives the first-level quantification.

The interiorized process is then encapsulated to form a single proposition which can be the object of an action. The action repeats the earlier one (alternating between existential and universal control) with propositions of this kind. Understanding second-level quantifications requires encapsulating and deencapsulating first-level quantifications and coordinating the process with the same process repeated with the other quantification. Repeating all this yet again leads to a third-level quantification.

Figure 6.9 shows an example of how we use computer activities to get students to make the constructions in the genetic decomposition described in Fig. 6.8.

The problem is to construct an understanding of a particular statement, which is a second-level quantification. Writing the code in line 3 is intended to get the subject to interiorize the process of a single-level existential quantification. Working with the procedure **Exi_1** helps students encapsulate this process, because the entire proposition is returned as the "answer" or result of the procedure. Embedding the proposition in **Exi_2** tends to help the subject go back and forth between the object (encapsulation) and the process (deencapsulation). The last line coordinates the new instantiation of the construction with the process/object constructed at the first level.

The results of this approach can be very encouraging. Students who have been through it display an ability to interpret multilevel quantifications, to use them to

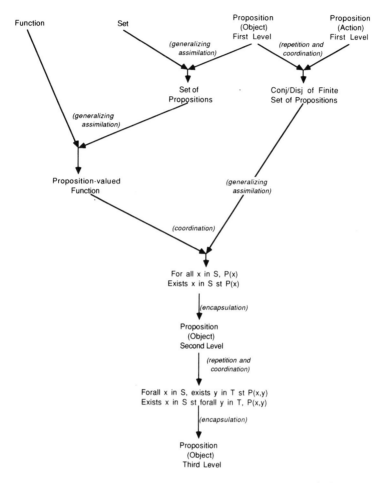

FIG. 6.8. Genetic decomposition of proposition-as-object (reflective abstractions in parentheses).

interpret a situation and even to reason about such statements (see Dubinsky, 1991, and Dubinsky, Elterman, & Gong, 1988 for details).

Functions

The concept of function may well be one of the most important ideas that students must learn in all of their studies of mathematics. Today there is a large and growing interest in this topic among researchers in mathematics education.

Most of the contributions made by these workers have been to point out some of the difficulties students have with functions. Schoenfeld, Smith, and Arcavi

For every integer a in the set $\{1, 2, 3, \ldots, 16\}$ the equation

$$ax \equiv 1 (\bmod\ 17)$$

has a solution in the set $\{1, 2, \ldots, 8\}$.

1. Evaluate $ax \equiv 1 (\bmod\ 17)$ for many values of **a** and **x** (possibly using the computer).

2. Write the following programs (fixed value of **a**) and run for several values of **x**.

```
P13 := func(x);
         return (13*x -1) mod 17 = 0;
       end;
```

3. Execute the following code.

```
exists x in {1..8} | P13(x);
```

4. Construct the following programs:

```
Exi_1 := func(P,S);
           return exists x in S | P(x);
         end;

Uni_1 := func(P,S);
           return forall x in S | P(x);
         end;
```

5. Construct the following:

```
PO := func(a,x);
        return (a*x-1 mod 17 = 0;
      end;
```

6. Construct the following:

```
Exi_2 := func(P,S);
           return func(u);
                    return exists v in S | P(u,v);
                  end;
         end;
```

7. Run the following code:

```
Uni_1((Exi_2(PO, {1..8}), {i..16}));
```

FIG. 6.9. Computer activities for quantification constructions.

	Prefunction	Action	Process	Unknown	Total Number of Students
Group A	38%	23	0	38	13
Group B	50%	17	3	30	30
Group C	35%	28	17	20	40
Group D	50%	18	9	23	22
Totals	40%	31	3	25	104

FIG. 6.10. Responses to "What is a function."

(1993) have shown how hard the concepts of slope and y-intercept can be for a single student. Goldenberg (1988) has pointed out the distortions that can arise relative to the graph of a function. It has been observed that many students come to calculus without the intuition that the limiting position of a secant is a tangent (Dubinsky, 1992). Only a few of them develop that intuition in their study of calculus.

I am in the midst of a long-term research project attempting to use the framework described in this chapter to study functions. I begin with an investigation of the process conception of function developed by mathematics majors with a concentration in education during their first two years at the university.

My initial findings, based on responses to the question "What is a function" (see Fig. 6.10) and performance on a task asking them to use functions to describe various mathematical and nonmathematical situations, suggest that most students fail to develop much of a process conception of function. They remain at

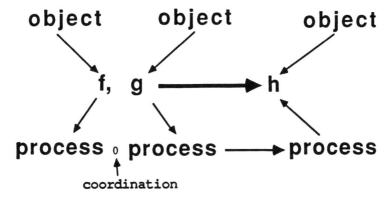

FIG. 6.11. Composition of functions.

the level of action in the sense that, for them, a function must have an explicit formula which they must be able to calculate. When faced with a two-column table of values, they search for a formula before they are willing to accept it as a function. They are not able to do much with functions other than calculate their formulas for individual values of the unknowns and substitute one formula in another.

I am working on a genetic decomposition of the function concept, and the reader may have noticed that the general description of objects and processes is very close to something that would make sense for functions. When the action of a function can be considered without an explicit algorithm, and when the totality of this action can be thought about, reversed, and composed with others, it is considered that it has been interiorized to a process. When the function can be an element of a set or the object in the various function operations in mathematics, then the process has been encapsulated to be an object. The end result is that a function can be thought of as an object or a process as need dictates.

As an example of what a genetic decomposition of the function concept may look like, consider Fig. 6.11 in which a cognitive description of the concept of composition of functions is suggested.

The way we would like students to think about this is that one takes two functions and forms their composition (if it is possible) to obtain a new function. Thus one is to imagine beginning with two objects and combining them somehow to get a new object. Before making the combination, each of the two objects must be deencapsulated to obtain processes. These processes can be linked or coordinated serially to form a new process, which is then encapsulated to obtain the resulting object.

I have designed a large collection of computer activities to induce students to make constructions like the ones just described. Figure 6.12 shows several computer representations of functions (as algorithm and set of ordered pairs), which help students to interiorize the process connected with a function. It also shows how one can construct sets of functions that help students encapsulate these processes.

I treat the question of multiple representations a little differently than do other researchers. Of course, I have students use computer tools, which let them compare several different representations of the same function. What is unique is that the students construct these tools themselves rather than only use them. Fig. 6.13 shows what they construct to convert back and forth between algorithm and set-of-ordered-pairs representations of functions.

In Fig. 6.14 I show how various computer operations (in particular, composition) with functions correspond very closely to the cognitive description suggested by the theory. Again, my hypothesis is that writing these programs results in students making useful mental constructions relative to the mathematical concepts involved.

This instructional treatment seems to have a powerful effect. Figures 6.15 and

The two functions *F* and *G* each having the domain {1, 2, 3, ... , 10} and defined by,

$$F = x^3 - 17$$
$$G(x) = \text{The integer part of } \sqrt{x+12}$$

can be represented on the computer as follows:

```
F  :=  func(x);
          return x**3-17;
       end;

G  :=  func(x);
          return fix(sqrt(x+12));
       end;
```

They can also be represented on the computer as sets of ordered pairs as follows:

```
F  :=  {[x, x**3-17] : x in 1..10};
G  :=  {[x, fix(sqrt(x+12)] : x in 1..10};
```

or

```
F  :=  {[1,-16], [2,9], [3,10], [4,47], [5,108], [6,199],
        [7,329], [8,495], [9,712], [10,983]};

G  :=  {[1,3], [2,3], [3,3], [4,4], [5,4], [6,4], [7,4],
        [8,4], [9,4], [10,4]};
```

One can then write

```
S  :=  {F};
T  :=  {F,G};
```

FIG. 6.12. Computer activities for constructing sets of functions.

```
conv1  :=  func(f,D);
              return {[x,f(x)] : x in D};
           end;

conv2  :=  func(S);
              return func(x);
                 if x in domain(S)
                    then return S(x);
                 end;
              end;
           end;
```

FIG. 6.13. Computer programs to change representations of functions.

Composition

```
comp := func(f,g);
            return func(x);
                        return f(g(x));
                    end;
        end;
```

Addition

```
addfun := func(f,g);
              return func(x);
                          return f(x) + g(x);
                      end;
          end;
```

Inverse

```
inv := func(f,D);
           return func(y);
                       return arb({x : x in D | f(x) = y});
                   end;
       end;
```

FIG. 6.14. Examples of function operations.

Suppose that `int_str` is an ISETL representation of a function that takes a positive integer and converts it to a string which is its name. Thus, one could have the following terminal session.

```
int_str(237);
"two hundred and thirty seven";
```

Now consider the following ISETL **func** represention of a function F.

```
F := func(n);
         if is_integer(n) and n > 0 then
             return {[x, int_str(x**n)] : x in [1..10]};
         end;
     end;
```

1. What is the value of $F(3)(2)(4)$? Explain how you got it.
2. Find a such that $F(4)(3)(a) = $ "t".
3. Find b such that $F(2)(b)(1) = $ "o".
4. Find c such that $F(c)(1)(2) = $ "n".

FIG. 6.15. Exam questions for function as process.

Let F be the set of all functions whose domain and range are the set of all real numbers. Let D be the operation that acts on a function, say f in F and transforms it to the function f' (the derivative). Let K be the operation that acts on a function, say f in F and transforms it to the function h where $h(x) = f(-2x)$.

1. Are D and K functions? Explain. If not, can you change them a little so they are?
2. If f is the function in F defined by $f(x) = x^3$, what is $K(f)(3)$?
3. Describe the inverse of K.

FIG. 6.16. Exam questions for function as object.

6.16 show the kinds of questions on which students (not expecially strong in mathematics—their math SATs average in the low 500s) exhibit success rates in the range of 50%–70%. More details about our latest results on functions can be found in Breidenbach et al. (1991).

CONCLUSION

I presented a general discussion of my research program with examples from several projects that have either been completed or are in progress. I have not said anything about results. In fact, I have a considerable amount of data describing the performance of students who have experienced the instructional treatments described here. These include indications of students' ability to solve the kind of problems mentioned, comparison with apparent knowledge before and after the treatment, and in-depth interviews used to model what may be going on in the students' minds with respect to these mathematical ideas.

This information can be found in the papers I have published reporting the individual investigations (Ayres, Davis, Dubinsky, & Lewin, 1988; Dubinsky, 1986, 1989, in press; Dubinsky, Hawks, & Nichols, 1989). All of these results are in the context of regular university courses. The overall story that the data seem to tell is that the approach being taken is extremely effective. Students appear to be very successful on problems more difficult than what is usually given in comparable courses. There is some indication that transfer of learning takes place. I also have had some experiences in which results are replicated by teachers other than the ones who developed the material, which is another kind of transfer.

These results suggest that it might indeed be possible to design coherent instructional treatments explicitly based on a theory of learning and using the computer. It also seems possible to do this in the context of regular college-level classes, and to produce course packages that can be used by other teachers.

Finally, it appears that when all this is done, significant improvement in student understanding and problem-solving abilities can be obtained.

I hasten to add that this work is just beginning, and there is a long way to go. Using my approach, it takes a long period (at least a year or two) of intensive work to develop a particular course. I am just beginning to learn how to get other teachers to use my material to replicate our results. Ultimate success, however, will not be possible until a "critical mass" of courses are developed and a reasonable number of teachers are implementing them. This will take a major effort by many people over a long period of time.

Before mounting such an effort, there are a number of issues that will have to be resolved. Many questions remain unanswered. What is the effect of class size? How important is the enthusiasm everyone shares from doing something new and can it be sustained? How much more effort does it take by an individual teacher to teach a course using these methods? Are there students whose reaction to computers is so negative that this approach will not work for them? How much does the efficacy of my approach depend on cultural features? Does it work as well for women as for men? How about minority groups? And finally, my approach asks students to pay the price of working harder for longer periods in order to learn more. Once in my courses, I find that they are willing and able to do this, at least for a semester or two. But given free choice, in the long run, will students decide that learning mathematics is worth this kind of effort, or will they look for less strenuous paths to take in their education?

In spite of these unanswered questions, my initial results suggest the possibility that developing an approach to teaching and learning mathematics that could work in a society willing to pay the price could be successful. Whether or not we are living in such a society remains to be seen.

REFERENCES

Archambault, R. D. (1964). *John Dewey on education: Selected writings.* New York: The Modern Library.

Ayres, T., Davis, G., Dubinsky, E., & Lewin, P. (1988). Computer experiences in learning composition of functions. *Journal for Research in Mathematics Education, 19*(3), 246–259.

Beth, E. W., & Piaget, J. (1966). *Mathematical epistemology and psychology* (W. Mays, Trans.). Dordrecht: Reidel. (Original published 1965)

Breidenbach, D., Dubinsky, E., Hawkes, J., & Nichols, D. (1991). Development of the process conception of function. *Educational Studies in Mathematics,* 247–285.

Dubinsky, E. (1986). Teaching mathematical induction I. *The Journal of Mathematical Behavior, 5,* 305–317.

Dubinsky, E. (1989). Teaching mathematical induction II. *The Journal of Mathematical Behavior, 8,* 285–304.

Dubinsky, E. (1991). Constructive aspects of reflective abstraction in advanced mathematical thinking. In L. P. Steffe (Ed.), *Epistemological foundations of mathematical experience.* New York: Springer-Verlag.

Dubinsky, E. (1992). A learning theory approach to calculus. In Z. A. Karian (Ed.), *Undergraduate mathematics education* (MAA Notes No. 24, pp. 48–55). Washington, DC: Mathematical Association of America.

Dubinsky, E. (in press). On learning quantification. In M. S. Arora (Ed.), *Mathematics education: The present state of the art.* UNESCO.

Dubinsky, E., Elterman, F., & Gong, C. (1988). The student's construction of quantification. *For the Learning of Mathematics, 8*(2), 44–51.

Dubinsky, E., Hawks, J., & Nichols, D. (1989). Development of the process conception of function in pre-service teachers in a discrete mathematics course. In C. Vergnaud (Ed.), *Proceedings of the 13th Annual Conference of the Psychology of Mathematics Education.* Paris, France.

Goldenberg, P. (1988). Mathematics, metaphors, and human factors: Mathematical, technical, and pedagogical challenges in the educational use of graphical representations of functions. *Journal of Mathematical Behavior, 7,* 135–173.

Piaget, J. (1964). Development and learning. *Journal of Research in Science Teaching, 2,* 177–186.

Schoenfeld A., Smith, J., & Arcavi, A. (1993). Learning. In R. Glaser (Ed.), *Advances in Instructional Psychology* (Vol. 4, pp. 55–175). Hillsdale, NJ: Lawrence Erlbaum Associates.

von Glasersfeld, E. (1985). Learning as a constructive activity. In *Proceedings of the 9th Annual Conference of the Psychology of Mathematics Education.*

Comments on Ed Dubinsky's Chapter

Ronald G. Wenger
The University of Delaware

Dubinsky's emphasis on research and development "as different parts of a single whole" is a practical position. I'd like to emphasize the "practicality" theme in these brief comments. Then I want to explore possible implications for the design of textbooks.

I share Ed's belief about the present "state of the art." There is great risk in attempting to translate the results of small experiments into instructional practice prematurely. During this period of major concern about mathematics education, the inclination to do this is very strong: There have, for example, been a large number of aggressive suggestions about how the algebra curriculum should change in the presence of computer algebra systems. However, like Ed, I consider the perspectives and methods that have resulted from "cognitive science" research (loosely defined) very powerful *both* practically and theoretically. The validity of the constructivist emphasis is supported not only by the research community but by the experience of many dedicated classroom teachers. But teachers need help from the research community on how to effectively implement good ideas and perspectives from the research.

Ed mentioned binary synthesis as a theme of his chapter. A form of binary synthesis which deserves special attention is the relationship between conceptual entities (Greeno, 1983) and their "organization" in the learner's memory. I'm using the term *conceptual* to suggest a network in which linking relationships are as prominent as the discrete pieces of information (Hiebert & Lefevre, 1986). Dubinsky discusses related themes using the terms *encapsulation* and *interiorization*. Certainly the process—object duality in his discussion—is designed to represent both the character of the "objects" the learner constructs and the kinds of associations or linkages between those objects (and their attributes) and others that have already been constructed. But I expect teachers and textbook authors would appreciate more guidance concerning what criteria to use in their choice of the entities or processes to be developed as systematically as the examples he provides.

Perhaps the most serious impediment to the design of more effective learning environments or instructional materials is the fact that, in general, we lack principled descriptions of the forms of understanding we seek to develop in students. Certainly textbook authors do not even attempt to describe the process aspects of learning mathematics emphasized in Ed's chapter. His "genetic decompositions" of various topics such as mathematical induction provide one way of thinking about such knowledge organization issues. The even more specific "global task analyses" (Wenger, 1987) are also an effort to provide more principled descriptions of tasks, most of them quite procedural, which permit us to better describe the kinds of "understanding" we desire for students in a particular domain. More paradigms for describing the forms of understanding desired are badly needed.

A practical question to consider is: What can this talented community of people, knowledgeable about the research on mathematics learning, say to textbook authors? To illustrate the urgency of my question, let me use "graphic thinking" and "thinking to think graphically" and the textbook treatment of it.

A few years ago in a committee meeting (called, presumptuously, the Knowledge Organization Committee) in the Math Center at Delaware, I had an insight that troubled me greatly. The insight was that there were virtually no problems in precalculus textbooks, which one had to "think graphically" to solve. There are, of course, huge numbers of problems in which the student is told to graph a function. Graphs appear everywhere in the textbooks. Yet, with few exceptions, these textbooks contain virtually no problems which: (a) contain no graphic cues, and (b) cannot be effectively solved unless they are represented graphically. The following is an elementary example: At the end of a precalculus course, ask how many solutions the equation $2 - x^2 = e^x$ has. Odds are that the students, "trained" by their experience, will try and fail to solve the problem analytically.

The purpose of this "graphic thinking" example is to illustrate several points. First, even on a topic for which the mathematics community is in total unanimity concerning its importance, the learning activities in which students are asked to engage are extraordinarily poorly thought out by textbook authors. Their attention is on the "encapsulation" of the skills with the implicit assumption that the learner will not only be able to use them when told to do so but will think to think graphically even if not told to do so. For that reason I find myself very interested in the "control" issues discussed by Schoenfeld in his problem-solving framework (Schoenfeld, 1985). I think Ed's "genetic decompositions" help make some of the components more explicit. But it is not quite so clear how his more general model deals with such control issues. Having students learn to use computer environments or languages such as ISETL to encapsulate procedures and to construct mathematical objects "feels right." But my instincts are that the more powerful and versatile a representational medium is in its own right (say ISETL), the more important it is to have the learner see more than one as an instantiation of an object or process, such as decomposition/composition of functions or recursion. Students' efforts to mediate differences between two or more quite different encapsulations of a concept are important even when the medium is as versatile as ISETL. Yet each such computing environment, whether ISETL or a computer algebra system, imposes some of its own structure on learners' construction and organization of their knowledge. The choice of that environment by a teacher will influence the way the learners encapsulate their knowledge.

Another assertion regarding instructional technologies is that each new computing environment not only influences the chapter of the learner's inferences but also fosters its own misconceptions. The work of Goldenberg (1988; Goldenberg & Kilman, 1988) provides a nice illustration of this point—the context of computer-based graphing software. That we are not yet as aware of the possible misconceptions of using environments such as ISETL should not lead us to be too complacent about the potential difficulties.

Schwartz made a useful distinction about kinds of software in one session: (a) software for doing things, and (b) software for learning how to do things. In some sense, Ed's instructional experiments with ISETL blends both these—with whatever advantages and disadvantages that implies.

In summary, I urge that we continue to work at developing more principled descriptions of some of those forms of understanding that have been notoriously difficult for students to learn. Dubinsky's work helps in that enterprise. If we do not start to use such descriptions as genetic decompositions or global task analyses to guide the development of tasks for students, we should not be surprised if textbooks continue to be permutations of their predecessors, and that students' understanding does not improve. Writing textbooks will remain an art. But there is much to be gained by having people from mathematics who have cognitive science perspectives collaborate more effectively with cognitive scientists who have strong mathematics interests. Such collaboration is likely to result in the design of better instructional materials.

REFERENCES

Greeno, J. G. (1983). Conceptual entities. In D. Gentner & A. L. Stevens (Eds.), *Mental models* (pp. 227–252). Hillsdale, NJ: Lawrence Erlbaum Associates.

Goldenberg, E. P. (1988). Mathematics, metaphors, and human factors: Mathematical, technical, and pedagogical challenges in the educational use of graphical representations of functions. *Journal of Mathematical Behavior, 7,* 135–173.

Goldenberg, E. P., & Kilman, M. (1988). *Metaphors for understanding graphs: What you see is what you see* (Tech. Rep. No. 88-22). Cambridge, MA: Harvard University, Educational Technology Center.

Hiebert, J., & Lefevre, P. (1986). Conceptual and procedural knowledge in mathematics: An introductory analysis. In J. Hiebert (Ed.), *Conceptual and procedural knowledge: The case of mathematics* (pp. 1–28). Hillsdale, NJ: Lawrence Erlbaum Associates.

Schoenfeld, A. H. (1985). *Mathematical problem solving.* Orlando, FL: Academic Press.

Wenger, R. H. (1987). Cognitive science and algebra learning. In A. Schoenfeld (Ed.), *Cognitive science and mathematics education* (pp. 217–252). Hillsdale, NJ: Lawrence Erlbaum Associates.

Comments on Ed Dubinsky's Chapter

Andrea A. diSessa
University of California, Berkeley

We all come to the discussion table with points of view that have been developed by our individual experiences and predilections. Points of view are not things that we "know" and can use simply to affirm or reject statements that arise from other points of view. But, neither should we hide our perspectives. I am a firm believer that science evolves by coordinating and mutually refining multiple points of view, and by developing new views when prior ones have fatal flaws. So, while my comments might appear in some respects to say, "Ed got this right and this wrong," my intention is rather to present an alternate point of view, hoping that confluences and divergences will help us all to see beyond our present ideas. Except for the rhetorical form that places my views as comments on his, Ed's point of view is also an implicit critique of mine on each issue where we differ.

WHERE WE AGREE

Let me start where Ed's and my points of view seem nearly entirely aligned.

Research and development should be synthesized, articulated, not "balanced" or opposed. I couldn't agree more. I have spent a great deal of time self-consciously working to create a personal research agenda that intertwines for optimal effect: making things (e.g., computer environments), working with students and teachers, and, in complementary manner, building theories and other "more researchy" activities.

To Ed's list of three arguments in favor of joining research and development, I would emphasize and add the following. First, I believe education will always be an area of complex design. As with building airplanes or designing buildings, there is plenty of need for science at the core. But designing for the complexity and multiple goals of real-world contexts always puts a stress on artful compromise, on invention, on having the right "material" with which to design (like computer technology in education), on understanding the particular properties of that material, and exploiting them. Those who expect education to be in any respect a simple applied field of general scientific principles, I believe, are simply mistaken. So, if we are to be educators, we will be designers as well as "pure researchers."

But do we need to be educators to be education researchers? I have two arguments that we should be. The first comes from my long-cultivated belief that any understanding arises from experience and from the intuitive knowledge that results from experience. We need to ground ourselves broadly and deeply in experiences of learning and watching others learn to begin the road to "high" science. Of course, we could do this without participating in educational design. But that seems likely to be an impoverished experience.

The second reason to join the fray of educational design is a sense for the scale and complexity of the scientific problems we have to face. This is akin to Ed's argument of practicality but emphasizes fundamental scientific aims. I believe that the cognition of learning and conceptual change is an effect of large and complex systems of human knowledge. In order to understand change, you need to understand a lot about the particular system of knowledge involved and the context in which that knowledge evolves. This requires big experiments that triangulate significantly on features of prior knowledge of students and on features of the experience we hope will foster learning. Every experiment needs to know about its context, and if contexts are complex, then experiments automatically become complex. If we add that the expense of big experiments probably should be amortized by some hope of practical effect, we put ourselves very much into the range of rich development, formative and other kinds of testing that *seem* more characteristic of engineering than science. We are more familiar with this effect in physics, where no one questions billion-dollar accelerators as part of experiments on fundamental physical principles, or where there is a mix of fundamental and "practical" orientation such as that in nuclear fusion research. We should get used to big experiments as a matter of fact in education as well.

Finally, let me add a most urgent plea along other lines Ed plotted, that we cannot expect any of our educational efforts to accumulate or converge without serious thought given to the creation and testing of theories. Even if we start from education as development or design, we should wind up doing *both* development and fundamental research for the sake of future designers and developers. Otherwise, we will not have defined our terms or specified the generality or limitations of our prescriptions sufficiently to help others design. The co-development of theory is a critical lack in too much contemporary education work, and it worries me when I wonder about the future of present "good ideas" that have inadequate theoretical foundations. I am happy to see researchers like Ed working on this and calling our attention to its importance. I have written extensively in other places about education, theory, and design (e.g., diSessa, 1991a, 1991b), and I refer readers to these for details.

The computer can be a powerful instructional instrument, especially when students have control of it as a representational medium. Ed's use of ISETL is similar to a mode of using computers in instruction that I recognize and strongly endorse.[1] I would schematize it in the following way. Think of computer programming as adding not just "another representation," but as a particular representational system that is so extremely flexible, so broadly applicable that it effectively counts as an entirely different language in which to express mathematical and scientific ideas. It is a different language not just as French differs from English, but in having a really different sort of expressive power that is characterized most easily by the language's capacity to build dynamic and interactive representations (think of a simulation as a simple example). Such representations allow reflection and consideration joined with action and feedback that is radically different from that afforded by "reading" or "hearing."

Ed and I differ a bit on what sort of breadth one should be striving for in choosing a programming language. My preferences run to languages that are aimed more broadly than mathematics (diSessa, Abelson, & Ploger, 1991), but the principles are roughly the same. Ed is also certainly right that his work differs, I judge profitably, from standard-form talk of "linked multiple representations," which is popular in mathematics education these days. But he may be a bit hasty in describing his work as "unique." I see some important facets of his work in the context, for example, of much that has come from the Logo community, in which programming is treated as a useful representational language for students to employ generally in expressing and exploring ideas in mathematics and science. Some recent work from our group in using programming as a representational system in biology and physics are presented in Ploger and Lay (1992) and Sherin, diSessa, and Hammer (1993).

Genetic decompositions are a good idea! I don't need to tell anyone that constructivism is an idea with wide positive repute these days. Yet, it is surprising how little this affects the practice of developing educational materials. Providing a genetic decomposition, an explicit representation of the fundamental constructions and paths of development in learning subject matter, captures the constructivist orientation in a difficult task that is well worth our efforts. I would emphasize a point that was implicit in Ed's descriptions: These analyses depend critically on empirical details. They are falsifiable with reference to data in ways

---

[1] I think Ed does not believe that any of the power of his approach is aptly characterized as I do here. What follows is my own interpretation.

that "abstract task analyses" or other a priori methods of generating a curricular decomposition are not. I take this to be a notably good feature of them.

A DIFFERENT POINT OF VIEW

Genetic decomposition is a good place to make a transition from agreeing with Ed to presenting some contrasting perspectives. As Ed made clear, his analyses depend on a particular theoretical frame. Interiorization, coordination, and encapsulation are theoretical terms that appear explicitly in his charts. My different point of view questions whether these are well-founded and sufficient terms in which to describe learning. In saying that, I should be held accountable for specific and detailed critiques. But, obviously, I cannot take the time to do that here. In any case, such a response would be a better companion to Ed's more extended presentations of his underlying theory. Instead, I will try to paint a picture in broad strokes of a different perspective on learning and point out where that picture seems at variance with what shows up in Ed's genetic decompositions.

Stroke 1: The content and forms of naive knowledge are critical pursuits. There are two brands of constructivism. Both say that students construct their new knowledge out of their past knowledge through interpretation of learning experiences. But one, version G, relies on *G*eneral, powerful learning principles (like reflective abstraction) and does not pay much attention to the specific schemata that form the "naive" interpretations out of which reflection (to name one mechanism) builds more adequate ones. The other version of constructivism, version S, believes that the naive state, as well as the expert state, needs thorough analysis, both in terms of the *S*pecific naive schemata that form the grist out of which better developed ideas evolve, and also, naturally, in terms of the theoretical categories that describe this thinking—what, exactly, constitutes "intuitive" thinking?[2]

Put succinctly (but perhaps flippantly), the genetic decompositions that Ed produces for us don't look much like kids to me.[3] That is, Ed is a constructivist of type G. He concentrates on general processes of development, like *encapsulation*, and general categories of mental constructs, like *processes* and *objects*. He is not so much concerned with naive knowledge and how different contexts of

---

[2]There are numerous versions of this distinction in the literature. Ackermann (1991) contrasts Piagetian developmental approaches (sometimes described as *structural*), which are G-type theories, with "differential" approaches. "Situated cognition" likely is an S approach.

[3]Ed says his empirical work, the data from his students, speaks clearly through these decompositions. Such a difference of opinion cannot be settled in a context such as this. I urge readers to formulate their own judgments. Consider his longer papers and evaluate how they represent naive knowledge in contrast to my accounts of that knowledge (e.g., diSessa, 1993).

naive knowledge might make the process of encapsulation and the resulting objects different from case to case.

In each of Ed's genetic decompositions, he lists the constructions to be made but does not describe the material out of which the construction is made. The latter is important to know if an instructor effectively draws out the productive ideas in students' prior conceptual systems to learn most effectively. The nature of that prior knowledge is as important for us as theoreticians to describe as encapsulated objects.

Because I have not done any work in Ed's areas of instruction, I am at a disadvantage here. But let me at least suggest what prior knowledge might become involved in learning one of the areas he describes—quantification. Start with the two statements he used to test quantification knowledge.

1. For every positive number a, there is a positive number x, which is less than a.
2. There is a positive number y, such that for every positive number b it is the case that y is less than b.

I believe that one pool of knowledge that may (perhaps must) be engaged and refined to create adequate understanding of these quantifications is knowledge of games, of rounds in games, of protagonists and antagonists, of finding the right specific strategy or ploy to win a round, or a strategy or ploy of sufficient breadth to win all necessary rounds. I am suggesting that instructors might implicitly or explicitly invoke this naive knowledge and work to refine it into "genuine" quantification knowledge.

To be a little more explicit, below is a rephrasing of the above two statements in terms of this game knowledge. I will try to stick mostly to the "game-level" knowledge to highlight it, and I will insert particular mathematical details only to help readers keep track of where games and the mathematics of quantification intersect:

1. There is a game of independent rounds, but I must win each round. The form of each round is simple. I (the protagonist) am given a number a, and I need to find an x following a particular specification involving a. If, given any a, I can find an appropriate x, I win. So, in particular, I have a lot of freedom to fiddle x for each a. That is, each round is independent.
2. There is a one game, with many challenges. I must find a certain number, y, that is "powerful enough" to meet all the challenges. The challenges are roughly like the game above, except my one powerful number must win the challenge for all possible numbers, b.

Described in terms of games, the difference between the two statements is fairly simple. In the first, there are independent rounds, with lots of freedom to attack each round with a different response. The second game requires commitment to one response for a collection of different challenges.

Let me defend against misinterpretations of this little example by elaborating, briefly, the theoretical frame behind it. I claim that much of our schooled learning relies on rather vast pools of intuitive knowledge acquired from experience. These pools contain many quite abstract and broadly useful elements. The learning task depends critically on the "cognitive ecology" of intuitive knowledge elements that are available to be used in instruction. The instructional task depends in large measure on marshaling the appropriate range of intuitive knowledge elements, in some cases "loosing" them from their habitual contexts of application, and, critically, making them work effectively together in the new (instructed) context. I claim that a concern for the details of the ecology cannot be replaced simply by listing the supposed general processes (encapsulation) that turn them into "real" mathematical knowledge. Too much important detail is lost in doing this.

In pointing rather vaguely in the direction of games, I do not mean to imply that the relevant intuitive abstractions are to be found only in students' familiarity with games. But, this is likely to be one context in which we can evoke relevant ideas. Second, I do not mean to be advocating the use of metaphors or other particular strategies of instruction. Intuitive ideas can be evoked by appropriate context without explicit mention, without analogies, and even without language, although language happens to be the form I used for convenience of exposition here. An alternative would be simply to "play out" the form of the game, and, as I said, I would certainly not rule out nongame contexts from which relevant intuitive abstractions might be drawn out. I also acknowledge again that to justify my "theory," I am obligated to describe not only the content of intuitive knowledge in greater detail, but also what constitutes an "intuitive abstraction." I've done this in other contexts, with respect to learning physics (diSessa, 1993). Finally, I emphasize that it is the form of analysis of naive knowledge and its role in learning, not particular details, that is important. I am not in a position to evaluate how much of a contribution "game knowledge" may provide in a full instructional treatment.

Stroke 2: Knowledge is diverse, and its evolution depends on fine details in this diversity. I suggested this in the earlier example, and elaborate it here. S-type constructivists are naturally inclined to believe that the particulars of the naive state (and of expert understanding too!) might be very different from one topic to be learned to another. They worry that reflective abstraction or encapsulation and deencapsulation might not be profitably defined in a context-independent way. They are very conscious that different disciplines, like mathematics and physics, have not only different content, but also different sorts of naive knowledge pools that contribute to them and probably are differing knowledge architectures in experts. Physics (say, Newtonian mechanics) is a domain that draws on rich everyday experiences, and it builds causal explanations. Mathematics may draw on everyday experiences, but schematized at a very high level of abstraction; it builds structures that can be seen as necessary, not contingent on "how the world happens to be." Will knowledge construction look the same for these areas? Ed's theoretical commitments point toward a "yes." Mine suggest caution, at least, in generalizing.

Stroke 3: The capacity to engage in certain kinds of activity is a form of knowledge and a generator of new knowledge. Tracking the evolution of student capabilities at extended, felt-to-be coherent and self-motivated activity is as important as tracking more familiar forms of knowledge. In recent years my concern for intuitive knowledge as a foundation for learning subject matter has expanded to another kind of knowledge that is systematically underestimated in importance in learning. This is the culturally and experientially developed capability and interest to engage in extended, personally meaningful activities.

I have watched my sons play, draw, read, interact with peers in ways that had them learn a lot about scientific subjects in surprising ways. What has struck me about this is how much these activities were really "childlike"—not in the deprecating sense of "infantile," but in the positive and appropriate sense of fitting into their world of interests, habits, social connections, and so on. I feel they were not being "little scientists" trying to collect and synthesize literature directed toward answering a top-level question, nor were they really doing "scientific experiments." Instead, they were collecting entertaining stories they could tell their friends and parents. They were drawing for the fun of drawing, doing thought experiments as a kind of fantasy rather than to answer a critical question.

These observations have led me to consider a second fundamental continuity with which we need to concern ourselves in bringing students along the road to mathematical and scientific competence. We need to track a continuity of activities and goals as well as a continuity of knowledge. We need to start where students are in terms of activities as well as where they are in terms of concepts. Now, of course, our goals might include students engaging in scientific activities such as doing experiments, and formulating theorems and sharp definitions. But these may be far down the road of development and not where we should start at all. In my own work I have advocated design, and, in particular, design of graphical representations as an early activity that can really engage children in sciencelike activities much better than many other supposedly scientific activities (diSessa, 1992).

When I look at Ed's genetic decompositions, I try to think about the activity context. The activity context seems to be "school, roughly as usual." Certainly, Ed does not highlight activities in his core descriptions of his work, in his genetic decompositions. This even seems also a bit ironic considering that Ed cited Dewey and Piaget about mathematics being an activity, not a static chunk of knowledge. In contrast, accounts by such contemporary researchers as Saxe

(1992) are much more attentive empirically and theoretically to the structure of activities and its implications for knowledge construction.

To be fair, Ed has students writing their own computer programs, which is certainly out of the ordinary, perhaps even radical, as far as school activities go. Another possible compensating factor is that Ed is working with students older than I work with, so maybe "school, roughly as usual" is a sufficiently good activity context in which to work. With our younger students, we are concerned that standard school sets are seriously restrictive of building good continuity in activities. And the theoretical questions remain, independent of whether an instructional activity adequately handles some issues that might be more problematic in other contexts.

I can put a point on this discussion by asking a question I have asked myself with some degree of concern. What is the function of (the mathematical concept of) "function"? What real intellectual power that students can perceive do they achieve by mastering "function"? It seems to me, in part based on the history of mathematics, that the contemporary concept of function serves esoteric functions having to do with "the most general possible formulation" and "encompassing all cases (including bizarre and ordinarily irrelevant ones)." Such goals are the pride of professional mathematicians but are very far from the activity structures that students ordinarily live. I do not excuse students for thinking that a function is an equation. But that definition is more operative in the students' school activity world of analytic exercises than the "better" definitions we try to instruct.

In contrast, to take a single example, I *can* describe some excellent functions of vectors. I have seen how enthusiastic students are in making three-dimensional shapes appear and move on a computer screen, and how vector arithmetic simplifies the conceptualization and computer implementation of exciting motion-oriented simulations. Of course, not every school subject can be as engrossing as others. But it seems to me we must take into account these factors seriously as theoreticians and as designers. There ought to be something interesting to say about how the student-perceived function of function might be improved by redesigning school activity structures.

EPILOGUE

I will close where I came in. I am not complaining that Ed is not concerned with activity structures so much as that he is not articulating his commitments and building theories about these. I am not focusing on naive knowledge to say that Ed's genetic decompositions are wrong or ineffective, but to say I do not think they are the whole story. My complaints are really a way to introduce another point of view about what may be important in learning, a view toward which I have some commitment. I am exercising my research prejudices on Ed's work so that we can all reflect on choices for future work that may be unarticulated if we politely suppress our differing points of view.

ACKNOWLEDGMENTS

This work was supported, in part, by grant number MDR-88-50363 from the National Science Foundation to the author. Views and opinions are those of the author and not necessarily those of the Foundation.

REFERENCES

Ackermann, E. (1991). From decontextualized to situated knowledge: Revisiting Piaget's water-level experiment. In I. Harel & S. Papert (Eds.), *Constructionism*. Norwood, NJ: Ablex.

diSessa, A. A. (1991a). Local sciences: Viewing the design of human-computer systems as cognitive science. In J. M. Carroll (Ed.), *Designing interaction: Psychology at the human-computer interface*. Cambridge, England: Cambridge University Press.

diSessa, A. A. (1991b). If we want to get ahead, we should get some theories. In R. B. Underhill (Ed.), *Proceedings of the Thirteenth Annual Meeting of the North American Chapter of the International Group for the Psychology of Mathematics Education* (Vol. 1). Blacksburg, VA: Virginia Tech.

diSessa, A. A. (1992). Images of learning. In E. De Corte, M. C. Linn, H. Mandl, & L. Verschaffel (Eds.), *Computer-based learning environments and problem solving*. Berlin: Springer-Verlag.

diSessa, A. A. (1993). Toward an epistemology of physics. *Cognition and Instruction, 10* (Nos. 2, 3), 105–225.

diSessa, A. A., Abelson, H., & Ploger, D. (1991). An overview of Boxer. *Journal of Mathematical Behavior, 10*(1), 3–15.

Ploger, D., & Lay, E. (1992). The structure of programs and molecules. *Journal of Educational Computing Research, 8*(3), 275–292.

Saxe, G. B. (1992). *Culture and cognitive development: Studies in mathematical understanding*. Hillsdale, NJ: Lawrence Erlbaum Associates.

Sherin, B., diSessa, A. A., & Hammer, D. M. (1993). Dynaturtle revisited: Learning physics through collaborative design of a computer model. *Journal of the Learning Sciences, 3*(2), 91–118.

7 The Role of Proof in Problem Solving

Susanna S. Epp
DePaul University

It is widely accepted that the kind of thinking done by mathematicians in their own work is distinctly different from the elegant deductive reasoning found in mathematics texts. In this era of public candor people freely admit things they might once have seen as compromising their dignity. When discussing the process of mathematical discovery, mathematicians now openly acknowledge making illogical leaps in arguments, wandering down blind alleys or around in circles and formulating guesses based on analogy or on examples that are hidden in the later, formalized exposition of their work. Enthusiasm for this more human view of mathematical thinking has led some to relegate proof to the position of an *ex post facto*, often unintuitive, somewhat pedantic justification for statements already known to be true to the intuition, a dull checking of final details. The view that intuitive understanding is separate from and precedes proof is sometimes given as a reason for presenting mathematics informally during the first two college years, leaving proof to junior and senior courses.

While I wholeheartedly agree that intuition is of major importance in mathematical discovery, I believe it is profoundly misleading to suggest that proof is mere formalism. Indeed, I believe that deductive reasoning occupies such a central position in mathematical problem solving that mathematicians are often unaware they are using it, taking their chains of inferences as much for granted as their breathing. This is not to imply that the discovery process is a straightforward linear progression from problem formulation to solution or that mathematicians do not become confused and sometimes make mistakes. I certainly do. But my experience working closely and interactively with students over many years has convinced me that the "illogical" thoughts of research mathematicians are different at least in number if not always in kind from those of most of their students. I propose that learning ways to help such students become better deductive reasoners deserves the attention of mathematics teachers and researchers in mathematics education, at least as much as do other aspects of mathematical thinking.

In this chapter I first examine how a relative novice might successfully approach three elementary questions in abstract mathematics. My account of the workings of such a person's mind is admittedly speculative, but I hope you will find that in its essentials it describes thought processes that are familiar. After examining approaches to the three questions, I consider problem solving in a more general context. Finally I explore consequences of this analysis for mathematics instruction and research into mathematics learning.

THREE QUESTIONS

I chose these particular questions because, although they are simple, they were not a standard part of the mathematics curriculum of 10 or 20 years ago, and so may be somewhat fresh. For the first question, recall that a binary relation on a set is a collection of ordered pairs of elements of the set. A binary relation R on a set A is called symmetric if, and only if, for all x and y in A, if (x,y) is in R, then (y,x) is in R. The question is this: If R and S are symmetric binary relations on a set A, is $R \cup S$ symmetric?

Now even if you know without thinking the answer to this question, try to imagine how one might approach it as a naïve undergraduate encountering the idea of binary relation for the first time. One might start empirically by considering examples of symmetric binary relations and checking whether their unions are symmetric. One might look at, say, $\{(a,b), (b,a)\} \cup \{(c,d), (d,c)\}$ or the union of the unit circle and the line $y = x$.[1] Because in both instances the union is symmetric, it would be natural to try to see if the result holds in general.

To consider the question in general, one might conjure up a kind of shadowy mental image of symmetric binary relations R and S and an ordered pair (x,y) of elements A satisfying the hypothesis that (x,y) is in $R \cup S$. In actual fact, one might not even think of the letters R, S, x, and y as attached to these objects; the objects might be imagined as nameless forms with blurred outlines. For instance, one might think of R and S as pools of things, one over here and the other over there, and (x,y) as a kind of general object with one shapeless form in a left position and another in a right position, imaging the fact that (x,y) is in $R \cup S$ as implying that this object lives in this pool over here or in that pool over there.

Then one might ask: Must the conclusion that (y,x) is in $R \cup S$ also be

---

[1] There are, of course, ways to imagine binary relations other than as static sets of ordered pairs. But, for simplicity, I am assuming that these alternate representations are yet to come in the experience of the person addressing the question.

7. THE ROLE OF PROOF IN PROBLEM SOLVING   259

satisfied? By definition of union, the conclusion means that (y,x) is in R or (y,x) is in S, or, pictorially, that (y,x) lives in this pool over here or in that one over there. But we are assuming that (x,y) lives either over here or over there. And if it lives over here, then when its elements are interchanged, the resulting ordered pair also lives over here by symmetry. Similarly, if (x,y) lives over there, so does (y,x). In either case, therefore, (y,x) lives in the union of the two pools, $R \cup S$, by definition of union. Hence, the answer to the question is yes: A union of symmetric relations is symmetric.

Note that I have radically slowed down a thinking process that might take place in only a fraction of a second. But even though the process might happen very quickly, some version of each step would be likely to occur at the time the result is discovered for the first time. I should emphasize that I have been describing only the thought process that might well precede a first discovery by a relative novice. Recalling a result or even rediscovering it is often a quite different process from discovering it in the first place.

My second example also concerns binary relations. A binary relation R on a set A is called transitive if and only if, for all x, y, and z in A, if (x,y) is in R and (y,z) is in R, then (x,z) is in R. The question is this: If R and S are both transitive binary relations on A, is $R \cup S$ transitive? As with the first example, join me in imagining what it would be like to address this question as a relative newcomer to the study of binary relations. A natural way to attempt to answer it would again be to conjure up a structure of vague mental images: transitive relations R and S and elements x, y, and z in A such that (x,y) is in $R \cup S$ and (y,z) is in $R \cup S$. But this time one would ask: must (x,z) be in $R \cup S$? Again, as for the first example, one would reformulate all or part of the hypothesis using the meanings of the terms. For (x,y) and (y,z) to be in $R \cup S$ means that either (x,y) is in R or (x,y) is in S and either (y,z) is in R or (y,z) is in S.

Now the conclusion says something about where (x,z) lives. Must (x,z) be in $R \cup S$, or, in other words, must (x,z) be in R or (x,z) be in S? To answer this question, one might imagine the various ways the hypothesis could be satisfied. One possibility is for both (x,y) and (y,z) to be in R. In that case, the transitivity of R implies that (x,z) is in R, and so (x,z) is in $R \cup S$, and the conclusion would be true. Similarly, in case both (x,y) and (y,z) is in S, the conclusion would follow readily. (Note, again, that one might go through this analysis entirely through a series of mental images without any names attached. One might think that if both ordered pairs were over here, then by transitivity the order pair consisting of the first element of the one and the second element of the other would also be over here. Similarly, if both ordered pairs were over there.) But there are still two more cases—when one of the ordered pairs is over here, and when the other is over there—and as one thinks about them one does not readily see a reason why if, for instance, (x,y) is in R and (y,z) is in S, the pair (x,z) must be in either R or S. Indeed, a lack of obvious reason for the conclusion to follow in this case would no doubt make one suspect that it does not. Now imagining the relations R and S and the elements x, y, and z in a shadowy, semiformed way in one's mind, one might ask oneself whether it is possible to bring these objects into concrete existence so that (x,y) is in R, (y,z) is in S, and (x,z) is not in either R or S. One might start to flip mentally through various possibilities noting that x, y, and z could be almost any elements as long as (x,z) is kept out of both R and S. To ensure this, one must arrange that (y,z) is not in R and (x,y) is not in S, because otherwise the transitivity of R and S would imply that (x,z) would be in R or S and thus in $R \cup S$. Once one has deduced this, one can come up easily with a counterexample and therefore conclude that there are transitive relations whose union is not transitive. So the answer to the question is no.

I should note that a person might begin to answer this question by considering concrete examples chosen at random and come across a counterexample immediately. But in the ordinary course of solving mathematical problems, one does not often happen on counterexamples purely by chance. Normally, as above, one performs some general analysis to find a set of conditions that a counterexample must satisfy.

My third example comes from the subject of graph theory. In the theory of graphs, a graph consists of finite collections of vertices and edges for which each edge is associated with either one or two vertices that are called its endpoints. The degree of a vertex is the number of end sections of edges that have the vertex as an endpoint (see Fig. 7.1). Imagine the response of a person fairly new to the subject of graph theory to the following question: Is there a graph with 9 vertices each of which has degree 3?

To discover the answer to this question, one might imagine a configuration of 9 vertices and various edges and mentally flip through various possibilities, checking whether it is possible for each vertex to have degree 3. Because complete enumeration is impractical, one would again have to think in a general way about the vertices and edges—the vertices would sit in one's imagination with the potential to be joined by edges in a large variety of different ways—but one would probably also test out a number of concrete possibilities on paper. After a longer or shorter period of frustration, one would probably come to a conjecture that there is no such graph. If one were very clever, one might notice that because each edge has two end sections, the sum of the degrees of all the vertices of a graph (the total degree of the graph) is always twice the number of edges of the graph. Or, less brilliantly but spurred by frustration, one might recall seeing this

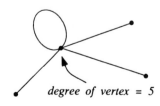

FIG. 7.1. The degree of a vertex.   *degree of vertex = 5*

7. THE ROLE OF PROOF IN PROBLEM SOLVING 261

theorem in class or as one glanced through the text on the way to the exercises. But, of course, this theorem implies that the total degree of a graph must be even. On the other hand, a graph with 9 vertices each of degree 3 would have a total degree of 27, an odd number. From this contradiction one would conclude that the answer to the question is no: There is no graph with 9 vertices each of degree 3.

PROBLEM SOLVING IN A MORE GENERAL CONTEXT

Now, consider the thinking discussed in these examples in a more general context. Problems in abstract mathematics are generally of two types:

1. What can be said about such and such? (The behavior of a function with a positive derivative on an interval, the number of zeros of a polynomial function, the total degree of a graph, the rationality or irrationality of $\sqrt{2}$, how to find the gcd of two integers, etc.)
2. Is it true or false that such and such? (If a function has a positive derivative throughout an open interval, then the function is strictly increasing on the interval; if f is a polynomial function of degree n, then f has n or fewer zeros; if a and b are integers, then $a/b \neq \sqrt{2}$; if one follows such and such a procedure for two integers, the number computed will be the gcd of these integers; etc.)

Questions of the second type can be put into the following form: True or false? For all (*elements in a domain*), if (*hypothesis*) then (*conclusion*).

The first type of question is crucial to the real development of mathematics as a field. But, as I've tried to suggest by the examples in (1) and (2) above, answering questions of the first type invariably involves answering questions of the second type as well. And it is questions of the second type that are most relevant to instruction in abstract mathematics at the college level.

The answer is either yes or no to a question of whether it is true or false that for all elements in some domain, if some hypothesis is true, then some conclusion is true (assuming the question is decidable), and regardless of which answer one finally decides on, one proceeds in a very similar way. One imagines one has elements that make the hypothesis true, and one asks oneself: Must these elements make the conclusion true also? What kind of elements does one imagine? It depends on whether one's tendency is to answer yes or no. (And in the process of working on the problem, one might change course several times.)

If one is tending toward an affirmative answer, then one generally thinks of the elements as particular but arbitrarily chosen. One might imagine them as shadowy images or generic objects that satisfy the hypothesis of the statement but nothing else; one makes no other assumptions about them. For instance, the first example I worked through shows that a union of *any* symmetric relations R

and S is symmetric. And to derive this result, one has to show that for *any* ordered pair (x,y) in $R \cup S$, the pair (y,x) is in $R \cup S$ also. Yet one works with only one R, one S, one x, and one y. What makes the argument valid is that this R, S, x, and y are generic: All that one supposes about them is that they satisfy the given conditions. They are as ready to take the form of any one set of elements of the sets from which they come as any other. For this reason, I call this method of reasoning "generalizing from the generic particular." In my opinion it is the most fundamentally characteristic part of mathematical thought. As Alfred North Whitehead once wrote, "Mathematics, as a science, commenced when first someone, probably a Greek, proved propositions about *any* things or about *some* things without specification of definite particular things" (1958, p. 7).

Other contributors to this volume have referred to this mode of thinking in a variety of ways. Schwartz designed his Geometric Supposer specifically to stimulate thinking in generic terms about geometric objects. A user can create a procedure on a token of a particular type, and this procedure can then be applied to any other token of that same type, allowing the user to riffle quickly through many different possibilities to make and test conjectures. White found it helpful when teaching youngsters about force to represent objects nonspecifically in order to foster thinking of their interactions in a generic way that could then be applied to a large variety of concrete instances. And Romberg used the idea of the generic to make his test example as general as possible to explore how to answer the problem of calculating the probability of winning a certain game.

Now return to the question of whether it is true or false that for all elements of some domain if some hypothesis is true then some conclusion is true, and imagine that one is tending to a negative answer. In this case one asks: Is it possible to find elements that satisfy the hypothesis but not the conclusion? Here one may not imagine elements that are arbitrarily chosen, but rather one might imagine sifting through the totality of all elements that satisfy the hypothesis in search of some with the additional property that they do not satisfy the conclusion. Even in this case, however, the mental image of the elements being sought has qualities of the generic. One often deduces as much as possible about the elements in general in order to narrow the search space before starting to look at concrete examples. And, in many cases, even when one considers concrete examples, one tries to make them as generic as possible.

Finally, suppose a direct proof of the given statement was sought in vain, and the attempt to find a counterexample also met with failure. The natural next thought would be: Maybe a counterexample cannot be found. Perhaps it is impossible to find a set of elements that makes the hypothesis of the statement true and the conclusion false. So, yet again, one might conjure up a mental image of somewhat shadowy objects, this time imagining those for which the hypothesis is true and the conclusion is false. Then one would explore this mental image, comparing different parts of it with each other and with one's general knowledge of the mathematical world to see if a contradiction results. If it does, one concludes that a counterexample does not exist, and hence the statement must be true.

What I've tried to convince you of in this discussion was referred to in a comment made almost in passing in Schoenfeld's chapter: That when a mathematician works on difficult problems, deductive reasoning is a fundamental part of the discovery process. Faced with a statement of the form, "For all (elements in a domain), if (hypothesis), then (conclusion)," whose truth or falsity is unknown, one reasons as follows: Suppose I have an element of the domain that satisfies the hypothesis. Must it also satisfy the conclusion? If the answer in all cases is yes, then the statement is true. If, on the other hand, there are some elements in the domain that satisfy the hypothesis and not the conclusion, then the statement is false. Yet, again, if there are no elements in the domain that satisfy the hypothesis and not the conclusion, then the statement is true. And that is it. Direct proof, disproof by counterexample, and proof by contradiction are three aspects of the same whole. One arrives at one or the other by a thoughtful examination of the given statement, knowing what it means for a statement of that form to be true or false. Proof and disproof are, therefore, indispensible tools, not just to check after the fact that one's illogical (or, better, alogical) intuitions have brought one to the correct answer to a mathematical question, but to be used side by side with intuition throughout the problem-solving process.

Deductive logic is in every nook and cranny of this process. In just the three examples discussed, I needed to use the facts that the negation of a universal statement is existential, that the negation of an existential statement is universal, and that if a property is true for all elements of a set, then it is true for each individual element of the set (universal instantiation). Also, I used the facts that to prove a universal statement true one proves it for particular but generic elements (generalizing from the generic particular), that the negation of "if p then q" is "p and not q," that the negation of "p or q" is "not p and not q," that the inference "p, therefore p or q" is valid, that a statement of the form "if p or q then r" is logically equivalent to one of the form "if p then r and if q then r," and also that if the negation of p implies a contradiction, then p is true, and so forth.

IMPLICATIONS FOR MATHEMATICS INSTRUCTION

Now what is the relation of this analysis to mathematics instruction? Anyone who has been in a mathematics classroom recently knows that deductive logic is not students' strong suit. Large numbers of students enter college with little intuitive feeling for the logical principles that are necessary to solve mathematical problems.

There are many reasons that this is the case. One is that although high school geometry is supposed to serve as a vehicle for teaching abstract thinking, it appears to be ineffective for the large majority of students. Perhaps the reason is that Euclidean geometry is so concrete. Because it is impossible to draw, say, a generic triangle, it takes a student with an unusually lively imagination to think of the particular one shown on a piece of paper in generic terms.

Another reason students have so much difficulty with deductive logic is that propositional and predicate logic are just a small part of the full range of logics that are used in everyday life. For instance, I regularly exhort my students that the negation of an if–then statement does not start with the word *if*. Yet in the logic of temporal statements, it may be appropriate to deny one if–then statement with another. (Statement: "If this program is input to the computer, it will run without error." Response after reading the program: "No, if this program is input to the computer, it will not run without error.") There are even a few mathematical situations in which it may seem natural to negate an if–then with an if–then. It recently occured to me to imagine how I would reply to the statement: "If x is a real number then $x^2 < 0$." I might well answer: "No. If x is a real number then $x^2 \not< 0$." Because the given statement is, in fact, universally false, it seems misleadingly weak to refute it by giving a single counterexample.

We mathematicians often unwittingly add to students' confusion by using language and logic in a misleadingly casual way. For instance, in ordinary speech people often express biconditional statements as simple conditionals (generally when one direction of the biconditional is regarded as obvious). For instance, both the parent who promises "If you eat your dinner then you'll get dessert" and the one who threatens "You will get dessert only if you eat your dinner" intend to communicate the message that eating dinner is a necessary and sufficient condition for getting dessert. Mathematicians regularly use the convention of implied biconditionality to write definitions in if–then rather than in if-and-only-if form. I think they do so with the laudable intention of avoiding intimidating formality. Yet both the *if* and the *only if* directions of a definition are equally important, and the naïve student who is presented with only one direction misses an important clue about how to use the term in practice. Another example is the habit mathematicians have of quantifying statements implicitly rather than explicitly, even though a correct interpretation of the quantification is crucial to determining whether a statement is true or false. Still another source of confusion is the tradition of giving the same letter several different meanings in a single sentence: If $f(x) = x^2$, graph $y = f(x)$ in the xy–plane and compute $(f(x + h) - f(x))/h$.

Some mathematicians have an attitude of "either you have it or you don't," that the ability to think abstractly is a gift that cannot be taught. Yet few people think that just because some children learn to read spontaneously we should stop teaching reading. For some reason it is common to feel impatient with anything less than perfect success in teaching logical thinking. Such impatience is curious, because we all accept a broad range of outcomes in the other topics we teach. Somehow, however, we feel a stronger negative emotional reaction when we have taught students, say, De Morgan's law, and they still do not use it correctly, than when we have yet again shown them how to add fractions, and they yet again do it wrong. But, really, which of these skills is going to benefit students more throughout their lives: being able to add fractions or having an intuitive sense for De Morgan's laws? Adding fractions comes up only in mathematical contexts, but De Morgan's laws are used in all fields involving analytical reasoning.

RELATION OF RESEARCH IN MATHEMATICS LEARNING TO MATHEMATICS INSTRUCTION

Extensive experience teaching a mathematics course that emphasizes reasoning has led me to believe that there is a sizeable group of students who can progress from being unable or only marginally able to cope with the demands of a technically rigorous scientific curriculum to a level at which they can survive and possibly even thrive. I very much agree with the view expressed at this conference that a major focus of our efforts should be to understand both the nature of our students' difficulties and also the strengths they bring with them in order to devise "mediating activities" to help them develop understanding of mathematical concepts. In this effort the informal observations of classroom teachers can be supplemented in important ways by the work of the researcher.

A small, but in my opinion striking, example of this kind of interaction is illustrated by the effect of Dubinsky's report in this volume about students' perception of what happens to a secant line through points $(x_1, f(x_1))$ and $(x_2, f(x_2))$ as x_2 is brought closer and closer to x_1. In his diagram, the secant lines approach a line with slope approximately equal to 1. Yet many students predicted that the secant lines would approach the vertical, and many others believed that they would get closer and closer to a horizontal line. Now for many years I have been skeptical of the view that most calculus students have an intuitive understanding of limits, even though they cannot handle the formal definition. But I took for granted that students understood the geometric image of the secants moving closer to the tangent in the way we would wish them to. Dubinsky's result was a revelation.

In past years a number of works by cognitive psychologists and mathematics educators have influenced my approaches to teaching and to developing various teaching materials. For instance, shortly after I started teaching a course that focused on proof as a process and discovered the severity of my students' problems with logic, I happened on work of a variety of cognitive psychologists (for a summary see Anderson, 1990) who had studied people's understanding of the logic of conditional and quantified statements. Reading this work helped me appreciate how typical my students' difficulties were. In the mathematical reasoning course I was developing, I was encouraged to devote more attention to devising new ways to help my students incorporate a knowledge of principles of abstract logic into their own thinking. And in the standard mathematics courses I

FIG. 7.2. Wason's experimental setup.

| E | K | 4 | 7 |

taught, I was stimulated to be more generous in explaining the logic as well as the substance of the mathematics I was presenting.

I was particularly struck by a study done by Johnson-Laird, Legrenzi, and Legrenzi that modified one originally done by Wason on how people interpret if–then statements (Wason & Johnson-Laird, 1972, chaps. 13 and 14). In the original set of experiments of Wason, subjects were presented with four cards. They were told that each card had a letter on one side and a number on the other (see Fig. 7.2). The cards shown face up were E, K, 4, and 7. They were then asked which cards they would have to turn over to check the following rule: If a card has a vowel on one side, then it has an even number on the other. The results were that only 4% gave the correct response, that both the E and the 7 would have to be turned over, 46% responded E and 4, 33% said E only, and 17% gave a variety of other responses. Hardly any subjects were aware that they had to turn over the 7, that a vowel on its opposite side would have shown that the rule was not obeyed.

In the modified study by Johnson-Laird, Legrenzi, and Legrenzi subjects were presented with four envelopes: two face down, one sealed and the other open, and two with the address side up, one with a 50 lire stamp and one with a 40 lire stamp (see Fig. 7.3). They were asked which envelopes they would have to turn over to check the rule: If a letter is sealed, then it has a 50 lire stamp. In this study 21 out of 24 subjects correctly indicated that both the sealed letter and the letter with the 40 lire stamp would have to be turned over to check the rule.

Another analysis of this result and further studies were done by Cheng, Holyoak, and colleagues (Cheng & Holyoak, 1985; Cheng, Holyoak, Nisbett, & Oliver, 1986). Their results suggest that the reason for the high success rate in the Legrenzi and Legrenzi study was that the set-up triggered a widely understood schema in the subjects' minds having to do with permission. When a rule demands that one act be performed to obtain permission for another, the schema tells us that in order to check whether the rule is being obeyed, one must check both that when the prerequisite act is performed permission is given, and also that the permission is not given without performance of the prerequisite act.

I draw two lessons from these results. First, there is hope for virtually all students. The potential for logically correct deductive reasoning is there if we can

FIG. 7.3. Johnson-Laird, Legrenzi, and Legrenzi's experimental setup.

just determine what mental images to hang it on. When we teach logic, we can appeal to some intuitive understanding on our students' part. This is good, because otherwise our effort might well be in vain. The second lesson is that context, motivation, and engagement with the substance of the subject are crucially important to the success of mathematics instruction. We need to stimulate our students' imaginations by giving them a great deal of experience with suggestive examples, and we should try to convince them by including frequent, realistic applications that what we are teaching is not abstract nonsense but is important and useful.

Other insights that seem to me to be particularly relevant to mathematics instruction are those that have come from research on the "society of mind" and the "social brain" (see, for example, Gazzaniga, 1985; Minsky, 1987). Both Minsky and Gazzaniga make the case that the brain functions as a community of different parts that often fail to communicate effectively with one another. This point of view is also supported by the work of diSessa (e.g., diSessa, 1983), which suggests that ideas or concepts are often tied to the contexts in which they are learned, and that students may have difficulty seeing their application in other contexts. For example, in a precalculus class I taught recently students asked me about a problem in which they were to find the domain of the function f defined by $f(x) = \log_{10}(3 - 4x)$. With help from the class I quickly reduced the problem to finding all real numbers x for which $3 - 4x > 0$, and then, knowing that we had covered inequalities just a few weeks before, I asked my students what to do next. Not a single student volunteered an answer—and this was a small, vocal class. It was not shyness that held them back, it was ignorance. My students' reaction was especially troubling, because I was well acquainted with Henkin's caution (Henkin, 1972) that students often learn a mechanical procedure to do "solve this inequality" problems without understanding that "solve this inequality" really means "find all real numbers for which the inequality is true," and I thought I had successfully communicated the equivalence of these phrases when I discussed inequalities in class. But then I looked at the section on inequalities in the course text and saw that every example and every exercise was stated in "solve that" form. For all the pleasure it had given me to feel I had said the right thing to my students in the classroom, their understanding was evidently shaped by their experiences with working exercises: Although they had a mechanical procedure to "solve" a linear inequality, they had not learned to find all values of the variable that make it true.

As I understand the lesson of the "society of mind" and "social brain" theories, it is that instruction needs to stimulate students to understand what they are learning from many different points of view, not to assume that understanding by one part of the brain will automatically connect with understanding by the other parts. As Minsky put it (Minsky, 1987, p. 193), teachers should help students create "robust, cross-connected webs" of mathematical knowledge. We need to provide learning activities that accustom students to a variety of ways of phrasing mathematical statements and to develop exercises that explore concepts from many different perspectives.

An additional, heartening lesson from these theories is that relatively minor changes in instructional strategy may have a major impact on overall effectiveness, because an understanding of the whole may be impeded by the lack of just a few small connections. I found, for example, that students had great difficulty writing formal versions of statements involving two quantifiers. One day when I was introducing the topic in class, I asked for a volunteer to formalize the statement, "Everyone is older than someone." When no one was able to answer correctly, it occurred to me to ask the students simply to finish the statement, "For all people x, there exists. . ." Not only were several students able to answer correctly, but the atmosphere also became noticeably livelier, and students became eager to try the more general problem. Ever since, I have included an exercise of this type to be done as homework and have observed a significant improvement in overall student performance.

As part of a research agenda, I would like to see more work in cognitive psychology and mathematics learning on issues of logic and language, especially as these apply to the activity of the mathematics classroom, and I would like to see wide dissemination of articles relating this research to practical questions of mathematics instruction. Students' fundamental problems of thinking are not likely to be solved by taking a single course. The more sensitivity all mathematics instructors have toward issues of logic and language, the better job they can do in all the courses they teach.

I end this chapter by describing a cartoon I'm particularly fond of. It shows a picture of the stage of an opera house, part of the main floor, and some box seats. Every seat is filled. On stage, backed by scenery, is the stage manager, all alone, dressed in a tuxedo, obviously making an announcement. The caption reads: "Is there any one here who can sing Siegfried?" What especially appeals to me about this cartoon is the touching naïveté of the stage manager's expecting to find a member of the audience capable of stepping into the role of Siegfried at a moment's notice. Though it is highly unlikely that a Gauss or an Archimedes will ever sit in our classroom just waiting to be called on, with thoughtful help from us our students can do better than we often give them credit for.

REFERENCES

Anderson, J. R. (1990). *Cognitive psychology and its implications* (3rd ed.). San Francisco: W. H. Freeman.
Cheng, P. W., & Holyoak, K. J. (1985). Pragmatic reasoning schemas. *Cognitive Psychology, 17*, 391–416.
Cheng, P. W., Holyoak, K. J., Nisbett, R. E., & Oliver, L. M. (1986). Pragmatic versus syntactic approaches to training deductive reasoning. *Cognitive Psychology, 18*, 293–328.
diSessa, A. (1983). Phenomenology and the evolution of intuition. In D. Gentner & A. Stevens (Eds.), *Mental models* (pp. 15–34). Hillsdale, NJ: Lawrence Erlbaum Associates.

Gazzaniga, M. S. (1985). *The social brain: Discovering the networks of the mind.* New York: Basic Books.

Henkin, L. (1972). Linguistic aspects of mathematics education. In W. E. Lamon (Ed.), *Learning and the nature of mathematics* (pp. 211–218). Chicago: Science Research Associates.

Minsky, M. (1987). *The society of mind.* London: Heinemann.

Wason, P. C., & Johnson-Laird, P. (1972). *Psychology of reasoning.* Cambridge, MA: Harvard University Press.

Whitehead, A. N. (1958). *An introduction to mathematics.* New York: Basic Books.

Comments on Susanna Epp's Chapter

James G. Greeno
Stanford University and the Institute for Research on Learning

We have reached widespread agreement about the primary goal of mathematics education. We want all students to be able to engage in authentic mathematical practices—to think and communicate mathematically about mathematical concepts and to use mathematics in their thinking and communication about other matters that are important in their academic and nonacademic lives.

This noble goal sets many problems. Some of the problems are political, some are organizational, some are practical at the level of classroom materials and activities of teaching and learning, and some are theoretical. The theoretical problems are important for two reasons. One reason is that better understanding of authentic mathematical practices of thinking, communicating, and sense-making will help in the political debates, the organizational reforms, and the changes in classroom practices that are needed. It is too easy at present to dismiss goals of education for mathematical practice on grounds that the goals are vague, romantic, and not grounded in rigorous, credible scientific theory. The other reason is that solving theoretical problems of mathematical reasoning and discourse will contribute significantly to cognitive science. Two central unsolved problems in cognitive science are the relation between intuitive and symbolic reasoning and the relation between individual and social processes in cognition. Both of these problems are involved crucially in the theory of mathematical activity, and therefore mathematical cognition provides an excellent developmental arena for fundamental cognitive research.

The topic of proof is particularly important for both our practical and theoretical concerns. The roles of proof in mathematical practice are not well understood. Regarding educational practice, I am alarmed by what appears to be a trend toward making proofs disappear from precollege mathematics education, and I believe that this could be remedied by a more adequate theoretical account of the epistemological significance of proof in mathematics. Regarding cognitive science, a theoretical account of interactions between intuition and reasoning with formal symbols that are involved in forming proofs would provide a significant scientific advance, as would an account of ways in which proofs function in mathematical discourse.

SIMULATIVE REASONING TO FORM PROOFS

Epp's chapter presents a provocative conjecture about exploratory reasoning in the process of forming proofs. According to Epp's conjecture, to consider a proposition that might be proved, "one would conjure up a kind of shadowy mental image" of mathematical objects that satisfy a hypothesis and explore this mental image to see (a) whether the conclusion must be satisfied, (b) whether a counterexample to the conclusion can be found, or (c) whether assuming the hypothesis and the negation of the conclusion leads to a contradiction.

Epp's conjecture accords in a general way with some recent thinking about logical reasoning in cognitive science, including studies that Epp cited, as well as some other work. Johnson-Laird (1983; Johnson-Laird & Byrne, 1991) has developed analyses of deductive reasoning based on hypotheses that people construct mental models that correspond to premises and draw conclusions based on properties of these representations. Barwise and Etchemendy (1991) are developing a rule-based inference system that uses information in a diagram in conjunction with statements in logical form in constructing proofs.

The scientific task of developing an empirically grounded theory of this kind of reasoning is in an early state, but there are some prospects that seem quite encouraging. One feature of this needed theory is a characterization of mental operations that can be applied to the objects in mental simulations. Johnson-Laird (1983) hypothesized mental tokens that represent generic objects that are associated with labels. Inferences about quantified propositions are made by examining the tokens to see whether relevant combinations of labels are present. Johnson-Laird and Byrne's (1991) analyses include a hypothesis that tokens have simulated spatial locations that support conclusions about relations that have not been stated. Barwise and Etchemendy's (1991) system includes construction of cases that are consistent with information in formulas and making inferences, including elimination of cases, that are justified by citing diagrams. Epp's conjecture requires operations on mental representations to infer whether sets contain elements with properties that are specified.

In all of these proposals, a key feature is that operations on objects in mental models simulate logical, spatial, or mathematical operations. To support valid mathematical reasoning, the objects in models must have properties and relations that correspond to the properties and relations of the mathematical objects that they simulate. The important requirement of mental models is that results of mental operations have properties and relations that correspond to the results of mathematical operations that the mental models simulate—that is, mental models preserve those aspects of the relational structure of the mathematical systems that the models represent that are needed for making inferences. This can be stated as a requirement that the behaviors of mental simulations are in accord with the constraints that apply in the represented domain, that is, the constraints of conceptual entities of mathematics as they are combined and transformed.

I have found it useful (Greeno, 1991) to think about reasoning with mental simulations as analogous to working in an environment such as a kitchen or woodworking shop, where there are various materials that can be combined and transformed into useful objects. The objects that are needed for mathematical reasoning are mental models, and the materials are conceptual resources that are used in constructing the models. With appropriate knowledge of the available conceptual materials and appropriate skill in constructing and operating models, the mathematical worker puts together and operates models in which the mental objects behave according to relevant mathematical constraints.

These general remarks apply to simulative reasoning in any situation that involves mathematics. The requirements for forming proofs are more specific. In considering a proposition as a candidate for a proof, Epp's conjecture corresponds to the following three requirements on simulative mental models:

1. To support construction of a direct proof, a mental model is constructed based on the hypothesis of the proposition, and, if required, transformations are performed by simulating mathematical operations on the simulated mathematical objects. The property stated in the conclusion of the proposition holds in the model without further specifications or transformations, and there are no indeterminate steps in the simulation that would generate contradictions to the conclusion.
2. To support the negation of a proof, a mental model is constructed based on the hypothesis of the proposition and required transformations, and the consequence of the proposition is contradicted by an example in the model or by an operation that is applied at an indeterminate step in the simulation.
3. To support an indirect proof, a mental model is constructed based on the hypothesis of the proposition and required transformations, along with the negation of the conclusion of the proposition, and the combination of these conditions cannot be satisfied because they include a contradiction.

As Epp noted, a key property of simulations for forming proofs is that their objects and operations can be taken as generic—that is, although they simulate specific objects and operations, their properties hold for the class of objects and operations that the proposition specifies.

STATING PROOFS

The ability to write proofs is a significant indicator of success in many high school geometry classes. An analysis of this ability is one of the solved problems in the cognitive science of mathematics education. Using the empirical and theoretical methods of recording and analyzing thinking-aloud protocols and writing computer-program simulations that Newell and Simon (1972) developed, processes of solving geometry proof exercises were analyzed successfully in the late 1970s (Greeno, 1978), and a simulation of learning the procedures of constructing statements of proofs was developed in the early 1980s (Anderson, 1982).

The theoretical problem that we solved corresponds to a learning problem that students have to solve in many school settings—they have to be able to write the statements and reasons of proofs. The simulation models are hypotheses about cognitive procedures and information structures that students need in order to construct correct proofs. They have components that are not explicit in most instruction, involving setting goals and planning. It should be possible to do a more thorough job of instructing students if these components are given explicit attention. This has been done in geometry in the form of a computer-based tutoring system (Anderson, Boyle, & Yost, 1985), and students in an evaluation study who worked with this tutoring system performed significantly better, by one to two grade levels, than students who received standard classroom instruction.

This development is promising from one point of view. Many students have difficulty in learning to write geometry proofs, and an instructional system that enables more students to learn this can improve their lives.

From another point of view, this development is problematic. The early versions of models of geometry problem solving provided quite a clear picture of a set of knowledge structures that are sufficient to produce correct proofs. We can examine these models to evaluate the significance of the knowledge they represent for students' mathematical understanding. In my judgment, these knowledge structures have a lot to do with problem-solving techniques, have less to do with geometry, and have practically nothing to do with proof. If this is what students learn in order to do proofs in school geometry, there is a serious question whether these activities of school geometry provide the kinds of learning experiences that we want students to have.

It is important to note that these models represent hypotheses that are sufficient to account for the data, and as is always the case, alternative hypotheses should be considered, especially if they present a more optimistic picture of the knowledge and understanding that students might acquire in learning to do proof exercises. Along this line, Koedinger and Anderson (1990) hypothesized processes that include students' recognition of familiar configurations in diagrams.

These processes represent cognitive interactions between propositional expressions that are written as proof statements and spatial properties and relations that are represented by diagrams. Models such as these could provide a richer basis for students' understanding principles of geometry that are represented in proofs. They may, in fact, address an issue that Epp referred to in saying, "The potential for logically correct deductive reasoning is there [in students' understanding] if we can just determine what mental images to hang it on." They do not, however, address questions of understanding roles of proof in mathematical epistemology and discourse.

PROOFS IN MATHEMATICAL DISCOURSE PRACTICES

Proofs play a crucial role in the practice of mathematics. They provide the primary means of establishing propositions as results. Proofs also can be a source of understanding relations among propositions and concepts. In this way, proofs in mathematics are analogous to observations and experiments in empirical science, which provide the evidentiary backing for knowledge claims. For students to learn mathematics without coming to appreciate the role of proof seems as impoverished as it would be for students to learn science without coming to appreciate the role of empirical evidence.

Fawcett's (1938) course in geometry is still our best example of teaching that focused on deductive reasoning as the means of establishing mathematical results. Fawcett's classroom discourse included discussions of alternative definitions and postulates with attention to their implications for ease of proving theorems about the defined objects. Each of Fawcett's students constructed her or his individual version of geometry, choosing from among alternative definitions and postulates and deciding what to include as theorems. Fawcett emphasized that formal deductive methods are not limited to the domain of mathematics. Considerable discussion was given to materials that were found outside of the classroom, including advertisements, political statements, and issues of enforcing ordinances, in which students discussed the adequacy of definitions and conclusions that depended on unstated assumptions.

We need a theory of semantics of proofs to clarify their role in facilitating understanding. An approach that seems promising uses the concept of *constraints*, developed in situation theory (Barwise & Perry, 1983; Devlin, 1991). Constraints are regularities involving types of situations, of the kind that are expressed by conditional statements. An important psychological concept is *attunement* to constraints, a general concept that applies in any situation involving skilled performance, in which the regularities of physical systems and other people's behavior are incorporated smoothly into a person's activity.

According to this idea, each statement of a proof refers to a constraint. A proof designates a sequence of constraints that link a set of conditions to a property or relation that necessarily holds in those conditions.

MENTAL MODELS AND CONSTRAINTS

Recall that I assume that a mental model works if operations on the objects of the model have results that simulate mathematical operations. Another way to say this is that operations in the model comply with the mathematical constraints that are relevant in the problem situation. Constructing mental models that behave according to certain constraints is one way in which someone's reasoning can be attuned to those constraints.

As an example, consider Epp's examples involving binary relations. The proofs depend on the presence or absence of ordered pairs in sets. Mental models for these problems are representations of sets that contain or do not contain various generic pairs that are critical to the proof. Reasoning that uses the models depends on the models' complying with constraints of set membership. If a person constructs a mental model that complies with the relevant constraints, then reasoning with that model can support a correct conclusion. Model-based reasoning that concludes correctly that the union of symmetric relations is symmetric depends on a construction of the union of sets that has all and only the members of those sets as members. If the operation of forming a union in someone's mental model did not have that property, that model might not support the conclusion. Model-based reasoning that concludes correctly that the union of transitive relations is not necessarily transitive depends on constructing a mental model that can have (x,y) in R and (y,z) in S, but not have (x,z) in either R or S. If someone constructed a restricted mental model in which all such related pairs were in one of the two sets, their model would support the incorrect conclusion that the union is transitive.

We can facilitate students' reasoning by providing conditions that afford their construction of mental models that include relevant constraints. I believe this is what happens in the situations studied by Cheng and Holyoak (1985). According to this hypothesis, when a logic problem is presented as a situation in which someone needs permission to enter a country, students are more likely to construct mental models that comply with the relevant constraints. In this case, it seems likely that incorrect solutions result from efforts to solve the problems without constructing mental models. In other cases, incorrect reasoning probably results from construction of mental models that follow incorrect constraints. Examples of this occur in reasoning about physical systems, such as inferring that an object dropped from a moving airplane would fall straight down. Yates et al. (1988) provided evidence that errors of this kind can result from students constructing mental simulations that follow constraints that are different from those that are understood in contemporary physics.

INTERACTIONS OF PROOF STATEMENTS AND MENTAL MODELS

These ideas suggest a conjecture about a relation between statements in a proof and properties of a mental model that supports the proof. The conjecture is that proof statements refer explicitly to constraints that are implicit in mental models.

According to this idea, learning to construct statements of proofs involves learning a set of representational conventions in which symbolic expressions (the statements of set theory, geometry, logic, or whatever) refer to constraints of the domain. If someone has learned to construct mental models that comply with those constraints, the statements of a proof correspond to general properties of those mental models. Constructing a proof statement includes forming symbolic expressions that represent constraints of the domain and arranging those expressions according to some conventions of the form of a proof. It seems likely, then, that learning conventions of representation involved in the construction of proof statements and learning to construct and operate on mental models with appropriate constraints can be complementary and mutually supporting aspects of learning to reason in a domain. Mental models can facilitate solution of proof problems, as Epp observed. It also seems likely that construction of proof statements can help students learn to construct mental models that simulate behavior of mathematical objects according to relevant constraints, especially if we teach in a way that makes the meanings of those statements evident to students.

CONCLUSION

I concur enthusiastically with Epp's proposal "that learning ways to help . . . students become better deductive reasoners deserves the attention of mathematics teachers and researchers in mathematics education, at least as much as to other aspects of mathematical thinking." I also agree with Epp's hypothesis that mental simulations play a key role in deductive reasoning that is productive for students' understanding and meaningful learning.

I think it is consistent with Epp's view to propose that instruction should focus more on interactions between the symbolic expressions of proof statements and the general properties of mental models. A reasonable way to do this is to provide representations of the mathematical objects that we want mental models to represent and to focus discussion on relations between proof statements and constraints that can be understood as properties of those representations. I expect that such an emphasis on interactions between formal and informal representations characterizes a great deal of successful mathematics teaching, in which presentation of proofs is accompanied by rich uses of diagrams.

On the other hand, discussions of proofs that neglect their relations to informal representations and mental models may lead some mathematics teachers to treat proofs as self-contained symbolic structures. It has led some cognitive scientists, such as Anderson and his associates (Anderson et al., 1985), to such a treatment.

Discussions of proofs that relate the symbolic expressions of statements and reasons to constraints could help students become attuned to the constraints and to their relevance in mathematical reasoning. Such discussions might be especially effective if the instructional resources included a combination of computational systems such as the Geometric Supposer (Schwartz, Yerushalmy, & Wilson, 1993), which supports exploration of invariant properties of geometric figures, and Anderson et al.'s (1985) Geometry Tutor, which presents graphical displays that show relations between steps of proofs and problem-solving goals that the steps achieve. Most importantly, instruction would need to emphasize how the steps of proofs represent invariant properties and relations.

ACKNOWLEDGMENTS

This chapter was supported by National Science Foundation Grant No. MDR-9053605. I thank Nancy Nersessian and Alan Schoenfeld for helpful comments on earlier drafts.

REFERENCES

Anderson, J. R. (1982). Acquisition of cognitive skill. *Psychological Review, 89,* 396–406.
Anderson, J. R., Boyle, C. F., & Yost, G. (1985). The geometry tutor. In A. Joshi (Ed.), *Proceedings of the Ninth International Joint Conference on Artificial Intelligence* (pp. 1–7). Los Altos, CA: Morgan Kaufman.
Barwise, J., & Etchemendy, J. (1991). Visual information and valid reasoning. In W. Zimmermann & S. Cunningham (Eds.), *Visualization in teaching and learning mathematics* (MAA notes No. 19, pp. 9–23). Washington, DC: Mathematical Association of America.
Barwise, J., & Perry, J. (1983). *Situations and attitudes.* Cambridge, MA: MIT Press/Bradford Books.
Cheng, P. W., & Holyoak, K. J. (1985). Pragmatic reasoning schemas. *Cognitive Psychology, 17,* 391–416.
Devlin, K. (1991). *Logic and information.* Cambridge, UK: Cambridge University Press.
Fawcett, H. P. (1938). *The nature of proof.* New York: Teachers College, Columbia University.
Greeno, J. G. (1978). A study of problem solving. In R. Glaser (Ed.), *Advances in instructional psychology* (Vol. 1). Hillsdale, NJ: Lawrence Erlbaum Associates.
Greeno, J. G. (1991). Number sense as situated knowing in a conceptual domain. *Journal for Research in Mathematics Education, 22,* 170–218.

Johnson-Laird, P. N. (1983). *Mental models: Towards a cognitive science of language, inference, and consciousness.* Cambridge, UK: Cambridge University Press.

Johnson-Laird, P. N., & Byrne, R. M. J. (1991). *Deduction.* Hillsdale, NJ: Lawrence Erlbaum Associates.

Koedinger, K. R., & Anderson, J. R. (1990). Abstract planning and perceptual chunks: Elements of expertise in geometry. *Cognitive Science, 14,* 511–550.

Newell, A., & Simon, H. A. (1972). *Human problem solving.* Englewood Cliffs, NJ: Prentice-Hall.

Schwartz, J., Yerushalmy, M., & Wilson, B. (Eds.). (1993). *The geometric supposer: What is it a case of?* Hillsdale, NJ: Lawrence Erlbaum Associates.

Yates, J., Bessman, M., Dunne, M., Jertson, D., Sly, K., & Wendelboe, B. (1988). Are conceptions of motion based on a naive theory or on prototypes? *Cognition, 29,* 251–275.

A Discussion of Susanna Epp's Chapter

John Addison

>**John Addison,** *University of California, Berkeley*
>**Susanna Epp,** *DePaul University*
>**Alan H. Schoenfeld,** *University of California, Berkeley*
>**Ed Dubinsky,** *Purdue University*
>**Robert Davis,** *Rutgers University*
>**Leon Henkin,** *University of California, Berkeley*

John Addison. Before the conference Alan told me that Susanna had gotten into this subject by writing a textbook on discrete mathematics, particularly for computer science students. I had the chance to look at that, and I sense from it where she is coming from. I like what I was able to read of the book very much.

Before I get to my main comments, Susanna, I do want to note that I didn't approach your problem about the union of transitive relationships the way that you did. Before you had gotten very far in your discussion, I just started thinking, what are some transitive relations? I thought of "less than" and "greater than" and took their union, and that solved the problem for me. So there are various ways we approach the problems, not just the one you mention.

I've always been interested in the question of what role logic should play in education. Let me say some very naive things. I have to tell you first that I'm a logician, so that colors everything I say. But, when I first grew up in mathematics, I thought my education was appalling. I began to think about it and find out what was going on, and I really wanted to get things cleaned up. Very soon the new mathematics appeared, and that looked very good to me. I was very excited about it. I thought that in short order they will have all of this bad stuff straightened out. Then the new math seems to have totally bombed. I'm not sure that I quite fully understand that. My impression, which maybe you can straighten me out on, is that there were two main things wrong with it: one, that the teachers in the schools were not prepared for it, and two, that the people that were introducing it didn't do enough to make it possible for it to be taught correctly. So you had a lot of horror examples of the use of set theory and the use of logic, which I think could have been avoided. I saw the new math as trying to do things correctly without too much pedantic emphasis on the correctness but to have everything correct in the background and to change the focus from rote learning to understanding. I'm still naive enough to think that those goals are where we should be going today in mathematics education. My feeling is that logic should be in probably all levels of education and come in a spiral fashion.

The kind of things that you are doing in your textbooks seems to me to be reasonable things for freshmen at Berkeley to be doing. I did ask myself the question, could you do some sort of course like that for the math majors? There you run again into political arguments. Certainly one big argument that would be put up against it immediately is: "We've got to get through all this calculus, because these students don't know whether they're going to be math majors or physics majors, and so there isn't time for that." Another one would be that mathematicians would look at this and say, "This is all trivial. We all know that if you have a universal sentence, then you have to begin by taking an arbitrary generic example . . ." and so on. In fact, I'm sure that many mathematicians on our faculty never explicitly thought through a lot of the kind of stuff that you've got in your book. But if I as a logician start pushing logic, people just think that's special pleading. So, I was pleased to hear that Susanna got her degree under Kaplansky in Algebra. She's really an outsider who can better plead this case. And she says she studied logic on her own after getting her thesis, and that she feels now that learning and understanding all of this logic made her understanding of mathematics better.

You would probably find mathematicians agreeing that this kind of approach might help the weakest students. I think they would say that all the brightest students would probably gather all of the logical principles by osmosis. I think that's not true. Some of that's true, but I know even in my own reasoning I am constantly using in very small steps principles of logic that I don't actually think through. I just know it's mechanical. For example, if you have an existential statement implying a universal statement or some universal existential statement, and that existential quantifier isn't appearing in the antecedent, I know that can come in the front as a universal quantifier. In all of my reasoning, I just use that automatically. I'm sure that if you ask some good mathematicians to respond to that quickly, they'd have to sit and think it through. So, I would be interested in seeing some research done that would try to prove your case and convince people that we should begin having sections of your kind of book at various levels in high school, a little more in college, and so on, and see if that's really doing any good. I do think that in terms of general literacy and understanding of mathematics, we should be working at all levels, and I'm certainly sure that the weakest levels of students would profit immensely from this.

Susanna Epp. Maybe I should just comment about this last point. In fact, quite a number of schools have courses where they are aiming to do this kind of thing. There are a number of "transitions to higher mathematics" courses, and we have the author of one such book here in this group, Steve Galovich. I'm very sensitive to the question of whether this is something the good students can profit from. I always hold my breath with my really good students. There have been a few cases where I've had some quite good students who've taken a casual attitude and acted as if they don't really feel that they need this very much. Often I find that they make major errors by the time of the final so their solutions look awkward and immature, but they are basically on the right track and generally get at least a B for the course. Overwhelmingly, the good students (and it's true we don't get the very best at DePaul, but the best we have are pretty good), really seem to like and appreciate the course. I take heart from that. I do know that there are similar courses in other places. There is certainly a very strongly perceived need in many different college environments for things that will help students learn to cope with more abstract ideas.

John Addison. Is the residue of the new math delaying that in high schools? If somebody tried to put a little more logic in high school, would people say, "Oh my god, here we go again!"?

Susanna Epp. I must say I have another hat. I have worked on the University of Chicago School Mathematics Project's Year 12 book. Quite a few of these ideas are embodied in the most recent version. But to my way of thinking Year 12 is really too late. Leon Henkin is a strong advocate for bring these ideas to younger students, and I agree wholeheartedly. I think that many issues should be raised in ninth-grade algebra, and I think there are things you can talk about in kindergarten. At a simple level, consider the idea of language being precise. Mathematics is the science of quantity. We talk about quantifiers, and we can talk carefully about "some" and "all" and "none" and "every" and "any" to children at a very young age to try to convey precision in the use of these terms.

Alan Schoenfeld. I'd like to pick up on one of the issues John mentioned, because it's central to this conference and my goals for it. The issue concerns the reasons that the new math ran into difficulties. I think that there are a number of fundamental reasons for its failure, and he bypassed a major one. I point it out, because it's been an implicit theme here, but it hasn't come to the fore.

There is no doubt that, at the elementary level at least, teachers were uncomfortable with and perhaps incompetent to teach the new material. That difficulty alone would have sabotaged the new math, and it was compounded by the fact that the mathematics appeared alien to parents as well as teachers: When Mommy and Daddy can't help the kids with their homework and don't see that it makes any sense, there will ultimately be hell to pay. There may well have been flaws in the material itself, certainly from the point of detailed mathematical subject matter analysis.

But there was another fundamental flaw in the materials: The psychology behind them was wrong. Even if everything else was just fine, there's a good chance the new math would have failed dismally, because kids would have had a hard time coming to grips with the notions in the way that they were presented. Bruner's maxim, "You can teach anything in an intellectually legitimate way to anyone at any age if you do it right," was one of the driving principles of curriculum development at the time. The way that new math implemented that idea at the elementary school level was to take complex abstractions and make them "grade appropriate" by reducing them to, and teaching them as, simple abstractions. And, the kids couldn't cope with the abstractions. That's a gross oversimplification, of course, but the fact is that there was really a mismatch between what the kids were capable of grasping psychologically and the mathematical structures they were asked to grapple with. The reason I harp on that point is that a central theme of this conference is that we need to get the two groups, mathematicians and psychologists, to work together. That you need to have the right mathematics, and to understand it deeply, is a given. But there's more, reflecting the cognitive perspective. To make things work you need more than just the mathematics; you need to look at the world from the student's point of view. And unless we have two-way communication, and both perspectives are represented, we run the risk of repeating the failures of the past.

John Addison. I was probably including some of that when I said that the people that were introducing it didn't do enough to make it possible for it to be taught correctly. Were you thinking about the fact that they were teaching second graders the axioms for rings?

Alan Schoenfeld. Yes, that was among the pathologies I was thinking of. With the proper collaborations of mathematicians, cognitive scientists, and mathematics educators, such debacles should be avoidable in the future.

Ed Dubinsky. I'd like to ask Susanna a question. You gave at the beginning of your talk a number of models of cognitive processes, explanations of how someone would think about one of the problems that you gave. You described the processes in some detail. You've heard from at least one person, John Addison, that your description didn't seem to match the way it worked for him. It's not automatic. So, my question is how did you come to think that is the way? What's your paradigm for deciding on that?

Susanna Epp. Let me start by asking John a question. That example came to your mind quickly, and I think that's the problem. That can happen with counterexamples. If it hadn't, did the paradigm seem reasonable to you? Would you have put it up in your head and played around with images like that?

John Addison. I think quite possibly. I think that if you've been studying transitive relations in the class and did it well, you would have studied a dozen transitive relations, all of the students would have those in their minds. So probably the first kind of thing I would have tried would be to have the students think about those. Now, if none of those examples worked, then maybe I would have them make the kind of logical analysis you suggest.

Ed Dubinsky. Susanna, what in the first place made you feel it was the way you described it? How did you come to that?

Susanna Epp. I know what you're getting at, and I think that it is partly my own way of looking at things, sometimes.

Ed Dubinsky. At least your impression of the way you look at things. You can't be sure.

Susanna Epp. Before I began teaching a course whose primary aim was to help students develop the ability to solve similar problems themselves, it would never have crossed my mind to analyze the three questions as I did in my paper. It was the give and take with my students over the many years of developing this course that led to the kind of analyses described in the chapter. Sometimes I ask students who have successfully solved a problem to share their thought processes with me and/or the rest of the class. Probably I should do this more often, but I find that such students often have difficulty expressing themselves coherently about such matters, presumably because they are not accustomed to engaging in this kind of introspection. So, more often, I spend a lot of time thinking about the kinds of difficulties students have in coming up with solutions to problems and try to figure out in detail what kinds of thought processes to suggest they use to overcome them. As these have become better and better enunciated over the years, student response has become more and more positive. In my own classes I get a lot more murmurs of agreement, smiles and nods, and (bottom line!) better performance. From students outside my classes and at other colleges and universities that use my book, I have also received a remarkably positive reaction to my approach.

Ed Dubinsky. You think it necessarily works the same way with your students?

Susanna Epp. This way of looking at the thought process that often precedes the discovery of a counterexample has come to me, in part, through analyzing the difficulties of students who have been unsuccessful. For example, I often ask students to determine the truth or falsity of the following statement: If a does not divide b, and a does not divide c, then a does not divide bc. What I find is that many students will start out by trying to prove the statement true and then run into trouble. So then they try to find a counterexample, but they proceed randomly, often trying only $a = 2$, $b = 3$, $c = 5$ and maybe two or three other cases, none of which does the job. Sometimes they make a logical error and find a "counterexample" that isn't, but more often they see that the statement is actually true in all of the cases that they have examined. They then conclude that the statement is probably true in general, but that they are just not clever or persistent enough to come up with a proof, so they stop working on the problem. Through the analysis of this and other cases, I have come to believe that when students are unsuccessful in discovering a counterexample, it is exactly because they search randomly rather than combining a search with a deductive process in which they focus imaginatively on the properties that a counterexample must possess.

Bob Davis. I'm very pleased Ed picked up on that issue, because it seems to me that gets into what this conference is all about. You thought of a solution by running through your list of possibilities. He thought of it by running through examples. I thought about it by trying to reinterpret it, namely to say, "What is this saying? This is saying that if two things are there, this third thing has to be. And that's true for this, that's true for this. Suppose one of the things is here, and the other one is over here, there's no obligation for the third one to be in the union."

That's literally how I thought about it. I think those are three different ways of thinking about it, and it seems to me that's what we're talking about, that although within mathematical notation things may often look the same, they are not necessarily thought about in the same ways.

Leon Henkin. I want to go back to thinking about the role of proof. One role is that proof is the way you establish results—that's how we get to know things. Now, I don't know what other ways there are. In the schools the predominant way is authority. We've unanimously rejected that. If we reject authority, and we want our students at every level to come to decisions from the material itself about what is going on in this domain, then it has got to be through proof. I don't mean formal proofs necessarily, but some such reasoning has to go down to kindergarten. There has to be this way by which kids can decide for themselves what is going on in whatever mathematical domain they're dealing with.

Now I have been concerned with teaching proofs for a long time. I think it's about 43 years, and it's only in about the last 7 that I have begun to divide my instruction into parts: helping students find the ideas for the proofs, and helping students to write up a proof after the ideas are found. Those are extremely different ways, and I believe that John Addison's point was simply to draw people's attention to these two different kinds of activities, both of which go into the final production of proofs. I also think there's an important place for Susanna's observations. At first when I heard them I thought, "this controverts what I heard about logic the very first day I every took a logic course," which was a lot longer than 42 years ago. It was stated by my logic professor Ernest Nagle at the philosophy department at Columbia. He said that although people think that logic is about thought, that isn't the case. It has to do rather with the form of sentences. I have never really gotten myself away from that distinction between thought and logic. There is an important goal for the kind of thinking that Susanna was emphasizing—that is, understanding what it is you're trying to establish. It's only in the last two years that I have been bringing into my courses an immense emphasis on how much, what large fraction of the time the students spend with the problem they have to spend finding out what really is wanted before they start trying to get the answer. I believe that what you were helping us seek was just those very first thought processes that students really have to go through to find out what those sentences mean that they are trying to establish. I think that those thought processes will only lead to the whole proof in problems that we would in the logic course call the routine problems. The really interesting ones—and we've all been emphasizing how important it is to get away from the routine—the really interesting ones are the ones in which the proof doesn't work that way. Yes, an important class of proofs do work that way. Of course, we should have some problems that students can get to feel confident about. Then we have to encounter problems in which the proof doesn't just come from the meanings of the terms involved. Then they need some new ideas, and there are no rules for furnishing those new ideas. That's what makes it interesting.

Susanna Epp. I'd like to comment briefly on that. That's absolutely right, and I definitely feel that the kind of thing that I'm trying to teach is the proof of really pretty routine statements. I've also been impressed as I've really gotten into this by how much, when I sit in the mathematics seminar, say the algebra seminar at DePaul, what a huge fraction of what I listen to is very routine. I kind of think that if I can teach students to do what we think of as the routine stuff, I'm taking them really far.

Ed Dubinsky. It seems to me that the problem is getting kids to understand that proof is necessary at all. Story: You get up in front of a fourth-grade class, and you do a little exercise like:

What's 4×4 . . . that's 16. What's 3×5 . . . that's 15. Good.
What's 5×5 . . . that's 25. What's 4×6 . . . that's 24. Good.

You do this for awhile and then you say, "Do you all see a pattern?" They say, "Yes, we all see a pattern." Then, "Does it always work? How many cases would you have to see before you believe that it always works?" Now they don't have the machinery for answering that question. That's not the point. They could understand quite naturally that there's something wrong with a scheme that's going to say 50 cases would convince me or 120 cases. They know there's something flawed there. Now the question I want to ask is linked to that. A major piece of work that proof does, and I suggest that this might be a way of getting kids to internalize the need for it, is that it allows you to deal with an infinite ensemble of cases which you can't possibly do. I mean if you had a finite number of cases, then you can do them a case at a time, and you're done. But if you have an infinite number of cases, then you need another way to do them all. Do you know if anybody had used that as a way of getting kids to internalize the need for a different way of going about things, other than simply argument by demonstration?

Susanna Epp. Dijkstra makes that point not just for infinite sets but for large finite sets. He talks about if you want to check the multiplication circuitry of a computer. What are you going to do, test every possible multiplication? It's absurd.

Alan Schoenfeld. Pólya deals with that issue extensively in his two-volume set *Mathematics and Plausible Reasoning*. He gives a collection of examples, in which patterns seem to hold for a *very* long time, and then fail. And I suspect that many faculty have their own favorite ways of making the point as well. For example, I give my students a series of problems where, after you've calculated enough cases, the answer is "obviously" 2 raised to some power:

- What is the sum of the coefficients of $(x + 1)^n$?
- How many subsets, including the whole set and the null set, does a set of n elements have?

and a few more.

For each problem, I ask what the answer is: "2^n." "Are you sure?" "Yes." "Do you want to prove it?" "No." "Humor me."

Then I give them the following well-known problem:

- Suppose you place n points on the circumference of a circle so that when you draw the line segments joining them pairwise, no three of the line segments intersect in a common point. What is the number of regions into which the line segments partition the circle?

The pattern the students get empirically is 1, 2, 4, 8, 16, at which point they're certain they know the answer. "Are you sure?" Yes. "Why don't you test one more case, and then wrote a proof?" It turns out that the next number in the sequence is 31—and after the students have checked their diagrams half a dozen times and are convinced that they haven't merely miscounted, we can begin to talk about why proof is necessary.

8
Classroom Instruction That Fosters Mathematical Thinking and Problem Solving: Connections Between Theory and Practice

Thomas A. Romberg
University of Wisconsin-Madison

> First, "knowing" mathematics is "doing" mathematics. A person gathers, discovers, or creates knowledge in the course of some activity having a purpose. (NCTM, 1989, p. 7)

During the past quarter century, the mathematical sciences education community has been struggling to redefine school mathematics and outline the features of a curriculum that would reflect current societal needs. The most recent and comprehensive set of goals are those prepared by the Commission on Standards for School Mathematics of the National Council of Teachers of Mathematics in 1989. Their report, *Curriculum and Evaluation Standards for School Mathematics* (NCTM, 1989), lists nine goals: four societal goals and five goals for students.

The four general social goals for education in the area of mathematics are:

1. *Mathematically literate workers.* The technologically demanding workplace of today and the future will require mathematical understanding and the ability to formulate and solve complex problems, often with others. "Businesses no longer seek workers with strong backs, clever hands, and 'shopkeeper' arithmetic skills" (p. 3).
2. *Lifelong learning.* Most workers will change jobs frequently and so need flexibility and problem-solving ability to enable them to "explore, create, accommodate to changed conditions, and actively create new knowledge over the course of their lives" (p. 4).
3. *Opportunity for all.* Because mathematics has become "a critical filter for employment and full participation in our society" (p. 4), it must be made accessible to all students, not just white males, the group that currently studies the most advanced mathematics.
4. *An informed electorate.* Because of the increasingly technical and complex nature of current issues, participation by citizens requires technical

knowledge and understanding, especially skills in reading and interpreting complex information.

These social goals require that students become *mathematically powerful*, a key phrase used by the *Standards* authors to describe desired outcomes of schooling. *Mathematical power* "denotes an individual's abilities to explore, conjecture, and reason logically, as well as the ability to use a variety of mathematical methods effectively to solve nonroutine problems" (NCTM, 1989, p. 5). The authors of the *Standards* go on to emphasize that mathematical literacy includes much more than familiarity with numbers and arithmetic: "To cope confidently with the demands of today's society, one must be able to grasp the implication of many mathematical concepts—for example, chance, logic, and graphs—that permeate daily news and routine decisions."

The authors then articulate the notion of mathematical literacy by proposing five general goals for students:

1. *Learning to value mathematics:* Understanding its evolution and its role in society and the sciences.
2. *Becoming confident of one's own ability:* Coming to trust one's own mathematical thinking, and having the ability to make sense of situations and solve problems.
3. *Becoming a mathematical problem solver:* Essential to becoming a productive citizen, which requires experience in solving a variety of extended and nonroutine problems.
4. *Learning to communicate mathematically:* Learning the signs, symbols, and terms of mathematics.
5. *Learning to reason mathematically:* Making conjectures, gathering evidence, and building mathematical arguments.

These goals imply that students should be exposed to numerous and varied interrelated experiences that encourage them to value the mathematical enterprise, to develop mathematical habits of mind, and to understand and appreciate the role of mathematics in human affairs; that they are encouraged to explore, to guess, and even to make errors so that they gain confidence in their ability to solve complex problems; that they read, write, and discuss mathematics; and that they conjecture, test, and build arguments about a conjecture's validity. The opportunity for all students to experience these components of mathematical training is at the heart of our vision of a quality mathematics program. The curriculum should be permeated with these goals and experiences such that they become commonplace in the lives of students. (p. 5)

These goals reflect a shift away from the traditional practice of summarizing desired mathematical outcomes as knowledge of skills, concepts, and applications to an emphasis on broader dispositions, attitudes, and beliefs about the nature of mathematical knowledge and about one's own mathematical thinking. The traditional skills, concepts, and applications are subsumed under the more general goals for problem solving and communication. Throughout the *Standards* document, the authors deemphasize the view that knowledge consists of distinct parts that should be treated separately. Rather, they emphasize providing students with experiences through which they can build rich connections among the various kinds of knowledge.

Treisman (1989) calls this a "romantic" vision of school mathematics. Nevertheless, there are a large number of mathematics teachers and other educators who are convinced that this set of goals captures their beliefs about both what should be taught and how instruction should be carried out. What these educators want is a collection of examples and a set of principles to use for the creation of a curriculum.

In this chapter, my purpose is to suggest some connections between the vision of mathematical power expressed in the *Standards* (NCTM, 1989) and what is known about mathematical thinking and problem solving. My objective is to present a set of principles about mathematical activities and how they are organized for teachers and students. In so doing, I hope that a picture of classroom instruction in typical situations becomes clear. My assumption is that teachers need to have that picture in mind. Without it, the vision of the *Standards* will never be attained. To build this argument, I first present an analogy; second, an example is given; then reflections from both are presented and used to develop the set of principles.

AN ANALOGY

The distinction between "doing" and "knowledge about" perhaps can be made clear by an analogy with other things people do (like play basketball, fly a plane, or learn to play a musical instrument). For example, music, like mathematics, has numerous branches categorized in a variety of ways (classical, jazz, rock, or instrumental, vocal); it has a sparse notational system for preserving information (notes, time-signatures, clefs), and theories that describe the structure of compositions (scales, patterns). However, no matter how many of the artifacts of music one has learned, it is not the same as "doing" music. It is only when one performs that one knows music. In a like manner, when students learn to play basketball, they are always aware that their goal is to play the game. What is being argued is that in mathematics, the "game" is to solve nonroutine problems. Basketball practice is important for skill development, learning strategies, and so forth. However, a coach would never get anyone to practice if they never played a game. Furthermore, practices are tailored to the needs of the team and the individuals. Today, in school mathematics all students practice skills, whether needed or not, spend almost no time learning strategies, and never get to play the game.

Obviously, it is important to learn some mathematical concepts and practice some procedures so that one is a reasonably skilled performer (like learning to read musical notation and practice scales or shoot freethrows), but it is also important for all students to have an opportunity to solve problems (perform), whatever their level of capability. Too often, students find arithmetic, algebra, and sometimes calculus, to be senseless, dull, and even intimidating. After all, who can enjoy routinely multiplying one 4-digit number by another or solving a system of simultaneous linear equations (analogous to forever playing scales on a piano)? As a result of such limited experiences, many students are prejudiced against the broader, more interesting aspects of mathematics. What is needed is a balance between "games" and "practices." This must involve specifying the principles for both the development of a collection of problem situations (games or compositions) and the concepts, skills, and strategies related to those problems, so that students have an opportunity to develop their "mathematical power."

AN EXAMPLE

To illustrate the use of problem situations, I chose the "fairground" problem from *Geometric Probability* (North Carolina School of Science and Mathematics, 1988):

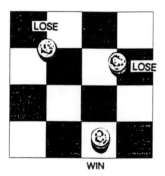

At a fair, players throw coins onto a board checkered with squares. If a coin touches a boundary, it's lost. If it rolls off the board, it's returned. But if it lies wholly within a square, the player wins his coin back plus a prize. Copyright (1988) by National Council of Teachers of Mathematics. Used with permission.

What is the probability of winning this game?

This problem encompasses several of the important features of mathematics instruction specified in the NCTM *Standards* (1989). First, it begins with a real problem situation. Second, to investigate the problem, it needs to be simplified so that an appropriate model can be constructed. In order to build a model, several concepts from different areas of mathematics can be employed (from probability, geometry, coordinate geometry, etc.). Third, several different mathematical procedures can be used (generation of random numbers, plotting points, simulation, etc.). These two features emphasize the interrelationship of concepts and procedures rather than their independence. And, finally, the problem leads to other interesting questions and the methods to other interesting problems.

To illustrate this, let me outline how a class of Grade 8 students actually solved this problem. First, they decided that it was really a geometric probability problem. They had been taught that probability can be understood from a geometric, as well as an analytical or counting, viewpoint. When faced with a question of probability, it is often possible to form a geometric model of the problem.

Second, the class decided to get acquainted with the problem by looking at a specific example: Let the radius of the coin be 3 cm and the side of the square be 10 cm. What geometric shapes represent the sample and event spaces? Consider the center of the coin as a dart. Where can the dart land and win? The center of the coin must be at least 3 cm from each side, otherwise, the edge of the coin will fall across the square. The sample space is the square with side 10 cm, while the winning event space is a square with side 4 cm. The ratio of these two areas is 16/100 = 16%. The probability of winning at this particular game is 16%.

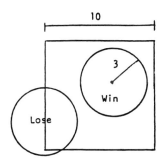
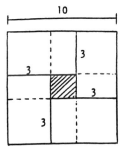

Third, after the students looked at this example, they examined the general problem: a square of side S and a coin of radius R. To win, the center of the coin must be at least R units from the side. The area of the square is S^2, while the area of the winning region is $(S - 2R)^2$. The theoretical probability of winning any such game is given by

$$P = \frac{(S - 2R)^2}{S^2}$$

where S is the length of the side, while R is the radius of the coin.

Fourth, they decided to examine the following question: If the square is 10 cm on a side, how small should the radius of the coin be if the player has just as good a chance of winning as losing? The equation above, which gives us the probability if winning, needs to be $1/2$. Solving the equation:

$$\frac{1}{2} = \frac{(10 - 2R)^2}{100} \Rightarrow 50 = (10 - 2R)^2$$
$$\Rightarrow \sqrt{50} = 10 - 2R$$
$$\Rightarrow R = \frac{10 - \sqrt{50}}{2}$$
$$\Rightarrow R = 1.46$$

shows that the radius of the coin must be approximately 1.46 cm. If the coin is larger than this, then the player has a better chance of losing than winning. If the coin is smaller, he or she has a better chance of winning than losing.

Fifth, some students constructed a board and empirically tested their results. Of course, they found the empirical probability to be different from (but close to) the theoretical probability.

Sixth, two students then wrote a computer problem to simulate the actual throwing of darts at a target (a Monte Carlo routine of plotting points at random on a computer screen and counting shots and hits). They used the program to study the relationship of their empirical results with the theoretical results, noting that as the number of shots increased, the difference between their empirical and theoretical results decreased.

Finally, they examined several related problems, such as:

1. Suppose that the squares are arranged in a checkerboard pattern, alternating red and white squares. If only the white squares win, what does this do to the probability of winning for a given size coin?
2. Suppose the pattern is made of congruent hexagons, rather than squares. If the side of a hexagon was 20 mm, while the coin had a radius of 7 mm, what is the probability of winning?
3. For a coin of radius 10 mm, how much better is a hexagon of side 25 mm than a square of side 25 mm?
4. For the original problem with squares, what is the relationship between S and R for which the probability is greater than $1/2$ of winning? What is the significance of this in terms of playing or not playing the game?

This example, worked on for several days by these students, illustrates the kinds of problem situations envisioned by the authors of the *Standards*. Such instructional activities should expect students to be active and should constantly extend the structure of the mathematics that they know by having them make, test, and validate conjectures.

REFLECTIONS

The analogy and the example have been presented to illustrate the vision of practice being proposed by the mathematical sciences education community. Keeping both in mind, I would like to comment briefly on the connections between this vision and what mathematicians do and how mathematics is learned.

Doing Mathematics

The *Standards* emphasize that students should learn "to do mathematics," and both the analogy and example were presented to illustrate how this might be accomplished. But, is this really what mathematicians do? To appreciate what it means "to do" mathematics, one must recognize that mathematicians argue among themselves about what mathematics is acceptable, what methods of proof are to be countenanced, and so forth. Kitcher (1988) claims that:

> Mathematical practice has five components: a language employed by the mathematicians whose practice it is, a set of statements accepted by those mathematicians, a set of questions that they regard as important and as currently unsolved, a set of reasonings that they use to justify the statements they accept, and a set of mathematical views embodying their ideas about how mathematics should be done, the ordering of mathematical disciplines, and so forth. (p. 299)

From this perspective, doing mathematics cannot be viewed as a mechanical performance or an activity that individuals engage in by solely following predetermined rules. Mathematical activity can be seen more as embodying the elements of an art or craft than as a purely technical discipline. This is not to say that mathematicians are free to do anything that comes to mind. As in all crafts, as Kitcher argues, there will be agreement, in a broad sense, about what procedures are to be followed and what is to be countenanced as acceptable work. These agreements arise from the day-to-day intercourse among mathematicians. Thus, a mathematician engages in mathematics as a member of a learned community that creates the context in which the individual mathematician works. The members of that community have a shared way of "seeing" mathematical activity. Their mutual discourse will reinforce preferred forms and a sense of appropriateness, elegance, and acceptable conceptual structures (King & Brownell, 1966). Furthermore, the community promotes and reinforces its own standards of acceptable work, and as Hagstrom (1965) suggests, a major characteristic of a mathematical/scientific community is the continued evolution of its standards.

However, recognizing that mathematics is a cultural product created by people does not specify what mathematicians do. In fact, because mathematicians are human beings, many of the things they do are quite uninteresting. What is important are its essential activities. Identifying and describing these features for educators has been the focus of the work of several scholars. Even though in each of the following examples, a different feature has been emphasized, all should be seen as different themes on essentially the same phenomena.

The most influential scholar in the field in the United States has been the noted mathematician, George Pólya. His writings stressed the importance of mathematical problem solving. His major books described, through examples, the way in which mathematicians examine nonroutine problems. These books are: *Mathematics and Plausible Reasoning* (1954), *How to Solve It* (1957), and *Mathematical Discovery* (1967). His emphasis has been on the elements of plausible reasoning that lead to the discovery of mathematical assertions. Pólya called this kind of reasoning *heuristics,* or the mental operations typically useful in the process of solving mathematical problems. His influence on mathematical education is reflected in such documents as *An Agenda for Action* (NCTM, 1980), and *Curriculum and Evaluation Standards for School Mathematics* (NCTM, 1989). In these (and many other) documents, problem solving is highlighted as the major activity of mathematicians. In fact, solving nonroutine problems is the central theme of the current reform movement in school mathematics in the United States.

Recently, Lynn Steen, past president of the Mathematical Association of America, has written several papers that have stressed mathematics as "the science of patterns." His writings include: *Mathematics Tomorrow* (1981), "Forces for Change in the Mathematics Curriculum" (1986), "Out From Underachievement" (1988a), "A Science of Patterns" (1988b), and "A "New Agenda" for Mathematics Education" (1988c). In addition, he was the principal author of the recent report of the Mathematical Sciences Education Board, *Everybody Counts* (1989) and is editor of two other MSEB reports, *Reshaping School Mathematics: A Philosophy and Framework of Curriculum* (1990a) and *On the Shoulders of Giants* (1990b). Steen's emphasis is on the fact that mathematics is at its roots an empirical science. As he states:

> The mathematician seeks patterns in number, in space, in science, in computers, and in imagination. Mathematical theories explain the relations among patterns; functions and maps, operators and morphisms bind one type of pattern to another to yield lasting mathematical structures. . . . Patterns suggest other patterns, often yielding patterns of patterns. In this way, mathematics follows its own logic, beginning with patterns from science and completing the portrait by adding all patterns that derive from the initial ones. (1988b, p. 616)

Thus, he sees that all students need to experience the search for patterns at all levels. Mathematics is not a fixed set of concepts and skills to be mastered, but an empirical science.

The most prolific and influential European scholar has been the Dutch mathematician, Hans Freudenthal. He was founding editor of the international journal *Educational Studies in Mathematics,* and his major books on mathematics educa-

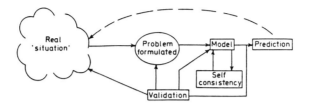

FIG. 8.1. The basic components in model building. Copyright (1976) by J. G. Andrews and R. R. McLone, and Butterworths-Heinemann, p. 3. Used with permission.

tion include: *Mathematics as an Educational Task* (1973), *Weeding and Sowing* (1978), and *Didactical Phenomenology of Mathematical Structures* (1984). Central to Freudenthal's work are "the strategies of mathematizing." Mathematizing involves representing relationships with a complex situation in such a way as to make it possible to put them into a quantitative relationship with one another. The basic components of mathematizing are shown in Fig. 8.1. The mathematician must first decide whether the variables and relationships between variables are important and which do not matter when confronting a complex problem situation. Second, a mathematical model is made, numbers assigned to variables, and numerical procedures used to make predictions. Then the results are examined. It is this way of approaching nonmathematical phenomena that has enabled researchers to make such wise use of computers as tools. The computer does not mathematize, but once a model is created for a situation and quantified, the computer can quickly perform extensive computations to produce predictions for complex situations. Freudenthal believes that learning these strategies should be the central focus of all mathematics curriculum. From this perspective, "Mathematics for all and everyone" is the motto he chose to define the work of the research and development institute he founded at the University of Utrecht. His argument is that all students can and must experience mathematics in this manner. The complexity of the problem situations and the sophistication of the mathematical models may vary, but all can build mathematical models.

Finally, as a mathematics education researcher and chair of the NCTM Commission on Standards for School Mathematics, I have presented still another complementary view. My works on the issue of "doing" mathematics include: "Activities Basic to Learning Mathematics" (1975), "A Common Curriculum for Mathematics" (1983), *School Mathematics: Options for the 1990s* (1984), and with Stewart, *The Monitoring of School Mathematics: Background Papers* (1987). In addition, I was the chair of the group that wrote the *Curriculum and Evaluation Standards for School Mathematics* (NCTM, 1989). In these works I argued that even with a superficial knowledge about mathematics, it is easy to recognize four related activities common to all of mathematics: abstracting, inventing, proving, and applying.

The abstractness of mathematics is easy to see. We operate with abstract numbers without worrying about how to relate them in each case to concrete objects. The process of abstracting is characteristic of each branch of mathematics. The concept of a whole number and of a geometric figure are only two of the earliest and most elementary concepts of mathematics. They have been followed by a mass of others, too numerous to describe, extending to such abstractions as complex numbers, functions, integrals, differentials, functionals, n-dimensional spaces, infinite-dimensional spaces, and so forth. These abstractions piled, as it were, on one another, have reached such a degree of generalization that they have apparently lost all connection with daily life, and the "ordinary mortal" understands nothing about them beyond the mere fact that all this is incomprehensible. In reality, of course, such is not at all the case. Although the concept of n-dimensional space is no doubt extremely abstract, it does have a completely real content, which is not difficult to understand.

Inventing involves creating a law or relationship. There are two aspects to all mathematical inventions: the conjecture (or guess) about a relationship, followed by the demonstration of the logical validity of that assertion. All mathematical ideas—even new abstractions—are inventions (like irrational numbers). Also, to assist them in the invention of their abstractions, mathematicians make constant use of theorems, mathematical models, methods, and physical analogues, and they have recourse to various completely concrete examples. These examples often serve as the actual source of the invention.

No proposition is considered as a mathematical product until it has been rigorously proved by a logical argument. The demand for a proof of a theorem pervades the whole of mathematics. We could measure the angles at the bases of a thousand isosceles triangles with extreme accuracy, but such a procedure would never provide us with a mathematical proof of the theorem that the base angles of an isosceles triangle are congruent. Mathematics demands that this result be deduced from the fundamental concepts of geometry, which are precisely formulated in the axioms. And so it is in every case. To prove a theorem means that the mathematician deduces it by a logical argument from the fundamental properties of the concepts related to that theorem. In this way not only the concepts, but also the methods, of mathematics are abstract and theoretical. The importance of building mathematical arguments about assertions cannot be underestimated. These arguments are central to all mathematical discourse (Putnam, Lampert, & Peterson, 1989).

Another characteristic feature of mathematics that all should realize is the exceptional breadth of its applications. In the first place we make constant use, almost every hour, in industry and in private and social life of the most varied concepts and results of mathematics, without thinking about them at all. For example, we use arithmetic to compute our expenses or geometry to describe the floor plan of an apartment. Second, modern technology would be impossible without mathematics. Scarcely any technical process could be carried through without building an abstract mathematical model as a basis for carrying out a sequence of more or less complicated calculations, and mathematics plays a very important role in the development of new branches of technology. Finally, it is true that every science, to a greater or lesser degree, makes essential use of mathematics. The progress of the sciences would have been completely impossible without mathematics. For this reason, the requirements of mechanics, astronomy, and physics have always exercised a direct and decisive influence on the development of mathematics. In other sciences (including the social sciences), mathematics plays a smaller role, but here, too, it finds important applications.

In summary, in each of the arguments mathematics is a rational human creation. It is a vast collection of ideas derived as a consequence of searching for solutions to social problems. The abstractions and inventions help us make sense of our world and ourselves. This is true whether one emphasizes problem solving, searching for patterns, mathematizing, abstracting, inventing, proving, or applying. These are merely descriptions of what mathematicians do. Note also that neither the acquisition of concepts nor of skills has been emphasized by these authors. This is not to imply they are unimportant. However, it does imply that their acquisition is vacuous unless they are used in "doing" mathematics.

The Learning of Mathematics

There are two parts to the assumption that "instruction should be developed from problem situations" (NCTM, 1989, p. 11). First, if students work on such problem situations, such work will enhance learning; and second, suitable activities will be found or created, which will "connect ideas and procedures both among different mathematical topics and with other content areas" (p. 11). The first part of this assumption seems warranted, the second is more problematic.

Psychologists claim that knowledge about related concepts, skills, and contexts is organized in a student's memory in "schema." These schema develop over long periods of time and by continual exposure to related contextual events. The development of schema promotes both problem-solving skill and recall of textual material. These schema appear to guide, organize, and direct both the search for a problem solution and the retrieval of expository details. For example, if students in Grade 1 have developed a "measurement schema" for measuring lengths with a variety of properties (units are arbitrary, common units are useful for communication, big units yield a small number and small units a big number, tools are useful, error is always involved), then it should be possible to create problem situations for new attributes (weight, area, volume, acceleration, value). Note, however, two things. First, it is important to know what knowledge students bring to such activities; second, one never knows all there is to know about any domain such as measurement. There will always be new attributes with special properties to be measured.

It is also argued that the encoding, comprehension, and retrieval of new information are aided when material is presented in a form that has structure and when the student is cognizant of that structure. In particular, these processes are facilitated when the information can be assimilated into existing schema of the learner. Thus, when information is presented in a familiar contextual setting, the transitions and the concepts and procedures are likely to be remembered. Some psychologists call this a generic story shell for the schema.

They also claim that although students appear to make use of cause-and-effect relations in encoding information from problem situations, they have difficulty with conditions regulating the use of specific mathematical procedures. Failure to recognize these conditions often results in the development of buggy algorithms. For example, students who say that $43 - 17 = 34$ probably have overgeneralized the statement "always take the smaller number from the larger."

It is also argued that problem-solving ability and encoding of information are enhanced when schema are interrelated and form a hierarchical arrangement analogous to the way knowledge is used. For example, if multiplication is often used to solve proportion problems, then the relationship between multiplication as repeated addition and proportion needs to be carefully developed and connected.

Finally, some scholars liken the process of conceptual change in individuals to that of the scientific community and draw on recent developments in the philosophy of science to gain insight into conceptual change. In order for the radical reorganization of central concepts to occur, it is claimed that four conditions must be met. (1) There must be some sense of dissatisfaction with the existing conceptual framework; (2) there must be alternate conceptions that are intelligible; (3) the alternate conceptions must be initially plausible; and (4) the alternate conceptions must be seen as fruitful, useful, or valuable. In fact, Nussbaum and Novick (1982) suggest a three-part instructional sequence designed to encourage students to make desired conceptual changes. They propose the use of an exposing event, which encourages students to use and explore their own conceptions in an effort to understand the event. This is followed by a discrepant event, which serves as an anomaly and produces cognitive conflict. It is hoped that this will lead the students to a state of dissatisfaction with current conceptions. A period of resolution follows, in which the alternative conceptions are made plausible and intelligible to students, and in which students are encouraged to make the desired conceptual shift.

These psychological views are important and give credence to the claim that instruction should be based on problem situations. However, what kind of activities are appropriate is not clear, because mathematical knowledge must develop simultaneously along two distinct dimensions. The first is a symbolic dimension and includes the signs, symbols, and rules for the use of those symbols in specific mathematical domains. The second is the set of problem situations for which those symbols are used and given meaning. de Lange (1987) portrayed the

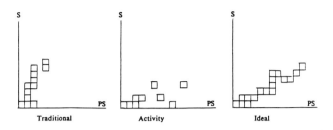

PS = Problem Situations
S = Symbols, Rules, Procedures

FIG. 8.2. Three types of instructional programs. Adapted from a model developed by de Lange (1987, p. 45).

curriculum engineering problem in a manner shown in Fig. 8.2. The first illustration shows the development of a mathematical domain in most traditional problems. The domain is first introduced via some problem situation, then most of the time is spent on mastering the use of the symbols with an occasional applied activity. The second illustration portrays an activity program that is rich in a variety of things for students to do but never has the students step back from the situations to develop the connections, symbols, and rules for the domain in a coherent manner. Ideally, what is being advocated is a program that does both, as in the third illustration.

Although there is no doubt that many interesting activities exist or can be created, "do they lead anywhere?" is a serious question. An activity approach could lead to "no mathematics at all." Keitel (1987) has argued that the problem is one of relevance. Too often a problem situation is judged to be relevant through the eyes of adults, not children. Also, that vision is undoubtedly a Western, middle-class, static vision. Concrete situations, by themselves no guarantee that students will see relevance to their worlds, will be relevant for all students, and will prepare them to deal with a changing, dynamic world. In response to this concern, D'Ambrosio (1987) argues that the "emphasis should be shifted to 'really real' situations" (p. 141). By this he means that projects of a global nature, "such as building a cabin, mapping a town, or assessing the water consumption of a community" (p. 141), should be taught, because they require modeling and problem solving. However, Greenberg (1987) stated: "It is hard for me to imagine that the system of mathematics education advocated by D'Ambrosio, ideal though it might be, could ever prepare our young for the future advancement and betterment of the socio-economic and political framework of society" (p. 217). In essence, to paraphrase Damerow and Westbury (1985): There may be a need to change the contents and conditions of the school mathematics curriculum so that all students can study more and somewhat different mathematics, but how to do this is currently elusive.

PRINCIPLES

The following five principles are a consequence of my consideration of the vision of mathematics instruction, the issues raised in this chapter, and my belief that an authentic school mathematics program can be developed. They are presented because learning by active participation implies that the program must be a program of activities from which knowledge or skills can be developed. However, simply developing a collection of interesting activities is not sufficient. The knowledge gained must lead somewhere. Thus, what is constructed by an individual depends to some extent on what is brought to the situation, where the current *activity* fits in a sequence leading toward a goal, and how it relates to mathematical knowledge. Thus, the suitability and effectiveness of selected learning activities is an empirical problem. It depends on both the student's prior knowledge and the student's expectations. These can only be determined by teachers. For this reason, curriculum development cannot be an exact science. The "intended" curriculum can only include our best guesses about what will both interest students and lead all toward development of mathematical power. At the same time, the "actual" curriculum depends on teacher choice, and the "achieved" curriculum depends on each student's interest and prior knowledge.

Principle 1. Conceptual Domains Should Be Specified

The mathematical domains that we wish to develop in students must be identified, and then a curriculum must be built around those conceptual domains (Vergnaud, 1983, called these conceptual fields). The domains should be selected because of their generality and ability to subsume more specialized components of the curriculum deemed desirable for the development of problem-solving ability and quantitative reasoning. Some of the domains that are reflected in the NCTM *Standards* are:

- counting and numeration (the assignment of numbers to sets)
- measurement (the assignment of numbers to continuous attributes)
- addition and subtraction of whole numbers
- multiplication and division of whole numbers
- common fractions
- decimal fractions
- ratio and proportion

- integers
- descriptive geometry
- motion geometry
- linear equations
- patterns, relations, and functions.

These domains should not be considered independent of each other. While it is true that each domain has some unique properties (signs, symbols, rules), I would rather think of them as petals of a flower whose center involves problem solving, communication, and reasoning and whose petals are all interconnected.

Principle 2. The Domains Should Be Segmented Into Curriculum Units, Each of Which Takes Two to Four Weeks to Teach and Each of Which Tells a Story

Students should be expected to construct meanings, interrelate concepts and skills, and use those meanings in a variety of problem situations. Each unit should be similar to a chapter in a Dickens' novel. It should introduce or reintroduce the characters to the reader, and there should be a new problem situation to be resolved that involves conflict, suspense, crisis, and resolution.

Principle 3. Students Should Be Exposed to the Major Conceptual Domains as They Arise Naturally in Problem Situations

Ideas are best introduced when students see a need or a reason for their use. Promoting the development of integrated schema requires an integrated curriculum. Problem situations may include the historic reasons for the development of a mathematical domain, the relationship of that domain to other domains, or the uses of that domain.

Principle 4. The Activities Within Each Unit Should Be Related to How Students Process Information

Each unit should provide review of prior concepts and skills and lay foundations for concepts and skills to be learned later. Activities used to teach algorithms should differ from those used to teach problem solving, and activities requiring assimilation should differ from those requiring accommodation. Furthermore, the method of instruction is likely to differ. Students might be addressed as a large group when being exposed to new information and work in small groups when inventing, proving, or applying. Assimilation may require exercises that involve little prior knowledge, while accommodation may demand a dissimilar array of problem situations involving varying cognitive structures. A higher degree of teacher-imposed structure and control may be desirable for lower level cognitive outcomes, while a greater degree of group autonomy may aid higher level cognitive outcomes.

Principle 5. Curriculum Units Should Always Be Considered as Problematic

All curriculum sequences need to be adapted and modified in light of what knowledge the students bring to the unit and the context in which instruction takes place. The difference between the intended and the actual curriculum should be apparent. What actually occurs will differ among classrooms. The program cannot be "teacher proof." Instead, the program should assist each teacher in making reasonable adaptations, so that the prior knowledge and interests of the students are taken into account in instruction.

CONCLUSION

The authors of the *Standards* have presented a vision of mathematics all students should have an opportunity to learn and of the way in which instruction should occur. Classrooms should be places where interesting problems are explored using important mathematical ideas. For example, in various classrooms, one could expect to see students recording measurements of real objects, collecting information and describing the properties of objects using statistics, or exploring the properties of a function by examining a graph. This vision sees students studying much of the same mathematics currently taught, but with quite a different emphasis. I can see how this picture is connected to what we know about problem solving and thinking. Thus, I am confident that such a mathematics program can be developed and implemented. The content that should be included in a school mathematics program can be developed and implemented. The content that should be included in a school mathematics program can be specified. Materials such as texts, courseware, and tests can be produced so that constructive learning will take place in classrooms. However, I harbor no illusions of immediate reform. This reform will only be accomplished by hard work.

ACKNOWLEDGMENTS

The research reported in this chapter was supported by the Office of Educational Research and Improvement of the U.S. Department of Education and by the Wisconsin Center for Education Research, School of Education, University of Wisconsin-Madison. The opinions expressed in this chapter are those of the author and do not necessarily reflect the view of the OERI or the Wisconsin Center for Education Research.

REFERENCES

Andrews, J. G., & McLone, R. R. (1976). *Mathematical modelling.* London/Boston: Butterworths.

D'Ambrosio, U. (1987). New fundamentals of mathematics for schools. In T. A. Romberg & D. M. Stewart (Eds.), *The monitoring of school mathematics: Background papers. Vol. 1: The monitoring project and mathematics curriculum* (pp. 135–148). Madison: Wisconsin Center for Education Research.

Damerow, P., & Westbury, I. (1985). Mathematics for all: Problems and implications. *Journal of Curriculum Studies, 17*(2), 175–186.

de Lange, J. (1987). *Mathematics, insight, and meaning.* Utrecht, The Netherlands: University of Utrecht.

Freudenthal, H. (1973). *Mathematics as an educational task.* Dordrecht, Netherlands: D. Reidel.

Freudenthal, H. (1978). *Weeding and sowing.* Dordrecht, Netherlands: D. Reidel.

Freudenthal, H. (1984). *Didactical phenomenology of mathematical structures.* Dordrecht, Netherlands: D. Reidel.

Greenberg, H. J. (1987). Mathematics education: A really, real, real world problem: Reactions to Chapters 6–10. In T. A. Romberg & D. M. Stewart (Eds.). *The monitoring of school mathematics: Background papers. Vol. 1: The monitoring project and mathematics curriculum* (pp. 213–221). Madison: Wisconsin Center for Education Research.

Hagstrom, W. O. (1965). *The scientific community.* New York: Basic Books.

Keitel, C. (1987). What are the goals of mathematics for all? *Journal of Curriculum Studies, 19*(5), 393–407.

King, A. R., Jr., & Brownell, J. A. (1966). *The curriculum and the disciplines of knowledge: A theory of curriculum practice.* New York: John Wiley & Sons.

Kitcher, P. (1988). Mathematical naturalism. In W. Aspray & P. Kitcher (Eds.), *History and philosophy of mathematics, Minnesota studies in the philosophy of science* (Vol. xi, pp. 293–325). Minneapolis: University of Minnesota Press.

Mathematical Sciences Education Board (MSEB). (1989). *Everybody counts: A report to the nation on the future of mathematics education.* Washington, DC: National Academy Press.

Mathematical Sciences Education Board (MSEB). (1990a). *Reshaping school mathematics.* Washington, DC: National Academy Press.

Mathematical Sciences Education Board (MSEB). (1990b). *On the shoulders of giants: A philosophy and framework of curriculum.* Washington, DC: National Academy Press.

National Council of Teachers of Mathematics (NCTM). (1980). *An agenda for action: Recommendations for school mathematics of the 1980s.* Reston, VA: Author.

National Council of Teachers of Mathematics (NCTM). (1989). *Curriculum and evaluation standards for school mathematics.* Reston, VA: Author.

North Carolina School of Science and Mathematics. (1988). *Geometric probability.* Reston, VA: National Council of Teachers of Mathematics.

Nussbaum, J., & Novick, S. (1982). Alternative frameworks, conceptual conflict, and accommodation: Toward a principled teaching strategy. *Instructional Science, 11,* 183–200.

Pólya, G. (1954). *Mathematics and plausible reasoning. Vol. 1: Induction and analogy in mathematics.* Princeton, NJ: Princeton University Press.

Pólya, G. (1957). *How to solve it.* New York: Doubleday Anchor Books.

Pólya, G. (1967). *Mathematical discovery.* New York: John Wiley & Sons.

Putnam, R. T., Lampert, M., & Peterson, P. L. (1989). *Alternative perspectives on knowing mathematics in elementary schools.* East Lansing: Michigan State University.

Romberg, T. A. (1975, October). Activities basic to learning mathematics: A perspective. *The NIE conference on basic mathematical skills and learning: Vol. 1: Contributed position papers.* Washington, DC: National Institute of Education.

Romberg, T. A. (1983). A common curriculum for mathematics. In G. D. Fenstermacher & J. I. Goodlad (Eds.), *Individual differences and the common curriculum* (pp. 121–159). Chicago: University of Chicago Press.

Romberg, T. A. (1984). *School mathematics: Options for the 1990s: Chairman's report of a conference.* Washington, DC: U.S. Government Printing Office.

Romberg, T. A., & Stewart, D. M. (1987). *The monitoring of school mathematics: Background papers. Vol. 1: The monitoring project and mathematics curriculum.* Madison: Wisconsin Center for Education Research.

Steen, L. A. (1981). *Mathematics tomorrow.* New York: Springer-Verlag.

Steen, L. A. (1986, Winter). Forces for change in the mathematics curriculum. *Wisconsin Teachers of Mathematics, XXXIV,* 3–7.

Steen, L. A. (1988a, Fall). Out from underachievement. *Issues in Science and Technology,* pp. 88–93.

Steen, L. A. (1988b). The science of patterns. *Science, 240,* 611–616.

Steen, L. A. (1988c). A "new agenda" for mathematics education. *Education Week, 28,* 21.

Treisman, U. (1989, November 30). Comments on the *Standards* made at the National Forum on Effective Urban and Metropolitan Schooling, New York.

Vergnaud, G. (1983). A classification of cognitive tasks and operations of thought involved in addition and subtraction problems. In T. P. Carpenter, J. M. Moser, & T. A. Romberg (Eds.), *Addition and subtraction: A cognitive perspective* (pp. 39–59). Hillsdale, NJ: Lawrence Erlbaum Associates.

Comments on Thomas Romberg's Chapter

Gaea Leinhardt
University of Pittsburgh

To set the stage for my reaction to Romberg's chapter, it is probably of some use for me to introduce myself and my perspective. I am an educational researcher who uses the tools of cognitive psychology and anthropology to study mathematics classrooms and the teachers in those classrooms. The teachers whom I study work in tough educational settings. Their students do well on standardized tests, on the quantity of future mathematics courses taken, and in terms of having a strong tendency to stay in school. These are not the classrooms of the university laboratory; these are not the suburban public schools; these are schools where it is hard to teach and harder still to learn. The majority of "my" teachers have lived through at least six major educational reforms in the past 25 years, and through their eyes I share a certain skepticism toward reform in general. However, through their eyes I also hold dear a vision of what education could be like and why education is so critically important for our children. With this as a context I turn to my major theme: Given the tremendous work of the National Council of Teachers of Mathematics (NCTM) in producing useful and usable standards, what are the cautions and hazards ahead that we ought to look out for? In this response I sketch six areas of potential concern.

BEWARE OF TURNING GOOD IDEAS INTO DOGMAS

The collection of activities and documents that I associate with the label "NCTM Standards" is very impressive. It includes years of work by hundreds of people all over the country. This has been a unique activity in American educational history, because it has been both well orchestrated and reasonably democratic. It is not hard to do either of these well. Getting top-flight scientists together to decide what should be is not a trivial task, but it is not impossible—witness the Woods Hole experience. Nor is it difficult to get large groups of people temporarily fired up about a particular reform. But it is hard to do both together. What the groups involved with the development and production of the NCTM Standards have done includes completely redefining the vision of mathematics education in terms of goals and processes. The effort is significant and is to be applauded.

There is, however, a fine line between building groundswells of popular support for a new positive reform and the implantation of a dogma. Dogmas are to be avoided. Dogmas can be recognized by the manner in which ideas that do not conform to the mainstream are dealt with. Are contradictory ideas ridiculed instead of answered? Are aspersions cast on the authors of such ideas rather than considering the merits and flaws represented by the concern? Is there any diversity represented among the leaders of the field, or is it all one big happy family? Is there a tendency for people to be purer than the purists? Stated differently, is anyone good enough other than blood relatives? Does the reform deal with the history of reforms that came before in ways other than simply tossing them on the garbage heap? Are failings and flaws treated as opportunities for improvement, or are they silenced because the "movement" cannot afford any dissension? We must be wary of "true believers," because real progress comes from reflecting analytically and critically on both success and failure.

WHAT ARE THE PARAMETERS OF A GOOD PROBLEM?

One cornerstone of the NCTM Standards is an anchoring of instruction to problems. These critical problems are supposed to help students build up repertoires of actively available information—knowledge that can be disentangled from the specific problem at a future time and used in a new situation. But what is a good problem? The family of parameters that make a good problem are not worked out well.

Words like "authentic" are often used to describe a good problem. But what makes a problem authentic? I would like to suggest that authenticity is like beauty, in the eye of the beholder. Surely, one cannot argue that the checkerboard problem Romberg cited is somehow more authentic than Lampert's problem of finding Descartes' discovery (linear graphs) (Lampert, 1988; Leinhardt, 1992). Or that determining how many baseball cards are needed to cover a wall is fundamentally more or less interesting than weird age problems. An authentic problem, it seems, must deal with some aspect of mathematics that is inherently important and not trivial. It can do that through both abstract and real-world issues. I personally do not find the grocery store inherently more interesting than the intersection of planes (Mathematical Association of America & Pólya, 1965).

Changing the setting or context of a problem alone does not necessarily improve its utility for teaching. Telling a good tale and then sticking in some mathematical ideas may not end up being more powerful than determining real mathematical subtleties and anchoring them to real situations. Solid empirical work is needed to discover what kinds of problems work, how they can be corrupted, and how they can be enhanced.

ISSUES OF COHERENCE AND ACCESS

The logic behind using core problems as defining aspects of the curriculum has a well-worked-out rationale. The intention is that students will situate important skills and strategies in a particular problem space, and that the use of problems will help to develop an intellectual coherence. The alternative (the one that is being replaced) is one in which students learn separate skills and concepts and are then expected to draw on these skills and concepts effectively when faced with problems. The fact that students are often unable to make use of their abstract knowledge, even when they have the necessary skills and concepts in place, is often referred to as the problem of inert knowledge.

The idea behind a problem-centered approach is that when students are taught to apply and invent procedures in the direct service of a meaningful task, they will both remember the procedure better, and they will be able to apply the procedure in the service of other tasks. The hope is that a kind of intellectual coherence will be achieved. I think this is very much worth a try, but if it does not work in the way we hope, we should realize that we are still leaving the students with a considerable task of transformation. Teaching problem- or context-based mathematics may beg both the transfer and transformation questions.

Suppose, for example, that a student is doing some type of problem that involves developing and then combining measurement units. In the course of this problem the student "invents" a process by which he or she adds fractions with unlike denominators. Now suppose the student, some months later, is given a different kind of problem, not at all in the measurement world, in which it might prove useful to combine fractions with unlike denominators. To do this the student must recognize the situation as having some similar properties to the earlier problem and realize that the procedure invented last time is "true" and will always work. He or she must detach the procedure, and then use the process correctly. Or, the student may invent a new procedure and go through the entire process again. The student still must access knowledge that has been constructed and stored in one way and use it in another unique way, and this is always hard. Making the initial construction and use more meaningful and less mindless may indeed be very helpful, but it does not, by any means, guarantee a particularly easy task when that original organization must be deconstructed and reaccessed.

BOOKS ARE THE FRIENDS OF IDEAS NOT THE ENEMIES

Let's have no book burnings here. The intense rejection of textbooks that appears to be a concomitant of the reform movement seems, to me, to be unfortunate. The consequence of rejecting textbooks is to reject scaffolding for the teacher, while at the same time scaffolding is applauded for the student. It is also to reject the visible trace provided by a written record, one that is inspectable both backward and forward, one that also helps to build coherence.

Many teachers rely too heavily on texts. However, this does not mean that they will be better teachers if we take texts away from them. Further, elimination of textbooks means that students will be completely dependent on the teacher as a source of all knowledge and as a source of curriculum. Even though the rhetoric behind the reform is to empower students with their own learning, there is a fundamental disempowering when the only curriculum resides inside the teacher's head. The curriculum then becomes hidden from public, parental, and student view. Rejecting textbooks suggests that somehow teachers will be better equipped to handle the difficult tasks of creating problems, teaching lessons, and designing tests than panels of individuals with expert consultants whose sole task is to design such texts. If teachers are overly dependent on texts, then we should teach teachers to use them and critique them better, not throw them away. Throwing them away is not necessarily the best action to take. If texts are limited and little more than collections of exercises with pictures in between, then they should be redesigned and expanded. If one textbook is too limiting, then students and teachers should be encouraged to use several.

WILL WE PRODUCE COMMUNITIES OF SCHOLARS OR LORD OF THE FLIES?

A considerable emphasis has been placed on the social construction of knowledge and the role of distributed knowledge in problem-solving groups. The vision supported by this emphasis is an admirable one—one in which groups of students with a shared set of goals work together and support each other in the quest for knowledge and the solution of complex and demanding problems. In the best view, students in these groups work well beyond each individual's competence and become involved in the entire problem. If things work this way, then this is undoubtedly an excellent course to follow. However, it takes a lot of work to teach teachers and students how to manage these dynamics. The nonfunctional version of "cooperative learning" has one or two students doing all of the work, because they are motivated by fear and grades. Or worse, that physically or socially powerful individuals in the class come to demand service from the "nerds" in the environment, service that often goes well beyond the particular problem on which the "group" was supposed to work. Intellectual problems do not simply become easier, more fun, and more engaging, because you tell a group of students to work on them. Getting groups to work together requires sophisticated social engineering not simple proximity.

THE POOL

My last caution is addressed to the mathematicians at our meeting. There is a certain awkwardness in calling for reform. Currently, mathematics programs keep very few of the incoming college freshmen who declare as math majors. Mathematics programs are comfortably situated in a buyer's market, and the top of the mathematical heap is doing quite well. The United States places last overall in international studies of the entire population, but frequently it places first at international mathematics competitions. My comments are an attempt to demonstrate that it is worthwhile both altruistically and selfishly to be worried about those in the society who are not mathematicians. Of course, the most convincing argument here would be if I could show that, of those who are deleted from the pool, there are individuals who would have made major contributions to mathematics as a field. I am unable to do this, but it is an issue worth considering. Instead I briefly describe the moments at which various cuts are made and suggest somewhat rhetorically that perhaps those tossed aside could benefit from and be of benefit to mathematics as a field.

In most schools in the United States the mathematics curriculum through the third grade is relatively homogeneous. That means the students are not tracked or laned for mathematics, and that they receive the same overall curriculum. The quick, interested student and the slower, bored student are either in the same class or are in equivalent classes. After the third grade the situation changes.

In the fourth and fifth grades both multidigit multiplication and long division have traditionally been introduced, while in fifth and sixth grades operations with fractions are introduced. These three sets of procedures take up a lot of time in the traditional curriculum. The conceptual underpinnings of multiplicative structures are often poorly taught and poorly understood by the teacher. This portion of the curriculum then becomes overwhelmingly procedurally driven. Students who do the "right" procedures quickly are considered good, while those that make careless errors or cannot follow the procedures are considered and come to believe that they are not very gifted mathematically.

The path for future mathematics tracking has begun. As currently constituted, middle school material (grades 6, 7, 8) can be done in 1, 2, or 3 years. Students are often divided into different groups depending on how well they got through multidigit multiplication, long division, and fractions and on how long they are expected to take getting through the next chunk of material. If the material is done in 1 year, the student may well take both first-year algebra and geometry before high school (sometimes a school will use the 2 years to teach 1 year of algebra more slowly). If the middle school material is covered in 2 years, the student will take algebra in eighth grade. If the middle school material is covered in 3 years, then the student will take algebra in the first year of high school or even possibly something else called consumer math or general mathematics. This something else is a stripped down and reformulated middle school mathematics curriculum. In many states that is all the math a student needs to take. In several states the students who take 1 year of general mathematics or 1 year of algebra in high school are eligible to, and have become, our elementary school teachers. These are the same teachers who in turn will teach the middle school mathematics curriculum.

Mathematics tracking occurs, formally and informally, at various points in the curriculum. Many of the decisions are made informally as early as third grade (based on skill tests and paper-and-pencil procedures); other more formal tracking contributes to the approximately 50% annual attrition rate in mathematics enrollment from grade 8 onward. By the end of high school there is a small pile of mathematical survivors and a large pile of discards. The people who end up deciding on grants, administering our government and our universities, who run our schools and teach our teachers, which pile have they landed in? Most frequently they have been in the group that did not accelerate through the middle school years before high school. In fact, many of our doctors and lawyers were in that same group as well. They do, however, take a fairly full load of high school mathematics. This means that when mathematicians wish to argue for a certain literacy in the public at large, they are speaking to people who have a very solid grasp of the mechanics of fractions but little understanding of them.

The point here is that even mathematicians exist in a larger context, in a world where decisions that affect their lives and the lives of their children are taking place. These decisions are made by people who have never had the opportunity to experience or enjoy the beauty, humor, and delight of mathematics itself. As long as the view of mathematics is that it is a largely biologically determined knowledge, an endowment that is unattainable, it will have no special place in our culture, and it will have to fend for itself among a somewhat hostile and resentful population. The NCTM *Standards* represent an attempt to improve that situation.

In sum, the work on which Romberg has reported is of immense value. It is of value for the students in the United States today and for our society as a whole tomorrow. The claims made and goals striven for are large as well as high; we need to recognize that they may not all be realized. However, it is likely that more of them will be realized if some concern is shown to issues of dogmatism, problem formulation, intellectual coherence, uses of texts, social dynamics, and the current pools from which mathematicians are and are not drawn.

REFERENCES

Lampert, M. [Teacher]. (1988, October). Lesson on graphing.

Leinhardt, G. (1992, April). *Standing to the side: A case of instructional dialogues on functions and graphing.* Paper presented at the 70th annual meeting of the National Council of Teachers of Mathematics, Nashville, TN.

Mathematical Association of America [Producer], & Pólya, G. [Instructor]. (1965). *Let us teach guessing* [Film]. Washington, DC: Producer. (See also McQuaide, J., Fienberg, J., & Leinhardt, G. [1991]. *Illustrated transcript of 1965 MAA/George Pólya film, "Let Us Teach Guessing"* [CLIP Tech. Rep. No. CLIP-91-01]. Pittsburgh, PA: University of Pittsburgh, Learning Research and Development Center.)

Comments on Thomas Romberg's Chapter

Robert B. Davis
Rutgers University

UNIVERSITY HIGH SCHOOL AND URBAN SCHOOL, USA

I approached Romberg's remarks in a state of considerable confusion. The confusion was not Tom's fault. For a decade I had the privilege of working with students at University High School, in Urbana, Illinois, which is a school for gifted students. Uni students complete, in just 3 years, all of the mathematics usually found in grades 7 through 12 (and therefore usually taking 6 years, not 3)–this is a pace that the Uni students themselves determined; no adult would have proposed it, I think, but the Uni students handle it easily. As a result, they have 2 years available for calculus, and their calculus course—which I had the pleasure of teaching—is a shared delight, a 5-day-a-week party, never a "course," and certainly never a burden. Some original mathematical work by Uni students appears from time to time in the *Journal of Mathematical Behavior* (see, e.g., Davis, 1987a, 1987b; Imrey, 1987a, 1987b; Pandharipande, 1985, 1987a, 1987b; Secrest, 1985—the work of these students is well worth looking at!). In a period of 4 years, three Nobel Prizes (in physics, medicine, and economics) were awarded to people who had graduated from Uni. At the time, Uni was graduating about 22 students in a typical year, so that we were approaching the point where a student graduating from Uni would have almost a 5% chance of winning a Nobel Prize. (The Uni program is described in Driscoll, 1986.)

I have now moved to Rutgers and have the opportunity of working with some high school students in a low-SES (mainly minority) urban area in a reasonably typical Northeastern city. This is not the first time I have worked in such an area—I did so in St. Louis 30 years ago—but these New Jersey students are different, different from Uni students, and also different from the African-American students I worked with 30 years ago. (They are also different from some 8-, 9-, and 10-year-olds—mainly African-American—with whom I work every Saturday morning.) I think that understanding these differences, and dealing with them in an appropriate way, should be a very high priority for U.S. education—no, for U.S. society. I do not understand the differences, nor know how one would deal with them properly, so I approach nearly every educational question in a state of great confusion. From this foggy perspective I did not look forward to viewing Romberg's vision; no one seemed to know what to do about the different students and different educational experiences that constitute the reality of U.S. schools, and I more or less assumed that Tom Romberg didn't, either.

ROMBERG'S VISION OF SCHOOL MATHEMATICS

I have now read Romberg's paper five times, and every time I read it I am more favorably impressed—more awestruck, I suppose—by how well Romberg *does* manage to deal with all of the kinds of students I have worked with in recent years. Not perfectly (I'll come to that later), but far better than I had thought possible.

We are, of course, seeing Romberg's vision in two, or possibly three, manifestations, since he is not only the author of the present remarks, but also the main author of the *Standards,* and possibly also the inspiration for many of the recent reports authored by others, because of his influential contributions to the national dialogue on the reform of school mathematics.

It is a vision of students working in small groups, working on interesting and challenging problems, which they are expected to come to understand, working toward some valuable goals (for example, "learning to value mathematics," "becoming confident of one's own ability," "becoming a mathematical problem solver," "learning to communicate mathematics," and "learning to reason mathematically"). It is a vision of students "doing mathematics," which Romberg carefully explains does not mean filling in the blanks on ditto sheets. "In mathematics," Romberg argues, "the 'game' is to solve nonroutine problems." As Romberg says, "Today in schools, all students practice skills, whether needed or not, spend almost no time learning strategies, and never get to play the game": an elegant description of a major part of the problem that we face in typical school programs today.

Elsewhere Romberg writes, "Mathematics cannot be viewed as a mechanical performance or an activity that individuals engage in by solely following predetermined rules. Mathematical activity can be seen more as embodying the elements of an art or craft than as a purely technical discipline."

I'm delighted that Romberg has taken such pains to spell out, very clearly, what he means by "doing mathematics," because every day I meet teachers, administrators, and parents who presume that "doing mathematics" means exactly and only the following things: listening attentively as someone tells you how to carry out some algorithm, practicing this algorithm, and working hard to memorize the key bits and pieces that (they wrongly believe) constitute "mathematics." Only this, and nothing more.

As I look at Romberg's vision of what school mathematics might be like—bearing in mind the various students with whom I have worked over the last three decades—I am deeply impressed by the correctness of Romberg's vision. I think that, if it is realized in practice, it can make things very much better for many students, probably for most students.

WHAT WE ARE ASKING OF TEACHERS

During the past year perhaps the main lesson I have been learning is how very much we ask of teachers, when we look at mathematics in the way Romberg suggests. This was not an entirely new idea to me, but I continue to discover by how large a margin I have underestimated the difficulties.

If mathematics means knowing a few standard facts and algorithms, then I, as a teacher, can know them, show them to students, supervise some drill and practice, and that is about it. That is relatively easy.

But suppose students are going to work with manipulable materials. Suppose they are to work on nonroutine problems. Suppose they are to talk with one another about what they are doing, discuss strategies they are inventing for dealing with some task, look for flaws in one another's suggestions, look for patterns, dig hard to find what things *mean*. . . This is a whole new ball game.

If you haven't studied this process carefully—which we do do, using videotaped recordings of teachers interacting with students—then I am almost certain that you, too, underestimate the difficulties that teachers will face. I give one example:

A group of fifth-grade boys, working with Dienes's MAB blocks, had in effect proved that $2 \times 3 = 60$. Now, of course, the boys knew this to be an incorrect result, and so, equally clearly, did the teachers. But if we care about mathematical *reasoning*, then we need to focus our attention on what has gone astray in this instance of mathematical reasoning. It is not enough to observe that the answer is incorrect.

In a 2-hour discussion, four very intelligent and very highly motivated teachers were not entirely able to agree on what, precisely, was wrong with the line of argument. The fact is that the boys had been working on a problem involving hundredths, written as decimals, something like

$$2.54 \times 3.91 = .$$

The students knew that, if they used the small "unit" cubes in the Dienes's blocks base-10 set as their unit in this problem, they would have no smaller blocks available to represent tenths and hundredths. They chose, therefore, to use the block that is sometimes called a "flat" (Davis, 1984), which is (approximately) 10 cm × 10 cm × 1 cm. They proclaimed that, for this particular problem, this flat would "be their unit." This is all very insightful and in fact almost unassailable—provided you are careful to keep in mind which attribute of the flat you have in mind (and somehow manage to deal with one or two other complications). The basic problem is that there is really no such thing as "a unit" as an absolute abstraction—only after we specify an attribute can something become "a unit." When a boy held up a flat and said, "This is our unit," was he specifying a unit of mass, or of weight, or of volume, or of area, or of length, or of acoustic absorbency? In fact, he was paying no attention to this question at all, so it is no mystery that he slipped into the error of changing his meaning without noticing that he had done so, and this mistake was too elusive for the teachers to catch.

Real mathematics is full of pitfalls like this. There is no real way to avoid them, and no short-cut that somehow "covers" all such difficulties. I strongly suspect that one of the reasons for the really poor school mathematics programs we so often see is that they do avoid these difficulties, by the draconian expedient of tossing out the "whole mathematics" baby with the admittedly hazard-rich bath water.

Clearly I think Romberg's goals are the right ones. Why stop teaching real mathematics, merely because it asks a lot from teachers. But let's make no mistake about the fact that it does. In our work at Rutgers, "teacher development" is one of our main concerns (the other is the careful detailed study of how humans think about mathematical situations). We hear people say, "Oh, the teachers will need a 3-day workshop in order to get the hang of all this." No, they won't—at least in the sense that, after a 3-day workshop, they will not know how to deal with many of the kinds of things that will arise when students plan and discuss real mathematics. We have been working in one school for 8 years now, and the teachers are not yet ready to deal with all such matters the way they—and we—feel they should. (I could argue from the other side and point to evidence that things are much improved in this school—it is one small school, at that!—but matters are still nowhere near satisfactory, and because the students and teachers really are thinking about real mathematical questions, it is not easy to conceal or ignore the problems. Neither the teachers, nor the parents, nor the administrators, nor the Rutgers staff would wish to go back to "the old way," but this school, rather like the former USSR, is trying to move toward a brighter future by sailing through some uncharted and hazardous seas.) Anybody who thinks that what is needed is "a 3-day workshop for the teachers" just has not got the picture!

WHAT WE ARE DOING TO STUDENTS

Having said that the Romberg vision will be hard to realize in practice, I now argue that what we need is even more extreme than Tom's proposals, valuable though they would be. I want to argue that we need to look much more closely at each individual student and try to devise programs and environments that will be beneficial to *this* student right at *this* particular point in his or her life.

In the case of the students at Uni High, I know of very few places in the United States where students are taken seriously as "younger colleagues" to the extent that they were at Uni. We enjoyed working together on mathematics. (If you buy a billiard table with a bumper shaped like an ellipse, and if you put a pool ball at one focus, and if you hit it rather hard, and if the ball rolls with no loss of energy—so that it keeps on rolling forever—will the path ever close up and start retracing itself? One of the 11th-grade students posed this problem, the class worked on it, and another student ultimately solved it.) This was not math as most people know it, these were not "classes" as most people think of them, and this was not "school" in its most familiar form. It was, however, fun, and at least as intellectually honest as anything I have ever done with any college students at any level.

Do some of the young people in our rich society deserve the opportunity to experience this? I think so, but I know that very few students will have that chance. Is this important? Well, three Nobel prizes may be telling us something . . .

Let me turn to students in some urban schools in the Northeast. I have watched students who come to class and put their heads down on the desks, making no pretense of working or of caring about whatever it was that the class was supposed to be doing. People say that these students are in ninth-grade algebra. This is not true. They are in a warehouse for teenagers, just as many senior citizens are in a warehouse for old people. Nor is their school day more promising in other subjects. In English they are supposed to be studying Macbeth, but in fact many of them cannot read, and a number of them are really unable to speak English. They say that they are studying "MacDeath," and they may well be largely correct. One of their teachers recently saw one of her students at work in a service station, putting gasoline and oil into automobiles. She was visibly shaken: "Michael is behaving like a normal human being! He never comes close to doing that in school!" This episode speaks to me very loudly. That school is allegedly there to *help* Michael, to give him a more-or-less ideal situation for growing up and learning about himself, other people, and the world in general. The gas station is there, really, to make money for the people who own it. Yet the evidence seems clear that it is the service station, and not the school, that is giving Michael the best chance to learn a more mature way of dealing with the world. It is probably the case that the gas station is also giving him a better chance to find out what mathematics is all about, and maybe also to learn about the subtleties of human relations, which he probably does *not* get from his struggles with the MacDeath which he despises, and from which he is learning far too much, most of it about his own inadequacy and the hopeless complexity and unappealing nature of the world of "learning."

The part of the *Standards* that worries me the most are the suggestions that all students in U.S. schools should be learning more or less the same mathematical content. Let me turn again to what I saw in Illinois: The city of Urbana not only contains University High School. It also contains a high school where, each year, a selected group of students start with an empty plot of land, purchase lumber and plumbing and shingles and nails, build an actual house, sell it in the local real estate market, and put the money back so that next year another group of students can go through the same process, building and selling that year's house.

Clearly, the city of Urbana has some serious alternatives; I would argue that each of these alternatives makes good sense. This availability of alternative pathways through education was once commonplace in the United States, but lately this seems to have changed. In the United States today we seem not to dare to speak of alternatives in education, because too many people assume that it will be black and Hispanic students who will build the houses, and Asian students who will do science and mathematics. White students will be classified as "other" (as in fact they are in New York City) and will learn neither carpentry nor science and mathematics. But, real as such problems are in some people's minds, they are in truth abstractions. I cannot see the world in terms of such immaterial ideas. I cannot stop myself from looking at actual students—actual individual human beings—and concluding that our schools are coming close to destroying many of these young people. Asking our youth to attend school every day, to sit and listen quietly to things they neither understand nor care about, to base their lives on the activity of listening and doing what they are told in ways they are told, in trying to remember a large number of meaningless bits and pieces of "knowledge," because they may need it someday (although they and their teachers doubt that this day will ever come), and—most devastating of all!—in learning a "base-level" education that never aspires to excellence—when we ask that of our children we are bringing them face to face with what Kozol has called "Death at an Early Age." There is under all of this a presumption that bland and vacuous education is appropriate for "the masses." Anything else would be undemocratic. Nowadays one says it more forcefully—it would deny the rights of minorities.

We need to rethink this. Working hard for excellence is not anti-black. I think the truly racist view is the one that says: If we work hard at teaching real mathematics, African-American and Hispanic students will necessarily do badly. If one discusses the program at Uni, some people hear that as anti-black. (Of course some of the successful and gifted students at Uni are black, and most of the students who build the house in Urbana are white.) This "anti-elitist" view argues that only a bland (or mediocre) school program can be nonracist. That, it seems to me, is the really racist point of view. Have African Americans shown themselves unable to compete? Certainly not in baseball. Nor in TV comedy. Nor in the command of the English language (read, for example, James Baldwin, or talk with Billy Taylor). Nor in command of the military (remember Colin Powell). It is significant success, not mediocrity, that has usually been the ticket out of the ghetto. (To see how the State of Oregon is attempting to reintroduce significant alternative paths into their public schools, see Celis, 1991.)

There is also a second issue here. Does our society truly dishonor carpenters? Is it a disgrace to work in a service station? John Gardner wrote that, as long as we honor every philosopher more than any plumber, "neither our pipes nor our theories will hold water." Two of the men I respect the most are the man who takes care of my heating and air conditioning (he owns his own company), and the man who built an addition onto my house (he also owns his own company). One of these men is black, one is white. They both are very good at their jobs. I think that commands respect.

I do not see how we can achieve really good education for young Americans as long as we insist that nearly all students must learn nearly the same thing. We have created schools that are ineffective, in part because they are *designed* to be ineffective. We need to take a fundamental look at our individual young people and make decisions that are appropriate to each individual one of them.

But I admit to being confused on the issue. What is it that a rich society owes to its citizens? Not, I think, warehousing under the pretext that they are "in ninth-grade algebra" or "learning Macbeth," but something far deeper, more valuable, more personal, more meaningful. When a student is bored, I do not think this is necessarily a failure of the student, nor of the teacher, but more often a failure of the curriculum. Why are we trying to force this student to learn how to solve algebraic equations right now? Why are we forcing him to struggle with Mac-Death? There is, of course, a quite serious theory that argues that a major task of the schools is to cause many citizens to come to view themselves as hopeless failures, so that they will give up on high personal ambition and become available as a labor force to serve anyone who is in a position to pay them. My goals for young people are exactly the opposite of this. They should *not* give up on themselves, they should *not* see their future as hopeless, they should *not* come to believe that the world is a collection of silly tasks that only a nerd could want to deal with. Todays students are learning far too much from our schools, and most of it is negative and destructive. Somehow we need to make this stop.

(Incidentally, I could recount a great many "horror stories" that many readers might identify as tales from low-SES minority areas, but in fact the teachers are white, the students are white, and the schools are in very affluent suburbs. Failing to provide students with effective opportunities to grow and to learn is ubiquitous in the United States at the end of the 20th century.)

The fact is that, at the present time, we are not giving students what they need. To do so, we will have to question nearly all of our basic assumptions—and I am not sure that we are ready to do this.

COGNITIVE SCIENCE

Some of these assumptions fall in the area of cognitive science. For brevity, I give only one example. Nearly every day we get more and more data from the neurosciences indicating that young children have a brain power not matched later in life. A compelling case can be made that children in grades 4 and 5 are better thinkers on mathematical questions than children older than that. (I argue this at some length in Davis, 1986, using both neurological data and clinical classroom data.) Language learning is more successful *before the age of 2* (Blakeslee, 1991). Now, if what we are building is a nationwide system for inexpensive babysitting, we should disregard this data. It only complicates our task.

But if we are serious about educating people in powerful ways, we are seeing more and more data that say: *Do it early*. A large amount of mathematics is best learned in elementary school. A large amount of foreign language learning should take place even earlier than that.

If we mean to create a highly effective set of educational arrangements to provide for all (or, realistically, nearly all) of our young people, we face a need for a truly major redesign effort. I think it would be worth it, I think we need it desperately, but I'm not sure that we're up to the task.

I believe that my vision of education has been clouded by my experience at Uni. I know what is possible, and I want it for far more students. The Uni community took all students seriously and expected great things—or at least very valuable contributions—from each of them. By contrast, most U.S. schools expect very little for most of their students, and I find this hard to accept. Become a physicist or a physician or a carpenter, but be good at it. Be serious about yourself. I cannot settle happily for anything less.

THE COMMON ASSUMPTIONS OF SCHOOLS

The major changes in education that I would like to see are probably visionary in the worst sense of that word, and Romberg is probably wise not to let their siren song lead him astray. One might describe his vision as one where we ignore Utopia and settle for challenging some of the basic assumptions of common school practice today. Even this will not be easy.

These common assumptions do not have the intriguing sound of the more exotic propositions that academics usually prefer to discuss, but—prosaic though they are—they have a very tight (and pernicious) hold on day-to-day school practice. I have just spent hours studying a high school that I admire—the faculty and administration are warm and caring people, and the students are getting about as good an education as one can usually expect in the United States today. That education nonetheless is shaped by the non-Rombergian assumption that mathematics does indeed consist of a few specific algorithms that must be memorized in a meaningless way—time and again one hears teachers say to their classes "Remember this" and "Remember that" and "Remember thus-and-such." "Remember to place the decimal point by . . . " "Remember that you must first rewrite the mixed number as an improper fraction." (The testing program focuses on precisely these same small bits of performance.)

You never hear from teachers any suggestion that there might be any other way to deal with the problem at hand (even though their students often invent such methods). You never encounter any evidence that the teachers are interested in how students think about any problem—nor, for that matter, in how one ought to think about the problem. Specific ways of dealing with specific problems are often explained to students by means of the most incredible mumbo-jumbo. Consider this instance: "Percent over 100 equals 'is' over 'of'." Or this: "The 24th letter of the alphabet, X, is really a cross, and when you see this, remember that you must *cross*-multiply!"

Underlying the instruction at this generally good school are many dubious propositions: There is one correct way to solve this kind of problem. This method must be told to students. No student could ever invent a way to solve this (much less invent a better way!). If we were to consider alternatives, that would confuse the students. Anyhow, there really are no alternative ways. Teaching means telling the students (one hardly ever hears a *student* talk, and in hours of observation no one saw a single instance of a student writing on the blackboard). In mathematics there is nothing to "understand," only meaningless procedures that must be memorized. Mathematics consists of a collection of separate rules; a good test determines which of these rules a student can use reliably, and which he or she cannot. Once we know which ones the student cannot use reliably, our work is cut out for us—we reteach these specific rules, in the same meaningless way that someone else must have done last year. The job of a student is to sit quietly, listen carefully, and memorize tenaciously.

This kind of background will never get a student to the level of the youngsters at Uni. Year after year of this kind of experience must be very hard to take, indeed!—no wonder so many students conclude that mathematics is something that only a nerd could love.

Challenging these assumptions will not be easy. At best it may be just barely possible. I suspect that Tom Romberg is wise, indeed, to pick this as his main target for further improvement in school mathematics. Hitting this target will be very difficult, but it may not be unrealistic to try for it.

As for me, I will probably continue to wish for more. I want a school where the development of every student is taken as seriously as the armed forces must have taken the preparation of the men and women who were trained for the war in the Persian Gulf. More peaceful purposes in most cases, I hope, but every bit as much seriousness. Not joylessness, and certainly not predatory testing—perhaps the key is to take seriously the way students think about mathematical situations, or even the fact that students do think. That may be the true essence of "seriousness." Of course, it goes against the traditions of American schools.

REFERENCES

Blakeslee, S. (1991, September 10). Brain yields new clues on its organization for language. *New York Times*.

Celis, W. (1991, July 24). Oregon to stress job training in restructuring high school. *New York Times*.

Davis, R. B. (1984). *Learning mathematics: The cognitive science approach to mathematics education*. Norwood, NJ: Ablex.

Davis, R. B. (1986). The convergence of cognitive science and mathematics education. *The Journal of Mathematical Behavior, 5*(3), 321–333.

Davis, R. B. (1987a). Mathematics as a performing art. *The Journal of Mathematical Behavior, 6*(2), 157–170.

Davis, R. B. (1987b). "Taking charge" as an ingredient in effective problem solving in mathematics. *The Journal of Mathematical Behavior, 6*(3), 341–351.

Driscoll, M. J. (1986). Teaching the mathematically talented: Three American programs. *The Journal of Mathematical Behavior, 5*(2), 143–156.

Imrey, L. (1987a). The key lemma in the elliptical billiard table problem. *The Journal of Mathematical Behavior, 6*(2), 241–246.

Imrey, L. (1987b). The Isis problem revisited. *The Journal of Mathematical Behavior, 6*(3), 381–382.

Pandharipande, R. (1985). The Isis problem. *The Journal of Mathematical Behavior, 4*(1), 101–104.

Pandharipande, R. (1987a). An interesting result concerning covariance. *The Journal of Mathematical Behavior, 6*(3), 369–371.

Pandharipande, R. (1987b). The problem of the deviant ball. *The Journal of Mathematical Behavior, 6*(3), 377–379.

Secrest, D. (1985). Regular polyhedra. *The Journal of Mathematical Behavior, 4*(1), 105–118.

Epilogue

Alan H. Schoenfeld
University of California, Berkeley

Following the exchange of ideas reflected in chapters 1 through 8, conference participants turned to a discussion of next steps. This set of mathematicians, cognitive scientists, and mathematics educators shared a common commitment to mathematics education, an understanding of the magnitude of the tasks we face, and an understanding that each of the communities present had valuable things to contribute—individually, of course, but much more effectively in synergy with the others. The question was how to move forward. What steps could the constituent disciplines take, individually and collectively, to advance both our understandings and the practical state of the art?

The discussions took place in three phases. The first phase consisted of reviewing issues raised prior to and during the conference, and brainstorming others for completeness. Following an evening of such conversations, members of the group noted that items on the list seemed to cluster naturally, and that those clusters might determine foci of interest. Participants broke into three groups to address the "big issues" in those clusters. The themes addressed by the three working groups were:

A. Issues related to the status of research in mathematics education; steps toward "community building" we might take to bring more people into the area, and to continue the kinds of efforts represented at this conference.
B. Issues related to language as a form of mathematical expression, epistemology, and learning.
C. Issues related to the "big picture" (the larger context of schooling, politics, and social issues within which R&D on learning and teaching are conducted), teacher preparation, implementation, and so on.

Discussions of these themes constituted the second phase.

In the third phase we reconvened as a group of the whole to receive and discuss the reports from the three theme groups. There was a great deal of overlap and consistency in the reports of the three groups, with some comments or recommendations made independently by all three. The distillation of the groups' work, reported later, was endorsed by acclamation at the conference and confirmed in subsequent exchanges.

GROUP A

There was general agreement that the endeavor in which we had been engaged—bringing together mathematicians, cognitive scientists, and mathematics educators to tackle the tough educational problems we face—is both important and difficult. Not least among the issues to be faced is the one of legitimizing the field of research in mathematics education, of convincing the mathematical community at large that there is a solid field that can offer useful insights into mathematical thinking, teaching, and learning at all levels, in particular at the undergraduate level. A variety of suggestions for community building were made and suggestions for promoting ongoing dialogue. These included, for example, having presentations and special sessions on research in undergraduate mathematics education at the national mathematics meetings, making connections with Mathematicians for Educational Reform, establishing close liaison with *UME Trends,* and exploring the possibility of having *UME Trends* evolve into a journal with a focus on research in undergraduate mathematics education. It was suggested that "demonstration projects"—existence proofs, so to speak—would provide tangible evidence that research has something to offer. It was also suggested that there could be a lovely joint effort by mathematicians and cognitive researchers: detailed studies of the structure of mathematical thinking, in which the two groups worked together to study the evolution of mathematical ideas in real time—for example, where a group of cognitive scientists worked with some mathematicians as they worked on some big problems, and they jointly elaborated the thinking and discovery processes involved.

GROUP B

This group considered a host of issues related to thinking and doing mathematics. The role of language was a central concern, and the notion that there may be different mathematical languages (a rich, multiply connected one for "doing," another rather formal one for "convincing") was a provocative idea put forth as worthy of serious exploration (note that this idea has clear instructional implications). There were extended discussions of learning, predicated on the fundamental assertion that students' partial understandings are critically important—that what everyone brings to mathematical situations (ofttimes incomplete, partially correct, and partially incorrect) is the base from which more complete understandings are constructed. That is, understanding "what's in someone's head" is vitally important, even if that something is wrong. There was a consensus that as a field we need (a) to develop better ways to characterize partial understandings, (b) to understand how knowledge evolves and is built on the structures one has, and (c) to figure out the kinds of activities (often collaborative) that connect with people's partial understandings and promote conceptual growth in reliable ways. The general sense was that such positive outcomes are much more likely to occur when students are engaged in the meaningful and purposeful use of mathematics, not merely the "imitation lessons" that predominate in much of today's instruction. Concept-specific studies were suggested as well (i.e., how do students learn X?) to add to our knowledge base. Group B noted that it is important for all of us to unpack our understandings and to recognize or admit that we often reconstruct our mathematical knowledge, rather than having it all accessible in crystalline form. The human side of this constructive activity is important and important for students to understand. Finally, to return to the theme of partial understandings, it was noted that all of the constituencies at the conference had partial understandings of the big issues we face, and that we can build on them together.

GROUP C

This group was concerned with the "big issue": arriving at a reasonable characterization of "quantitative thinking" and worrying about how we can create a nationwide context that results in a quantitatively literate populace. Among the points of agreement in this group:

1. Assessment is of major importance, because tests drive both curricula and the public's perception of what is important. It was suggested that our joint communities might propose our own assessment measures focusing on what is important, using them as a lever for change.
2. The issue of teacher preparation is critical, and it needs to be addressed. Aspects of this problem include: (a) the mathematical preparation of teachers, especially elementary school teachers; and (b) college mathematics instruction, because prospective teachers have university faculty as their last "models" of what to do in mathematics classrooms.
3. The role of technology needs to be carefully explored. Not much time was spent in discussions of specifics, but it was generally agreed that this is an important area that needs to be examined closely.
4. We need to focus on critical junctures in the pipeline. The demographics are frightening: we lose $1/2$ of the students from the pipeline from eighth

grade on, a disproportionate percentage of them are minorities and women. But the process of attrition may have begun even earlier, when students decide mathematics is not for them and stay enrolled only because they have to.
5. We worried about broad social perceptions of mathematics—that it's for nerds, has no use in the real world, and so on. Only somewhat whimsically, the idea of sound bites and TV commercials for mathematics—testimonials from Bill Cosby, New York mayor David Dinkins, and others who have mathematics in their backgrounds, were suggested. More down to earth, the idea of strengthening mathematics' public relations apparatus was endorsed.

In the brief interval of time since the conference, there has been significant progress on some fronts. In particular, with regard to Group A's agenda, there has been increased recognition of the importance of mathematics education at the university level and a move toward the solidification of the community. The Mathematical Association of America (MAA) has established a standing Committee on Research on Undergraduate Mathematics Education, which is about to be jointly sponsored by the American Mathematical Society (AMS). The AMS has established a Committee on Education for the first time in its 100-year history. The Research Advisory Committee of the National Council of Teachers of Mathematics (NCTM) has actively sought collaboration with its analogues in the other two societies, and the leadership of all three societies has sought to foster communication and collaboration on common projects. Invited talks on both research and practice in mathematics education, both exceptionally rare through the 1980s, are now common events at the annual joint meetings of the mathematics societies; special sessions on both research and classroom instruction draw huge crowds at the joint meetings. In 1994, the first annual volume in the series *Research in Collegiate Mathematics Education,* sponsored by the Joint Policy Board in Mathematics, will appear; if the series of volumes flourishes, it has the potential to evolve into a regularly published journal.

Less vivid progress has been made on the agenda outlined by Group B, but that is not surprising: Deep research issues will take time to work out, and the zeitgeist (at least in the mathematics education community) is generally supportive of the kinds of work suggested. Progress on issues related to Group C's agenda has been uneven. There has been a major NSF initiative on assessment, and one can expect to see the fruits of that work in the next few years. We have made less progress on issues of teacher preparation and technology, though they are receiving some attention. It is not clear that we have, as suggested earlier, begun to focus on critical junctions in the pipeline; it is clear that there is a *long* way to go before mathematics is favorably perceived by the public.

Nonetheless, such changes cumulatively represent significant progress in the brief amount of time since the conference. One can only hope that the progress will continue and that this volume will help, at least in some small way.

Author Index

A

Abelson, H., 250, 256
Ackermann, E., 251, 256
Addison, J., 279-286
Alexanderson, G. L., 19, 26, 29
Anderson, J. R., 265, 268, 273, 277, 278
Andrews, J. G., 295, 303
Arcavi, A., 235, 243
Archambault, R. D., 224, 242
Artigue, M., 176, 191
Ayers, T., 118, 152, 170, 241, 242

B

Balacheff, N., 61, 69
Banchoff, T., 65, 69
Baron, M. E., 114, 152
Barwise, J., 131, 153, 271, 274, 277
Benacerraf, P., 54, 69
Bessman, M., 275, 278
Beth, E. W., 228, 242
Bishop, E., 131, 153
Blakeslee, S., 319, 321
Bochner, S., 99, 102, 153
Borba, M., 182, 191
Boyer, C., 87, 89-90, 95, 97-99, 103, 105-108, 111-112, 118, 120, 123, 127-128, 130, 153
Boyle, C. F., 273, 277

Breidenbach, D., 118, 134, 153, 170, 241, 242
Bridger, M., 131, 153
Bridges, D., 131, 153
Brown, J. S., 61, 69
Brown, S., 45, 51
Brownell, J. A., 293, 303
Burton, D., 115, 153
Burton, L., 63, 69
Butler, M., 61, 69
Byrne, R. M. J., 271, 278

C

Cajori, F., 78, 80, 115, 124, 153
California Department of Education, 54, 57, 72
Carpenter, T. P., 56, 69
Celis, W., 318, 321
Cheng, P. W., 266, 268, 275, 277
Cipra, B., 133, 153
Clagett, M., 87, 90-97, 104, 153
Cohen, M. S., 193-208, 199, 206, 208
Collins, A., 61, 69
Confrey, J., 136, 149, 153, 172-192
Conley, M. R., 205-206, 208
Cornu, B., 119, 153, 157

D

D'Ambrosio, U., 299, 303

Damerow, P., 299, 303
Davis, G., 118, 152, 170, 241, 242
Davis, R. B., 312-321
de Lange, J., 298, 299, 303
Dennis, D., 180, 183, 192
Devaney, R., 131, 133
Devlin, K., 274, 277
diSessa, A. A., 94, 97, 153, 248-256, 267, 268
Douglas, R. G., 209-220
Dreyfus, T., 123, 153
Driscoll, M. J., 312, 321
Dubinsky, E., 118-119, 134, 152, 153, 157-171, 221-243
Dugas, R., 95, 153
Dunne, M., 275, 278

E

Edwards, C. H., 87, 100, 106-107, 110, 112-114, 116-117, 119-120, 122, 127, 153, 175-176, 179, 192
Eliot, T. S., 77, 153
Elterman, F., 235, 243
Epp, S. S., 257-269
Errecalde, P., 100, 156
Etchemendy, J., 271, 277

F

Fawcett, H. P., 61, 69, 274, 277
Flores, F., 151, 156
Fowler, D. H., 179, 192
Fraser, C., 88, 122-123, 153
Freudenthal, H., 295, 303
Frid, S., 131, 153

G

Garcia, R., 84-85, 102, 106, 110, 124, 134, 155, 158, 159, 171
Gardner, H., 85, 153

Gaughan, E. D., 199, 205-206, 208
Gazzaniga, M. S., 267, 269
Gillespie, C., 110, 153
Gleason, A. M., 19, 29
Goldenberg, E. P., 237, 243, 246
Gong, C., 235, 243
Grabiner, J., 85, 124, 153
Grattan-Guinness, I., 124, 127, 154
Greenberg, H. J., 299, 303
Greeno, J., 118, 148, 154, 244, 246, 270-278
Greenwood, R. E., 19, 29

H

Hagstrom, W. O., 293, 303
Hall, R., 148, 154
Halmos, P., 30
Hammer, D. M., 94, 97, 153, 250, 256
Harel, G., 118, 154
Hawks, J., 118, 134, 153, 170, 241, 242, 243
Henkin, L., 71-75, 267, 269
Hiebert, J., 244, 247
Hoffman, K., 53-55, 69
Holyoke, K. J., 266, 268, 275, 277

I, J

Imrey, L., 312, 321
Jertson, D., 275, 278
Johnson-Laird, P. N., 271, 266, 269, 278
Joseph, G. G., 172, 192

K

Kaput, J., 77-156, 158-170, 173-174, 176-180, 184-185, 187-188, 190
Karplus, E., 135, 154

Karplus, R., 135, 154
Keisler, H. J., 83, 131, 154
Keitel, C., 299, 303
Kelley, L. M., 19, 29
Kilman, M., 246
King, A. R., Jr., 293, 303
Kitcher, P., 54, 69, 293, 303
Klein, J., 111, 154
Kline, M., 79, 97, 109-111, 118, 120
Klosinski, L. F., 19, 26, 29
Knoebel, R. A., 193-208
Koedinger, K. P., 273, 278
Koestler, A., 28, 29
Kuhn, T. S., 59, 69
Kurtz, D. S., 193-208

L

Lakatos, I., 53-54, 59, 69, 173, 176, 192
Lampert, M., 61, 69, 296, 303, 306, 311
Larson, L. C., 19, 26, 29-38
Lave, J., 61, 69
Lay, E., 250, 256
Lefevre, P., 244, 247
Leinhardt, G., 101, 155, 305-311
Lewin, P., 118, 152, 170, 241, 242
Lindquist, M. M., 56, 69
Linn, M. C., 8-17
Luce, R. D., 95

M

Mahoney, M., 102-103, 108-109
Mamona-Downs, J., 119
Mason, J., 63, 152, 155
Mathematical Association of America, 306
Mathematical Sciences Education Board, 294

Matthews, W., 56, 69
McLeone, R. R., 295, 303
Mertzbach, U., 87, 103, 153
Minsky, M., 267, 269
Mokros, J., 148, 155
Morrison, F., 131, 152, 155
Moss, L., 131, 153

N

Narens, L., 95, 155
National Council of Teachers of Mathematics, 54, 287-290, 294, 295, 297
National Research Council, 48, 54
Nemirovsky, R., 134, 136, 148, 155, 156
Newell, A., 273, 278
Newman, S., 61, 69
Nichols, D., 118, 134, 153, 170, 241, 242, 243
Nisbett, R. E., 266, 268
North Carolina School of Science and Mathematics, 290
Novick, S., 298, 303
Nussbaum, J., 298, 303

O, P

Ogonowsky, M., 134, 155
Oliver, L. M., 266, 268
Olkin, I., 39-51
Pandharipande, R., 312, 321
Paulos, J. A., 74, 75
Pea, R., 8-17, 113, 155
Peitgen, H., 131, 155
Pengelley, D. J., 193-208
Perry, J., 274, 277
Peterson, P. L., 296, 303
Piaget, J., 84-85, 102, 106, 110, 124, 134, 155, 158-159, 171, 175, 192, 224, 228, 242, 243

Piliero, S., 190, 191
Ploger, D., 250, 256
Pólya, G., 44, 51, 53, 69, 207-208, 294, 303, 306
Popper, K. R., 59, 70
Priestly, W., 86, 155
Putnam Brochure, 19, 20
Putnam, H., 54, 69
Putnam, R. T., 296, 303

R

Reznick, B., 19-29
Reznick, S., 28, 29
Richardson, C., 135, 155
Richter, P., 131, 155
Rizzuti, J., 186, 190-192
Robinson, A., 83, 122, 130, 155
Romberg, T. A., 287-304
Rubin, A., 134, 148, 155
Russell, S.-J., 134, 155

S

Salomon, G., 101, 155
Sawyer, W. W., 136-137, 155
Saxe, G. B., 254, 256
Schoenfeld, A. H., 39-51, 53-70, 88, 155, 207-208, 235, 243, 245, 247, 323-326
Schwartz, J. L., 1-7, 71-75, 148-149, 277, 278
Schwingendorf, K., 119, 153
Secrest, D., 312, 321
Sfard, A., 118, 155
Sherin, B., 94, 97, 153, 250, 256
Sierpinska, A., 157, 171
Silver, E. A., 56, 69
Simon, H. A., 273, 278
Skemp, R., 132, 155
Sly, K., 275, 278

Smith, E., 172-192
Smith, J., 235, 243
Smith, R., 135, 155
Smith, S., 61, 69
Stacey, K., 63, 69
Starr, S., 133, 155
Steen, L. A., 54-55, 70, 79, 155, 175, 294, 303
Stein, M., 101, 155
Steussey, C. L., 205-206, 208
Stewart, D. M., 295, 304
Strang, G., 135-136, 155, 156
Struik, D., 87, 95, 111, 156

T

Tall, D., 119, 131, 149, 156
Thompson, P., 134, 148, 156
Tierney, C., 134, 155, 156
Tinker, R., 148, 155
Tolstoy, L., 166, 171
Treisman, U., 289, 304
Tucker, A., 149, 156

U, V

Unguru, S., 172, 179, 192
Van der Waerden, B. L., 88, 156
Vergnaud, G., 100, 156, 300, 304
Vinner, S., 123, 153, 156
von Glasersfeld, E., 228, 243

W

Walter, M., 45, 51
Wason, P. C., 266, 269
Waterhouse, W. C., 28, 29
Webster's New Universal Unabridged Dictionary, 57
Wendelboe, B., 275, 279
Wenger, E., 61, 69
Wenger, R., 244-247

West, M., 88, 135, 154
Westbury, I., 299, 303
White, B. Y., 209-220
Whitehead, A. N., 262, 269
Williams, S., 123, 151, 156
Wilson, B., 277, 278
Winograd, T., 151, 156

Y

Yates, J., 275, 278
Yerushalmy, M., 2, 148-149, 277, 278
Yost, G., 273, 274
Young, G., 137, 156
Youschkevitch, A., 122, 125-127, 156

Z

Zaslavsky, O., 101, 155
Zia, L., 135, 156

Subject Index

abstract mathematics
  form of problems in, 261
  solving problems in, 261-263
abstracting as part of mathematical practice, 296
abstraction, 175-7, 190-191
action notation system, 101, 167-168
activity context, 253-255
algebra, 182
  and geometry, relation between, 179-180
  relation to abstraction, 175-176
  role of in 17th century, 108-110
  16th and 17th century, 101-103
Algebraic Proposers, 8
algebraic reasoning in the 18th century, 120-124
applications as part of mathematical practice, 296-297
assessment, 3, 325
  of project-based learning, 212-214
  role of in educational reform, 1
assumptions of school practice, 319-320
authenticity in problems, 306-307
authority
  classroom, 4
  mathematical, 61-62
  teacher, 62-65, 71

biconditional statements, 264
binary relations
  hypothetical student's approach to problems in, 258-260
  mental models for, 275
calculus
  as action notation system, 119-120
  18th century, 120-124
  geometric and algebraic reasoning in development of, 116-118
  history of, 83-132
  methods for curricular change in, 214-216
  Oresme's influence on development of, 97-99
  role of motion imagery in development of, 111-113
  role of notation in development of, 113-115
  root aspects of, 83, 86, 160-163, 176
  standard approach to, stability of, 78-79
  student research projects in coursework, connection with, 212
    criteria for, 209-210
    design of, 211
    effect on number of math majors, 216-217

(continued)
   effect on students, 201-206, 216-217
   examples of, 194-198, 202, 204, 208
   grades of students, 204-205
   grading of, 206
   helping students with, 206-207, 219-220
   how assigned, 199-200
   how to create, 196-197, 199
   pedagogical roles of, 210-211
   schools using, 200
   sequencing of, 211-212
   student evaluation, 205
   student learning in, assessment of, 212-214
   student-designed, 217-218
   teaching assistants and, 200-201, 206
case studies, methodology for, 40
classroom atmosphere, methods for changing, 11-15
cognitive science, 270
   implications of for instruction, 265-267, 319
community building in mathematics education, 324, 326
competition, 32-33
Computer as Lab Partner Project, 10
computer program.
   See *software*.
conceptual change, 298
conceptual domains, curriculum design and, 300-301
conceptual entities, 244
constituencies of mathematics education, 40
constraints, proof and, 274-277
constructivism, 227-229, 251
context, 253-255, 267

continuity, development of, 104-106, 127, 129-130
continuous variation
   representations of, 90-97
   the Greeks' understandings of, 87-88, 184
   13th–15th century understandings of, 88-97
control, 245
cooperative learning, 308-309
counterexamples, 262, 283-284
covariation, 186-190
curricular artifacts, 4-5
   design of, 10, 13-14, See also *curriculum design* and *software*.
   impact of on curriculum, 16-17
curricular change, 71-73
   methods for, in calculus, 214-216
curriculum, 309-310
   calculus, 133-135, 163-164, 174, See also *calculus*.
      technology-based, 136-152, 164-165
curriculum design. See also *curricular artifacts*.
   conceptual domains and, 300-301
   history and, relation between, 160-163, 165
   implications of history for, 118-119, 132-133, 165, 174
   principles of, 300-302
   problem situations and, 301-302
   skills and, 301
decimals, student work on, 314-315
deencapsulation, 230
definite integral, 129
Dienes's blocks, 315
differential, 128
display notation system, 101
dogmas, 306

SUBJECT INDEX 335

educational reform
  role of assessment in, 1
  social networks for, 5-6, 12
  strategies for, 3, 15
empirical science, mathematics as, 57-58, 294
encapsulation, 229, 251-252
epistemological obstacle, 157
epistemology. See also *mathematical epistemology*.
  relation to pedagogy, 53, 225
"epistemology of multiple representations," 178, 183
everyday and mathematical language, differences in, 264
form of problems in abstract mathematics, 261
Function Probe, 172, 182-183
function, 235, 237-41, 255
  as object, 237-238
  as process, 237-238
  in 18th and 19th centuries, 124-127
gender, 47-49
genetic decomposition, 229, 250-251, 254-255
  of mathematical induction, 232
  of quantification, 235
Geometric Supposer, 2-4, 262, 277
geometry, 2
  algebra and, relation between, 179-180
  problem solving, models of, 273-274
  proof, 273
Geometry Tutor, 277
goals for mathematics education, 287-289
  skills and concepts, role of, 288-289
good problems, 306-307

graph theory, hypothetical student's approach to problem in, 260-261
graphic thinking, 245
groups (cooperative), 308-309
heuristics, 42
  in mathematical practice, 294
historical and individual development
  Kaput's perspective, 159-160
  parallels between, 84-85
  Piaget and Garcia's perspective, 158-159
historical work, conduct and role of, 172-174
history
  curriculum design and. See *curriculum design and history*.
  of calculus, 83-132
  of mathematics, 179
  role of in mathematics education, 157-160
hypothetical student's approach to problems in
  binary relations, 258-260
  graph theory, 260-261
  if–then statements, 264
individual and historical development. See *historical and individual development*.
inequalities (mathematical), 267
infinitesimals, 105-108, 128
instruction. See also *mathematics instruction*.
  assumptions underlying, 320
  implications of cognitive science for, 265-267, 319
  implications of psychological research for, 297-298
  implications of research on mental models for, 275-277
  use of problem situations in, 298

interiorization, 229
intuitive thinking, proof and, 257-258
ISETL, 222, 246
    activities using, 232-233, 236, 239-240
language
    mathematical and everyday, differences in, 264
    role of in mathematics, 324-325
limit, 128, 162-163
literacy, mathematical, 288
logic
    in mathematics education, 279-281
    role of in problem solving, 263
MathBikes, 147
MathCars, 182
    use in calculus curriculum, 137-147, 149-151
mathematical and everyday language, differences in, 264
mathematical authority, 61-62
mathematical community, 61-62
mathematical epistemology, 53-61, 71, 73, 75, 178, See also *epistemology*.
mathematical induction, 230-233
mathematical literacy, 288
mathematical models, 295
mathematical power, 288
mathematical practice, 293
    abstracting as part of, 296
    applications as part of, 296-297
    as an empirical science, 57-58, 294
    as model building, 295
    as problem solving, 294
    heuristics as part of, 294
    inventing as part of, 296
    proving as part of, 296
    role of proof in, 270-271, 274

mathematical truth, 59-60, 71-72, 74
mathematics
    as a social activity, 58, 60
    beliefs about, 225
    classroom experience of related to beliefs about, 75
    classrooms, *Standards'* vision of, 302, 313
    different varieties of, 50, 65
    "doing," 293-297
    "doing" vs. "knowledge about," 289-290
    history of, 179
    role of language in, 324-325
    social perceptions of, 326
    student beliefs about, 57
    teachers', administrators', parents' conceptions of, 314, 320
mathematics education
    constituencies of, 40
    goals for, 287-289
        role of skills and concepts in, 288-289
    goals of, 47-48, 270
    logic in, 279-281
    role of history in, 157-160
    role of research and development in, 223, 248-250
    status of research in, 324, 326
    technology in, 325
mathematics instruction. See also *instruction*.
    implications from "social brain" for, 267-268
    implications from "society of mind" for, 267-268
    implications of cognitive science for, 265-267, 319
mathematics learning, beliefs about, 224

mathematics majors, 48, 216-217
mathematics pipeline, 325-326
mean value theorem, 95
mental models, 271-272
  for binary relations, 275
  implications of research on for instruction, 275-277
  proofs and, 272, 276-277
model building as part of mathematical practice, 295
models of geometry problem solving, 273-274
natural exponential, development of, 120-122, 188-189
new math, 279, 281-282
nonstandard analysis, 122, 130-131
notation, algebraic, 182
notation systems. See also *notations* and *representations*.
  action vs. display, 101, 181-182
  referential constraints, 103-104
  relation between two, 81-82
  17th century development of, 101-104
notations
  "building meaning" from, 81, 162
  framework for thinking about, 80-83
  producing or modifying, 81
  relation to conceptions, 80, 163, 166-167, 169-170, 174
object, 229, 251-252
  function as, 237-238
Oresme
  characterization of variable motion, 95-97
  influence on the development of calculus, 97-99
  representations of continous variation, 91-97
  view of ratio, 186
part (of a magnitude), 186
patterns
  proof and, 285-286
  search for as part of mathematical practice, 294
pedagogy, relation to epistemology, 53, 225
pipeline in mathematics, 325-326
problem aesthetics, 32, 41, 44, 46
problem situation, 290
  example of, 290
  student work on example of, 291-292
  use of in instruction, 298
problem situations in curriculum design, 301-302
problem solving, 40-41, 42
  in abstract mathematics, 261-263
  in geometry, models of, 273-274
  in groups, 308-309
  in mathematical practice, 294
  role of deductive logic in, 263
  role of proof in, 263
problem-centered curriculum, 306-307
  skills and concepts in, 307
problem-solving course, Schoenfeld's, 42, 61-68, 73-74
  proof, showing necessity of, 286
  Pythagorean Theorem in, 66-67
  teacher authority in, 62-65
problems
  authenticity in, 306-307
  examples of
    function, to build the concept of, 239-241
    induction, 231
    to build the concept of, 232-233
    quantification, 236

in abstract mathematics, form of, 261
in binary relations, hypothetical student's approach to, 258-260
in graph theory, hypothetical student's approach to, 260-261
Putnam
  examples of solutions, 34-38
  examples of, 33-34
  good, 23-24, 27, 41, 46
  writing, 21-22, 24-27
process, 229, 251-252
  function as, 237-238
programming language, 250, See also *software*.
"progressive absolutism," 177, 183-184, 190
project-based learning, assessment of, 212-214
projects, student. See *student research projects* and *calculus, student research projects in*.
proof, 284-286
  constraints and, 274-277
  geometry, 273
  in mathematical practice, role of, 270-271, 274
  in problem solving, role of, 263
  intuitive thinking and, relation of, 257-258
  mental models and, 272, 276-277
  patterns and, 285-286
  showing necessity of, 285-286
  situation theory and, 274-275
proving in mathematical practice, 296
psychogenesis. See *individual development*.
psychological research, implications of for instruction, 297-298

Putnam exam
  good problems, 23-24, 27, 41, 46
  meaning of success on, 22-23, 40, 47, 49, 50
  role of in undergraduate mathematics, 30-33, 48-49
  scores, 40-41
  writing problems for, 21-22, 24-27
Putnam problems
  examples of, 33-34
  examples of solutions for, 34-38
quantification, 233-235, 252
quantifiers, 268
ratio, 184-186
reflective abstractions, 229
representation, 250, See also *representations*.
  of magnitudes by line segments, 99-101, 184
    14th and 15th century, 100, 184
    student, 100-101, 184
representations, 82, 180, See also *notation systems*, *notations*, and *symbols*.
  of continuous variation, Oresme's, 90-97
research agenda, 268
research in mathematics education
  role of, 223, 248-250
  status of, 324, 326
research paradigm, 8, 226-227
rigor
  17th century relaxation of, 109
  18th and 19th century evolution of, 127-130
role of deductive logic in problem solving, 263
role of proof, 284-286
  in problem solving, 263

school practice, assumptions of, 319-320
secant lines, students' perception of, 265
situation theory and proof, 274-275
skills and concepts
    in problem-centered curriculum, 307
    role of in goals for mathematics education, 288-289
"social brain," implications for mathematics instruction from, 267-268
social networks for educational reform, 5-6
    Geometric Supposer Society, 6, 12
social perceptions of mathematics, 326
"society of mind," implications for mathematics instruction from, 267-268
software, 246, See also *curricular artifacts*.
    Algebraic Proposers, 8
    design of, 9
    Function Probe, 172, 182-183
    Geometric Supposer, 2-4, 262, 277
    Geometry Tutor, 277
    ISETL, 222, 246
        activities using, 232-233, 236, 239-240
    MathBikes, 147
    MathCars, 182
        use in calculus curriculum, 137-147, 149-151

solving problems in abstract mathematics, 261-263
Standards
    demands on teachers, 314-315
    vision of mathematics classrooms, 302, 313
student research projects, 194, See also *calculus, student research projects in*.
student work
    on a problem situation, 291-292
    on decimals, 314-315
students
    attitudes of, 316-317
    perception of secant lines, 265
    treatment of, 316-319
symbols, 298-299, See also *notation systems*, *notations* and *representations*.
tables, 183
teacher authority, 62-65, 71
teacher development, 315
teacher preparation, 325
teachers
    conceptions of mathematics, 314, 320
    Standards' demands on, 314-315
    textbooks, and, 308
technology in mathematics education, 325, See also *curricular artifacts* and *software*.
textbooks, 245-246, 308
tracking, 309-310
Trojan mouse, 3, 15, 73
undergraduate mathematics, role of Putnam exam in, 30-33, 48-49

QA 11 .A1 M2767 1994
Mathematical thinking and
 problem solving